Additional Praise for Jon R. Stone's *The Essential Max Müller*

"This is an impressive collection of essays by one of the great minds in modern history. Jon Stone has done all of us—specialists and general readers alike—a great service by compiling and introducing them."—Peter van der Veer, University of Amsterdam, author of *Imperial Encounters: Religion and Modernity in India and Britain.*

———·———

"One will find in this selection of Müller's lectures and essays a wonderful introduction to the work of a major nineteenth-century student of culture and founder of the scientific study of religion. Stone makes available here Müller's reflections on a wide range of topics that captured his interest over nearly forty years—from fetishism to nihilism to monotheism and from fables and myths to the perception of the infinite to the scientific study of the traditions of thought and practice embodying those perceptions—which illuminate the nineteenth-century landscape of the early social sciences. The anthology therefore provides a rich and complex account of Müller's pervasive influence on scholarship in the fields of language, thought, mythology, and religion, yet maintains a unity of focus in the attention the lectures and essays chosen for inclusion give to the emergence of the scientific study of religion, to which, Stone rightly points out, Müller made a lasting contribution. This volume, therefore, will be of special interest to scholars and students in the field of religious studies and to those more broadly interested in the emergence of the social sciences generally."—Donald Wiebe, Professor, Philosophy of Religion Trinity College, University of Toronto, and author of *The Politics of Religious Studies.*

———·———

"Who was Friedrich Max Müller? The Claude Lévi-Strauss or Mircea Eliade of the nineteenth century, or a throwback to eighteenth century Deism? A lone, strong voice against the racist Orientalism of British colonialism—the best friend India ever had—or a subtle propagandist for German Aryanism and American Anglo-Saxonism? Pious Christian, or the first 'advance man' for invasion of Asian religions into the West? The model for Reverend Casaubon of George Eliot's *Middlemarch,* or the hottest ticket on the late-Victorian lecture circuit? *The Essential Max Müller,* Professor Jon R. Stone's anthology of the principal works of Max Müller, will help answer some—if not all—of these tantalizing questions about an equally intriguing giant of the nineteenth-century intellectual scene.

Stone has performed the entirely admirable service of making accessible for the first time—and in a single convenient volume—a shrewdly selected and painstakingly edited collection of choice pieces from Max Müller's voluminous *oeuvre*—most of which are currently out-of-print or very difficult to obtain.

Along with his readable and informative introduction to Max Müller's life and times, this collection goes a long way toward explaining Müller's great vogue among students of comparative mythology and religion, the origins of language as well as his popularity among the informed reading public of his time. Here included are not only Müller's classic statements about the nature of the ancient religions of India, the philosophy of myth, or how to go about the study of religion, but also fascinating peeks into Müller's own mystic spirituality and personal religious vision, as well as his attempts to wrestle with his allegiance to his own Christian faith over against the claims made upon him by the nobility of the religions of India."—Professor Ivan Strenski, Holstein Family and Community Professor of Religious Studies, University of California, Riverside, and author of *Contesting Sacrifice*.

Photo courtesy of The Warden and Fellows of All Souls College, Oxford.
Reprinted with permission.

THE ESSENTIAL MAX MÜLLER

Also by Jon R. Stone

The Craft of Religious Studies
Expecting Armageddon: Essential Readings in Failed Prophecy
A Guide to the End of the World
Latin for the Illiterati
More Latin for the Illiterati
On the Boundaries of American Evangelicalism
Prime-Time Religion: An Encyclopedia of Religious Broadcasting

The Essential Max Müller
On Language, Mythology, and Religion

Edited by
Jon R. Stone

palgrave
macmillan

THE ESSENTIAL MAX MÜLLER
© Jon R. Stone, 2002

First published 2002 by
PALGRAVE MACMILLAN™
175 Fifth Avenue, New York, N.Y. 10010 and
Houndmills, Basingstoke, Hampshire, England RG21 6XS
Companies and representatives throughout the world

PALGRAVE MACMILLAN is the global academic imprint of the Palgrave Macmillan division of St. Martin's Press, LLC and of Palgrave Macmillan Ltd. Macmillan® is a registered trademark in the United States, United Kingdom and other countries. Palgrave is a registered trademark in the European Union and other countries.

ISBN 0–312–29308–9 hardback
ISBN 0–312–29309–7 paperback

Library of Congress Cataloging-in-Publication Data
Müller, F. Max (Friedrich Max), 1823–1900.
    The essential Max Müller: on language, myth, and religion/
Jon R. Stone, editor.
        p. cm.
    Includes bibliographical references and index.
    ISBN 0–312–29308–9 – ISBN 0–312–29309–7 (pbk.)
        1. Religion. 2. Mythology. 3. Language and languages—Religious aspects.
I. Stone, Jon R., 1959– II. Title.

BL50.M785 2002
200—dc21   2002025825

A catalogue record for this book is available from the British Library.

Design by Newgen Imaging Systems (P) Ltd., Chennai, India.

First edition: November, 2002
10 9 8 7 6 5 4 3 2 1

Printed in the United States of America.

In Memoriam

Ninian Smart
(1927–2001)

My Teacher

Contents

PREFACE

"A classic," humorist Mark Twain is credited as saying, "is a book that everyone talks about but no one has read."[1] Were Twain alive today he might have said the same of the scholarship of F. Max Müller (1823–1900), the famed German-born Oxford professor of comparative philology and Vedic studies.[2] A prolific writer and a highly popular lecturer, Müller enchanted the literate public of Victorian England with his learned essays and addresses on subjects as varied and exotic as Asian mythology, Western folklore, comparative linguistics, the philosophy of language and thought, and the origins and historical development of the world's religions. Though having never visited India himself, Müller supported the Indian nationalist struggle for equality with its British rulers. G.W. Trompf points out that while it had been through Müller's published writings that the English public first became introduced to the ancient wisdom of India (*ex Oriente lux,* light from the East), at the same time "his work was one of the finest symbols of urbane, liberal Oxonian scholarship during the hey-day of the British Raj" (1978: 1). Johannes Voigt notes that Müller's opinions regarding British–Indian affairs had "more in common with those of the leading Indians of his days [*sic*] than with those of the majority of Englishmen" (1967: 43), with the result that he was more revered in India than in either Britain or his native Germany—and remains so to this day. It is not surprising, therefore, to learn that his Oxford home became a place of pilgrimage for Indian gurus and pandits who traveled westward to see this enlightened sage, a man whom many Hindus called *mahatman,* great soul.[3]

Max Müller represented the best—and at times the worst—of nineteenth-century intellectual life. His work in the origins and growth of language, mythology, and religion typified Victorian armchair scholarship: bold, adventurous, pioneering, sometimes triumphalistic, but always convinced of its social and cultural superiority. To be sure, there is much to admire, much to despise, and much to be embarrassed by, in the antiquated scholarship of the Victorian era as a whole. But as a pivotal period in the history of human ideas, the historical and intellectual import of its scholarly literature should not be ignored by historians or summarily dismissed by present-day researchers as utterly worthless. Rather, it should be read and understood within its own social and cultural context.[4] In the case of the voluminous and, at the time, influential writings of Friedrich Max Müller, this observation proves no less true.

Max Müller was a strong, loud, and sometimes lone voice in many of the major intellectual and political debates of his time, especially over the implications of the reigning Indo-European (or Aryan) linguistic theories on the legitimacy of British colonial rule.[5] Though a religious man, Müller was one of only a few scholars who disagreed with Darwin's theory of the "descent" of man from apes on purely scholarly, not religious, grounds.[6] Today, Max Müller is largely forgotten. His death in 1900 seems to have confined him to that era, unlike his contemporaries, such as Darwin, Spencer, Kierkegaard, Schopenhauer, and Nietzsche, whose works have been revived or at least revisited.[7] If Müller is read by anyone, it is generally in excerpted form or in very brief passages, passages often unrepresentative of Müller's larger works or excerpts wrested from his masterfully laid out and profusely illustrated essays.[8] This present anthology, containing some of Müller's best-known essays and public lectures, is meant therefore to correct this unfortunate oversight. It should be pointed out, however, that the intention behind this anthology is not to defend Max Müller's findings, based as they were on the now-dated linguistic scholarship of his time, but, like the friendly voice that encouraged St. Augustine with the words *tolle lege,* the intent here is to invite students, scholars, and those beyond the ivory tower to "take and read." While most scholarship sits idly on dusty library shelves and is opened only occasionally by specialists, those who take and read herein will discover in the works of Max Müller some of the enchantment, as well as a little of the discomfort, that the nineteenth-century English public felt upon encountering exotic religions and mythologies whose spiritual outlook and aspirations differed little—in essence—from their own.

It is folly, one might say, to judge the value of a person's life's work, even the most idiosyncratic or arcane, without first reading a good portion of it. Indeed, it will surprise many readers how much there is that is of historical and philosophical interest in Müller's ambitious efforts to trace "scientifically" the development of human thought in terms of the artifacts of language, mythology, and religion. Moreover, by comparison to many of his starched and stodgy contemporaries, Müller's keen use of language and his expressive style were fluid, felicitous, and highly accessible.[9] To be sure, some critics have called his style beguiling, as if in exhibiting his skillful and effortless use of language Max Müller were somehow trying to smooth over gaps in his argument. Perhaps, on some points, this observation is true. But to acknowledge gaps or ambiguities in his work does not mean that there is nothing of worth—no useful insights—in what Müller thought or in what he wrote.

Müller himself knew the limits of a scholar's work, especially of one who had lived over three-quarters of a century. As he observed in his unfinished *Autobiography,* published the year after his death, "Another disadvantage from which the aged scholar suffers is that he is blamed for not having known in his youth what has been discovered in his old age, and still violently assailed for opinions he may have uttered fifty years ago" (1901c: 159). The specific essay he had in mind was his letter on the Turanian languages published in 1854, an outmoded linguistic analysis of non-Indo-European and non-Semitic languages that, as Müller complained, "continues to be criticized as if it had been published last year." Moreover, he continued, "though I have again and again protested that I could not possibly have known in 1854 what has been discovered since ... , everybody who writes ... seems to be most anxious to show that in 1894 he knew more than I did in 1854. No astronomer is blamed for not having known the planet Neptune before its discovery in 1846, or for having been wrong in accounting for the irregularities of Saturn. But let that pass; I only share the fate of others who have lived too long" (1901c: 160). Of course, one must approach such works with their obsolescence in mind, much the same way one should read Marx, Nietzsche, or Freud. But to say this is not to say that there is no contemporary value in Victorian scholarship, only that the reader needs to bear in mind the times in which this scholarship was undertaken, the specific questions the scholar sought to answer, and the types of resources at his or her command.[10]

All human knowledge, however antiquated, possesses some value for the present. True, such knowledge may not yield pure gold, but there might still be a few nuggets one can mine, specks of gold ore that can be used to enrich one's knowledge of human history and of human ideas. It is perhaps ironic that what Max Müller had observed in his "Lecture on the Vedas" became true of his own scholarship. For while he thought the Vedic hymns "tedious, low, commonplace," he still believed that "hidden in this rubbish there are precious stones."[11] In the same way, one will find some "precious stones" hidden in the writings of Max Müller, but only if that person is willing to mine them.

The "Essential" Max Müller: An Anthology of his Writings

Compiling an anthology of the essential writings of F. Max Müller has been no mean task. While the aim throughout has been to include Müller's best-known and most often cited essays and addresses, page limitations have restricted the number of selections to fewer than twenty. As a result, those that have been included represent a mere sampling of his voluminous output, but a sampling, nevertheless, that presents to the reader the range of Müller's research interests in the origins of language, mythology, and religion. In addition, in view of Müller's wide-ranging interests in the comparative study of religion, mythology, folklore, linguistics, metaphysics, and human cognition, it is hoped that the selections in this "essential Max Müller" will be of interest to scholars and students in fields as diverse as religious studies, philosophy, anthropology, early linguistic theory, and the history of Western ideas.

There were a number of problems I encountered in editing this collection of essays that need to be mentioned. The first has to do with the problem of multiple editions of and revisions to his catalogue of works. For instance, there are two published versions of Müller's famous 1870 "Lectures on the Science of Religion," an original edition, first published in 1872 under the title *Lectures on the Science of Religion,* and an expanded edition, published in 1873, that Müller retitled *Introduction to the Science of Religion* (as a point of interest, the latter edition was dedicated to Ralph Waldo Emerson). Further complicating matters, each version ran through several printings in Britain and the United States. Worse still, with each printing, Müller suggested corrections and revisions. In absence, therefore, of a *definitive* edition, for the selection included in this current anthology, "Lecture One," I have decided to use the 1872 edition, which is closest to the actual lecture his audiences would have heard him give. It is shorter, "edgier," and less circumspect than Müller's revised and expanded versions.

With respect to other essays in this anthology whose originals were not available to me, I have had to content myself with using Müller's later and sometimes final versions, such as those essays he himself had selected for his *Chips from a German*

Workshop, which by 1881 had grown from two to five volumes, as well as those he republished in his two volumes of *Selected Essays* (1881). Additionally, the three chapters from Müller's *Lectures on the Origin and Growth of Religion* (1878) reprinted here are from his new edition, published in 1882. For this new edition, Müller updated some of his sources as well as tightened up his prose.

A second set of problems I encountered were numerous stylistic and mechanical incongruities. Müller was sometimes inconsistent in his spelling, in English transliterations of foreign words and phrases, and in his use of accent and stress marks. In addition, at least by modern standards, Müller made awkward use of commas, colons, and dashes and tended to write highly complex and overly long sentences and paragraphs. Many of the inconsistencies, of course, can be accounted for in the stylistic differences between his several British and American publishers. But his awkward use of punctuation was probably idiosyncratic. Though, for the reader's benefit, I have attempted to bring some consistency in both spelling and punctuation and have sought to reduce and simplify other *res extraneae,* in the end, it seemed inappropriate to "restyle" Müller's essays to fit modern tastes. For one thing, I did not want to dilute the nineteenth-century "flavour" of Müller's writings; and, for another thing, because a large amount of his published work had been written for lecture audiences, retaining most of the original accent and punctuation marks may preserve for the reader Müller's own speaking style; that is, it may allow the reader to "hear" his voice—which, according to contemporary reports was clear, passionate, erudite, and engaging. For instance, as Nirad Chaudhuri relates, after presenting a lecture on the Science of Language in the Council Chamber at which Queen Victoria and the royal family attended, Müller wrote to his wife that the Queen "listened very attentively, and did not knit at all, though the work was brought" (1974: 185; see also, p. 187). When his lectures are read aloud, Müller's punctuation does indeed add variety to the pacing of his phrases and underscores their aural intensity. What is more, his indulgent use of commas and semicolons lends greater coherence to his long but carefully constructed sentences.

A final set of problems I faced in editing this collection was what to do with Müller's lengthy appendices, his block quotations in Greek, Latin, French, and Italian, as well as Middle English, and the many incomplete references and obscure citations in his footnotes. In these instances, as before, it seemed easiest to leave the essays in virtually the same form in which Müller himself had published them. One exception was Müller's inclusion of Latin and Greek quotes in some of his footnotes; another was Müller's lengthy notes and added appendices. With respect to Greek and Latin quotations, if Müller had already quoted their corresponding English translations in the main body of the text, then I took the liberty of deleting them as redundant. If no translation was provided, and the quotes were important to the flow of Müller's argument, then I let them stand both in the text or in the footnote. In some instances, I have inserted translations, either by myself or by much more adept colleagues.

With regard to my decision to delete Müller's extraneous notes and appendices: because many of his notes and appendices were added after he had revised his lectures for publication or had updated previously published works—material often added for illustrative purposes—it seemed prudent, given space constraints, to leave them out. The essay most affected by these emendations is "On the Migration of Fables," which Müller had published in two separate essay collections, one of which does not include the added notes and appendices.

A Few Words of Acknowledgment

It is fitting that this anthology of Max Müller's essays, lectures, and addresses should be published by Palgrave, the division of St. Martin's Press recently renamed in honor of the English poet Sir Francis Palgrave, editor of *The Golden Treasury* and a close friend and long-time supporter of Müller. It was to Palgrave that Müller dedicated the third volume of his *Chips from a German Workshop* with these words: "in grateful remembrance of kind help given to me in my first attempts at writing in English, and as a memorial of many years of faithful friendship." In a similar spirit, I would like to express grateful remembrance of kind help from Gayatri Patnaik, the religion editor at Palgrave, whose enthusiastic support for this project, despite its great expense of time and effort, was unwavering. I would also like to thank Amanda Johnson, Alan Bradshaw, and Donna Cherry, my production editor, for keeping the book on schedule (Alan had already discovered from an earlier project how notorious I am for wanting to tinker with sentence structure long after the copyeditor has cried "basta!," "enough!").

In addition to the editorial staff at Palgrave/Macmillan, I would like to say a word of thanks to friends and colleagues who have shown or have continued to show interest in my many and varied research projects. A few of these include: Catherine Albanese, Edward Amend, Sondra Bacharach, Eileen Barker, John and Carrie Birmingham, Mitch Breitwieser, Jeffrey Brodd, Michael Burdick, Juan Campo, the late Walter Capps, Wendy Doniger, Arthur Droge, David Eng, Robert Goldman, Allan Grapard, Giles Gunn, Phillip Hammond, Richard Hecht, Ron Hendel, Benton Johnson, Anne Kilmer, Gary Laderman, Gerald Larson, Eric Mazur, the late Robert Michaelsen, Birger Pearson, Clark Roof, Cybelle Shattuck, Deborah Sills, Frits Staal, Ivan Strenski, Brian C. Wilson, Bryan R. Wilson (who also helped me secure a portrait of Müller from All Souls College, Oxford), and Roy Zyla. In addition, I would like to thank Philippe Duhart and Elijah Jenkins for their help in proofreading the text and in compiling the index.

The solemn dedication that follows the title page stands as a belated tribute to my beloved teacher, the late Ninian Smart, who was the first to introduce me to the comparative study of religions. During my five years of graduate studies at the University of California, Santa Barbara, Ninian served as my sometime advisor, while I served as his sometime teaching assistant. My favorite memory of those years is of Ninian and me returning to campus after his lecture in the I.V. Theatre. He would ride his rickety old bicycle while I walked slowly beside him, nervously praying that he would not fall over (Ninian's very slow pedal speed—"Ninianissimo," I dubbed it—defied the laws of gravity). Along the way, we would stop by the Isla Vista newsstand to check cricket scores and the local surf report. Years later, when I visited India and the United Kingdom, Ninian's discourses on the rules and strategies of cricket came in quite handy. Indeed, in his scholarship and in his personality, Ninian embodied cricket's emphasis on fair play. Until his sudden passing in January 2001,

Ninian continued to guide my scholarly career. In fact, he took a great interest and delight in the success of all his students. It saddens me deeply that our dear Ninian is not here to accept this dedication.

"O mihi præteritos referat si Iuppiter annos!"—Virgil
("If only Jupiter would restore to me those bygone years!")

Jon R. Stone
January 2002

Notes

(For citations, see the Reference section at the end of the Introduction)

1. Twain's actual remark, set in the form of an aphorism, is found in volume one of his *Puddin'head Wilson's New Calendar*. It reads: " 'Classic.' A book which people praise and don't read."

2. There appears to be some uncertainty among scholars, biographers, and bibliographers about the proper designation of Müller's surname. It is sometimes written as and cataloged under "Müller" or "Mueller" and at other times hyphenated as and cataloged under "Max-Müller" or "Max-Mueller." Louis Henry Jordan made a similar observation in his overview of Müller's scholarship (1905: 150, note 1). Müller himself preferred the designation "Max Müller." However, for the sake of simplicity, his surname will be given both here and below as "Müller," with "Max Müller" being sometimes given to avoid the tedium of repetition.

3. See Voigt 1967: 47–48 and Chaudhuri 1974, pt. 3, chapters 3–5 *in passim*, and Rothermund 1986: 52. According to G. W. Trompf, in recognition of his support for Indian equality with the English, Müller's Hindu friends bestowed upon him the Sanskrit epithet "moksa mulara" (1978: 4). Many decades later, in 1957, the New Delhi branch of the Goethe Institute was named the Max Mueller Bhavan ("house") in his honor (see Voigt 1967: 85 and Chaudhuri 1974: 1–2).

4. Here one might cite, among many others, the valuable research of Michel Foucault (1973) in *Les Mots et Les Choses* (published in English as *The Order of Things: An Archaeology of Human Sciences*), and of Umberto Eco (1995) in *The Search for the Perfect Language*. One example from the philosophy of religion is Peter Byrne's (1989) lucid study, *Natural Religion and the Nature of Religion: The Legacy of Deism*.

5. See Rothermund 1986: 47–49, 52–53. Admittedly, from a post-colonial perspective, Müller's liberal position may appear only a few shades different from his pro-imperialist contemporaries. However, his political and especially his religious views were quite controversial for their time, sometimes bordering upon heresy. Later biographers suggest that, together with his German origins and Broad Church tolerance, sometimes leaning toward universalism, it was Müller's liberal politics that had ultimately cost him the 1860 election to Oxford's Boden Chair in Sanskrit, a defeat that gnawed at him for the remainder of his life (see Voigt 1967: ix and 69–70; Chaudhuri 1974: 221–231; Leopold 1974: 582, and van der Veer 2001: 108–110; see also Stephen and Lee 1917: 1024–1025). Some of these same biographers have wondered aloud that had it not been for Müller's obvious German origins—the umlaut over the u and his provincial German accent—his works might have been given a more generous hearing among scholars during the late nineteenth and early twentieth centuries, a period that witnessed growing anti-German sentiments within English, French, and, to some extent, American universities—a point touched upon by Voigt (1967: 83–84).

6. See, for instance, Müller's *Lectures on Mr. Darwin's Philosophy of Language* (1873b). In his 1884 essay discussing his *Sacred Books of the East* project (*S.B.E.*), entitled "Forgotten

Bibles," published just two years after Darwin's death, Müller reaffirmed his position that "the barrier [between man and beast] of language remains as unshaken as ever, and renders every attempt at deriving man genealogically from any known or unknown ape, for the present at least, impossible, or, at all events, unscientific" (1901b: 24–25).

7. Marcel Detienne makes a similar observation (1981: 95).

8. See, for example, the excerpts in Feldman and Richardson (1972: 481–487) and Waardenburg (1973, vol. 1: 85–95). In many overviews and assessments of his work, Müller's ideas are usually reduced to a few pithy quotes, such as "mythology is a disease of language," "nomina [name] becomes numina [spirit]," and "he who knows one [religion] knows none." This reduction of Müller's work to a few catchy slogans mischaracterizes his views as much as it betrays a lack of familiarity with his research programme as a whole. One noticeable exception is Peter Byrne (1989: 183–196); another is Ivan Strenski (1996a and 1996b).

9. See Wilson 1948: 88–89.

10. After this anthology had gone to press, an essay on Max Müller's scholarship, by Norman Girardot (2002), appeared in the February 2002 issue of *History of Religions*. Its author's baseless and bizarre portrayal of Max Müller is worth discussing at length. Girardot begins his essay by erroneously referring to Müller as "Frederick Maximilian Müller," apparently after an 1875 "Men of the Day" caricature of Müller in London's *Vanity Fair*. Girardot's essay, which has its merits, concerns Müller's role in the emergence of the "scientific expert" in the late nineteenth century—at least as it relates to the development of a comparative or "scientific" study of religion. But his discussion rarely moves beyond caricature, and, along the way, Girardot also takes a few gratuitous swipes at Stanley Fish, the late Carl Sagan, and the late Mircea Eliade. But, not only is Girardot's critique of Müller unnecessarily harsh, it truly borders upon character assassination. To be fair to Girardot, one can hardly doubt that the nineteenth-century break from as ecclesiastically controlled a discipline as religion would require of lay scholars a certain amount of "social networking, professional prestige, business acumen, and political skills" (p. 215), all of which Müller seems to have possessed in good measure. But Girardot is critical, not of Müller's scholarship, but of his supposed entrepreneurialship, of which Girardot greatly disapproves. But that is to put it mildly. For, in his essay, Girardot portrays Müller as an academic charlatan: a money-grubbing, behind-the-scenes operator whose only goal is to secure for himself profit and fame at the expense of others.

In constructing his picture of Dorian Gray, Girardot imputes to Müller all kinds of dark motives and deceptive behaviors, all, it should be noted, without offering credible support for these charges—except that Girardot believes that Müller must have somehow cheated his man, James Legge (1815–1897), the elder Oxonian Confucian scholar who had signed on to Müller's *Sacred Books of the East* project. Apparently, for his contributions to the series, Legge had contracted for less money than Girardot feels Legge should have signed for. According to Girardot, Müller, the editor, earned more from sales of the *Sacred Books* than Legge, a contributor. How shocking! (see pp. 226–229, footnotes 26–30). In other words, Girardot blames Müller for Legge's own apparent deficiencies in "social networking, professional prestige, business acumen, and political skills."

But this is not all. Here is another of Girardot's many baseless and bizarre charges against Müller: "Müller preferred the national fame and substantial remuneration that came from published books and accepting well-publicized lectureships over the tedium and obscurity of routine teaching. His inclination in these matters was particularly one-sided when it was associated with the special notability of inaugurating a new, and heavily endowed, invitational lectureship devoted to an approach to religion largely identified with his own reputation as the preeminent comparative scholar in the world" (p. 222, note 17; see also, pp. 225, 227–228, 231, 242–243, 244, etc.). Elsewhere, Müller is described as a man of "self-cultivated fame" (p. 215), as speaking in a "typical bombastic crescendo" (p. 218), as expressing "not so sublimated anti-Catholic bias" (p. 219), as feigning "self-styled" editorial courage (p. 227), as offering "always slipping and sliding arguments" (p. 243), and as holding "romantic and still reverent 'ding-dong' notions of primordial linguistic resonance" (p. 246, note 74)—all without corroborating support and all without serving any useful purpose other than to make sport of a dead man. It is unfortunate that, despite its stated thesis and the importance of its subject, Girardot's essay never rises above the level of a scurrilous screed. Hence the dictum: "mortuo leoni et lepores insultant" (even rabbits [will] strike a dead lion).

11. "Lecture on the Vedas" in *Chips from a German Workshop* (1895, vol. 1: 26).

INTRODUCTION

In a letter written during the last year of his life, Max Müller confessed to his son, Wilhelm: "I am prepared to go; it would be strange if I were not, with such a long life behind me, and most of it devoted to religious and philosophical questions." He then added, with a poetic air, "And having lived this long life so full of light, having been led so kindly by a fatherly hand through all storms and struggles, why should I be afraid when I have to make the last step?" (quoted in Chaudhuri 1974: 382). At age seventy-five, Friedrich Max Müller was dying, barely strong enough to care for himself, let alone continue the heavy demands placed upon him by his private correspondence. Müller's illness had lasted throughout the summer of 1899 and by September his doctors had given up hope for a full recovery. However, in early 1900, the world-renowned Oxford don regained enough of his strength to write one last controversial article—on the Boer War—a pro-British essay that outraged his German compatriots. After the controversy surrounding his article had died down, Müller himself lingered on for only a few months more, finally succumbing to his illness on October 28, 1900 (Chaudhuri 1974: 252–255, 368–370).

In that same letter, Müller looked back over his long productive life and felt certain that his scholarship would be "carried on by others, by stronger and younger men." "I have laid a foundation that will last," Müller then declared, "and though people don't see the blocks buried in the river, it is on these unseen blocks that the bridge rests" (quoted in Chaudhuri 1974: 382). These "unseen blocks" were in reference to his work in the comparative study, or "sciences," of language, mythology, religion, and thought, as he had come to call these new academic fields, which he himself had either created, modified, or extended as a result of his highly ambitious research programme.

A few years before, in the Preface to his two-volume tome, *Contributions to the Science of Mythology* (1897), Müller outlined the logic behind the four sciences he had championed, the same logic behind the numerous books and volumes of essays and addresses he had written and published during his fifty-year career at Oxford. To quote him at length:

> I have hesitated for a long time before making up my mind to publish these two volumes on the Science of Mythology. I was sorry, no doubt, that I should have to leave this gap in the work of my life as I had planned it many years ago, namely an exposition, however imperfect, of the four Sciences of Language, Mythology, Religion, and Thought, following each other in natural succession, and comprehending the whole

sphere of activity of the human mind from the earliest period within the reach of our knowledge to the present day.

There is nothing more ancient in the world than language. The history of man begins, not with rude flints, rock temples or pyramids, but with language.

The second stage is represented by myths as the first attempts at translating the phenomena of nature into thought.

The third stage is that of religion or the recognition of moral powers, and in the end of One Moral Power behind and above all nature.

The fourth and last is philosophy, or a critique of the powers of reason in their legitimate working on the data of experience. (Preface to vol. 1, 1897: v)

Then, in a corresponding footnote, Müller listed his major works to show how each fitted into three parts of his four-part project, philosophy apparently being subsumed under the category of thought:

I. The Science of Language, two vols., 1864, last edition, 1891.
II. The Science of Religion:
    1. Introduction to the Science of Religion, 1870.
    2. The Origin and Growth of Religion, 1878.
    3. Natural Religion, 1888.
    4. Physical Religion, 1890.
    5. Anthropological Religion, 1891.
    6. Theosophy, or Psychological Religion, 1892.
III. The Science of Thought, one vol., 1887; and
    Translation of Kant's Critique of Pure Reason, 1881; last edition, 1896. (see Preface to vol. 1, 1897: vi, note 1)

Müller's sketchy outline did not include all of his published works but only what seemed to him to be his lasting contribution to the scientific study of human cognition.

Interestingly, the "science" upon which Müller is best known and upon which he has received the most scathing criticisms—both in his lifetime and into the present—is the one about which he wrote the least.[1] Noticeably missing from his sketch is the science of mythology, again, the area upon which much of his notoriety rests. As Müller explained, "neither time nor strength was left to me for doing what I had been allowed to do for the other three sciences" (1897: v–vi). Still, Müller's book-length essay, "Comparative Mythology" (1856 [1909]), together with his *Contributions to the Science of Mythology*—and several chapters in his five-volume set of collected essays and addresses that he had titled *Chips from a German Workshop* (1895)—formed for Müller the fourth—actually the second—part of his grand scheme.

Over forty years had elapsed since the publication in 1856 of this, his most famous essay, before Müller once more devoted himself to putting down on paper his thoughts—this time his culminating thoughts—on the Science of Mythology. Although Müller's *Contributions* reiterates much that is already found in "Comparative Mythology" and therefore might be viewed as merely an updating and reworking of that essay and of his famous *Lectures on the Science of Language* (see Dorson 1968: 166), Müller expressed confidence that this final treatise would be the crowning achievement of his life's work, aimed as it was at tracing the historical growth of language, mythology, religion, and thought back to its origins.

And indeed, by the end of his life, Müller appeared to have achieved the lasting success for which he had hoped. After his death, an early tribute, published in the *Dictionary of National Biography*, acclaimed Müller as "one of the most talented and versatile scholars of the nineteenth century." Still further: "Though much in his works and methods may be superseded, the great stimulating influence his writings have exercised in many fields will give him a strong claim to the gratitude of posterity" (Stephen and Lee 1917: 1027). Müller was likewise praised in an obituary published in *The Times* of London, which read: "We lose in him one of the most brilliant and prolific writers of our time; one whose voice has charmed several generations of Englishmen; who was a great scholar ... unsurpassed as a scientific expositor, possessing ... a power of breathing human interest into dry bones, a curiously sympathetic intelligence and a rare mixture of the talents of the poet and the savant" (quoted in Voigt 1967: 81). Though more balanced in his praise of Müller, Robert Brown, a younger contemporary, nevertheless observed in 1898 that "however we may differ with him in detail, [Müller] has done many a doughty deed, illumined many a dark spot, vastly widened the bounds of our knowledge, placed his views before the world in due completeness, and, if the translation of the *Rig-Veda* be an achievement reserved for the twentieth century, sung his song to the last stanza. And it is upon his great contributions to human knowledge and to human thought ... that Prof. Müller's permanent fame will securely rest" (1898: 1–2).

Yet, for all the confidence that Müller and his contemporary admirers expressed in his having laid a solid foundation upon which others might build, it was not long before substructural cracks in his system began to show. For instance, in one of the first overviews of his new field, the Science of Religion, published scarcely five years after Müller's death, Louis Henry Jordan gave a not-so-flattering opinion of Max Müller's contribution to the comparative study of religion: "judged by the rigid standards of that Science which [Müller] did so much to inaugurate, a good deal of his own work seems to-day incomplete and strangely defective." Jordan continued:

> He was a singularly strenuous student. During his long and active life he certainly never spared himself. But he attempted to be an investigator in far too many departments. When unexpectedly he found a door that gave admission to some new and inviting domain, he could not resist the temptation to enter it. ... In view of the various discoveries he made, it is not in one's heart to reproach him. ... As a teacher, author, lecturer, and investigator, Professor Max Müller had always an overwhelming amount of work on hand; and, to very much of it, he was able to devote only such fragmentary leisure as he could manage to command. It was for this reason that he never really found time to apply himself, with resolute and persistent purpose, to the promotion of Comparative Religion. (1905: 153–154; cf., Jespersen 1922: 86)

It was not in Jordan's interest to discredit the work of so great a sage, but his candid assessment did at least point out one of the greatest failings of this Victorian scholar. Said at length, Müller's aim to settle questions surrounding the origins and growth of religion, mythology, and human thought through a "scientific"—that is, comparative and historical—examination of language, was ever beyond his reach, and this, one must note, despite the rigor of Müller's comparative method and the great store of facts that he had unearthed while excavating the "ruins" of classical and archaic languages in search of links tying the prehistoric proto-Aryan past to the Victorian present.

Further on in his assessment of Müller's career, Jordan relaxed slightly his negative critique and generously conceded to Müller a lasting contribution to the comparative study of religion:

> It may seem to some that such praise is faint ... but no words that have been used seek to obscure or minimise the fact that the assistance which Professor Max Müller did actually lend to this Science possessed a value that was greater than any one to-day can fully realise. Moreover ... if some of the books which cost Professor Max Müller great labour are unquestionably doomed to be superseded and presently forgotten,—it can never be remembered without gratitude that it was he who secured for us the translation of *The Sacred Books of the East,* and thus rendered accessible for all time a simply inexhaustible quarry of untold riches. This single achievement is sufficient to render his memory immortal. (1905: 155–156)

Most later critics of Max Müller's scholarship have tended to follow Jordan's double-edged assessment, crediting Müller with having rendered "his memory immortal" through his painstaking efforts at publishing definitive translations of the world's sacred texts that became the fifty-volume *Sacred Books of the East,* five of which Müller himself had contributed, while at the same time dismissing the whole body of his scholarly work as "incomplete and strangely defective." Except in India, where, even a century after his passing, Müller is still regarded as a friend of the Indian people, few scholars who are now engaged in the history of religions or the history of Western thought, have cared to read more than a few excerpts of Müller's voluminous writings.[2] His published work remains largely unread and, hence, largely unknown.

Another Look at Max Müller

It is perhaps an interesting twist that at the height of his fame as a public scholar, Max Müller offered a similar critique of the comparative work of Sir William Jones, the eighteenth-century British justice in Calcutta who, in 1786, had published the first paper on the linguistic similarities that linked the so-called Indo-European or Aryan family of languages. Indeed, Jones's discovery formed one of the several assumptions upon which so much of Müller's comparative work had been based. Müller found fault with Jones over several "superficial comparisons" that Jones had later made between the mythologies of India and other nations that proved to be "without any scientific value." "It is not pleasant to have to find fault with a man possessed of such genius, taste, and learning as Sir W. Jones," Müller wrote, "but no one who is acquainted with the history of these researches will be surprised at my words. It is the fate of all pioneers, not only to be left behind in the assault which they had planned, but to find that many of their approaches were made in a false direction, and had to be abandoned. But as the authority of their names continues to sway the public at large, and is apt to mislead even painstaking students and to entail upon them repeated disappointments, it is necessary that those who know should speak out, even at the risk of being considered harsh or presumptuous" (Müller, "False Analogies," 1881, vol. 2: 448–449).

The eclipse of Müller's renown as a scholar followed that of Jones, whose ground-breaking work on classical Indian law and language is almost completely forgotten. But in his scholarship, flawed though it might be, Müller did not devote himself to mining one area alone, as Jones had done. And not all of the conclusions Müller drew or alluded to rose or fell on late-eighteenth- and early-nineteenth-century theories regarding the origins of the Indo-European or Aryan languages—though, of course, it did not help Müller's cause that he was sometimes ambiguous when writing about the linguistic, as opposed to the racial, aspects of the Aryan family of languages.[3] More careful critics—and few there are—are keen to the subtle distinctions and nuances in Müller's work on this and other points. The late British anthropologist E. E. Evans-Pritchard, for instance, referred to Müller as "a linguist of quite exceptional ability, one of the leading Sanskritists of his time, and in general a man of great erudition," who "for a time had some influence on anthropological thought," an influence that he outlived (1965: 21, 23; cf., Stocking 1987: 305). Evans-Pritchard also noted sympathetically that Müller "has been most unjustly decried," especially over his early view of mythology as a "disease of language," "a pithy but unfortunate expression which later Müller tried to explain away but never quite lived down" (1965: 21, 22).[4] At the same time, though sympathetic, Evans-Pritchard's comment that, in Müller's view, "language exercises a tyranny over thought, and thought is always struggling against it, but in vain" (22) is not altogether accurate. More correctly, Müller held that language is an *artifact* in which one can examine the development of human thought—philosophical, religious, and mythological—as a geologist might examine the various layers of rock and sediment. For Müller, mythology represented an earlier period or strata in the evolution of human thought and, as such, was thus viewed by him as a vestige or faint residue of the past that still impressed itself on the thought and language of the present.

"Mueller today," writes Peter Byrne, "still appears to receive greatest attention as the author of a theory of myth which rests upon the notion that mythology is a disease or corruption of language." With Evans-Pritchard, Byrne likewise sees this as an "unfortunate expression" that later critics mistook for the whole thrust of his work. "But," as Byrne argues, "the general discredit into which his theory of myth has fallen does not detract from the main thing upon which his claim to be the founder of the science of religion rests—that is, his systematic account and defence of the aims, methods, and presuppositions of the historical and comparative study of religion. Mueller was unmatched in his day, and his apology did much to gain the discipline respectability and attention" (1989: 183).

Whatever his influence on the budding disciplines of anthropology and comparative religion, Müller was foremost a philologist who believed that, in the assumed linguistic link between India the ancient and Europe the modern, he had found the key to unlocking the theretofore unknown origins of mythology and religion, at least among peoples from these linguistic backgrounds. And, as alluded to above, Müller's view of the role of the philologist, to quote Tomoko Masuzawa, involved "a kind of linguistic geology, or archaeology of language, whose first and foremost task is to gather up and sort out the odd fragments of obliterated language formations of the past which are half-buried in the languages of today" (1993: 64).

It was in his linguistic workshop that Müller compared dialects and traced back the origins of words in order to discover, layer by layer, the development of human thought and language, and thence to explain changes in religious perceptions as evidenced by mythological conceptions of nature and natural phenomena. As Müller himself affirmed in his posthumously published *Autobiography,* "it would not be easy for those who have hereafter to report on our labours to discover the red thread from our first stammerings to our latest murmurings. It might be said that in my own case the thread that connects all my labours is very visible, namely, the thread that connects the origin of thought and languages with the origin of mythology and religion. Everything I have done was, no doubt, subordinate to these four great problems" (1901c: 2–3). Elsewhere, in his book *Biographies of Words and the Home of the Aryas,* Müller argued, "there is one thing which, as everybody can see, will follow by necessity from the admission of the inseparableness of language and thought, and that is that all thoughts which ever passed through the mind of men must have found their first embodiment, and their permanent embalmment, in words. If then we want to study the history of the human mind in its earliest phases, where can we hope to find more authentic, more accurate, more complete documents than in the annals of language?" (1888: ix).

Max Müller employed a number of metaphors to describe the approach he took to his work, including the indirect references to geology and archaeology in the passages quoted above. These metaphors and allusions, together with his acceptance of the reigning theories of the tripartite division and evolutionary development of human languages and peoples, closely describe how Müller saw his task as a comparative philologist—i.e., as that of the scientist—and also underscore the methodological assumptions he made in his attempts at uncovering the defining stages in the evolutionary growth of language, mythology, religion, and thought.[5]

To understand how Müller came by his approach to these four "sciences," and to appreciate better his influence in shaping some of the intellectual and, to a lesser extent, political and nationalistic discourse of his day, it will be helpful to trace briefly his intellectual biography. A sketch of Müller's intellectual portrait will also assist the reader in understanding the essays and public lectures contained in this anthology. The purpose here will not be to present a detailed account, or even a critique, of Müller's work in his four chosen "fields," but simply to introduce his thought as it had developed within the intellectual context of his time. By all accounts, Müller lived an exceptional life. But, as will be seen, while Müller was a creative interpreter of the materials at his disposal, who breathed "human interest into dry bones," he was nevertheless a man confined by the scholarly and scientific assumptions of his day.

Max Müller: An Intellectual Portrait

Friedrich Max Müller was born on December 6, 1823, at Dessau, in the small German duchy of Anhalt-Dessau some fifty miles southwest of Berlin. His father, Wilhelm Müller (1794–1827), who died when Friedrich was not yet four, was a distinguished young Romantic poet whose fame rested on his *Griechen Lieder*, a set of poems written in support of the Greek nationalist struggles against the Turks. Alternately called the "Greek Müller" (Griechen-Müller) and the "Byron of Germany," Wilhelm came to be highly regarded within the artistic circles in and around the neighboring city of Leipzig. In 1823 and again in 1827, a great honor was paid him when Franz Schubert composed a pair of song cycles—*Die Schöne Müllerin* and *Winterreise*—that immortalized two of Wilhelm Müller's best sets of poems.

Adelheide Müller (1799?–1883), Wilhelm's wife and Friedrich's mother, was the eldest daughter of Ludwig von Basedow, who had been the duke's chief minister. Though retired, Präsident von Basedow remained an influential voice in the politics of the region. As one might expect, Wilhelm's sudden death at age thirty-three meant that Max Müller's early life was spent almost exclusively in the company of his maternal grandparents, whose courtly lifestyle he observed and on some occasions imbibed (see Müller 1901c: 46–96 and Chaudhuri 1974: 13–27). However, after his grandfather's death in 1835, Müller's mother decided that it would be best for him, now age twelve, to be sent to boarding school in Leipzig. As Müller himself related, his mother "had judged rightly that it was best for me to be with other boys and under the supervision of a man. I had been somewhat spoiled by her passionate love, and also by her passionate severity in correcting the ordinary naughtiness of a boy" (1901c: 97). On Easter Sunday, 1836, Müller arrived in Leipzig to begin his middle-year studies at the Nicolai-Schule, where once Leibniz had been a student. While in Leipzig, he lived in the house of a Professor Carus, an old friend of his father, and was soon drawn into the swirl of social and musical activities around which the Carus home was a center.

During his first years in Leipzig, Max Müller became interested in studying poetry and especially music, for which he showed some early talent (incidentally, the German composer Carl Maria von Weber had been his godfather). Leipzig in the 1830s and 1840s had become one of the foremost centers of music in Germany, boasting the presence of the composers Felix Mendelssohn, who, as conductor of the Gewandhaus Orchestra, helped revive the music of Bach; Robert Schumann, who served briefly as Mendelssohn's associate; and Franz Liszt, the virtuosic Hungarian pianist. Müller had met Mendelssohn through Carus and found himself attending many of the concerts and other musical festivities Mendelssohn had arranged for the people of Leipzig. Himself an accomplished pianist and singer, Müller sometimes sang with the orchestra and often took part in the evening parties hosted by Professor and Madame Carus. For a time, Müller had considered a career in music but, having been dissuaded by Mendelssohn from taking that difficult path, he finally decided to pursue the life of a scholar (1901c: 98–111; Smith 1917: 1023, and Chaudhuri 1974: 28–32). Müller remained at the Nicolai-Schule until 1841, when, after securing a scholarship, he matriculated into the University of Leipzig.

The devotion of Müller's mother toward her only son continued. She and Müller's older sister, Auguste, moved to Leipzig to keep house for him while he studied and lived the life of a university student. But Müller's college days were not all work; there was time enough for amusement, as Müller and some of his fellow classmates made merry with drink and played pranks at the expense of the townsfolk. Even a brief jail term for wearing the lapel ribbon of a politically out-of-favor student club did not dampen Müller's bon vivant lifestyle. As Max Müller wistfully recalled, "I cannot say that either the disgrace or the discomfort of my two day's durance vile weighed much with me, as my friends were allowed free access to me, and came and drank beer and smoked cigars in my cell—of course at my expense" (Müller 1901c: 118). Müller even engaged in the favorite pastime of German students—dueling—having himself fought three sword fights, one of which to defend the honor of a classics professor (Müller 1901c: 119; Chaudhuri 1974: 34–35). The aim in dueling was to wound slightly, not to kill or seriously disfigure, and many Germans, both soldiers and statesmen, displayed their dueling scars with great pride.

These were politically sensitive times in the German states, especially in Austria and Prussia, which held administrative sway over Leipzig and Dessau. This was the period when Karl Marx and other revolutionaries were becoming active and when Austrian and Prussian officials were ruthlessly stamping out political and ideological dissent. Indeed, governments throughout Europe began enacting measures aimed at quelling the growing spirit of nationalism and, later, open rebellion among the masses. "There were some wild spirits among us who fretted at the narrow-minded policy which went by the name of the Metternich system," Müller observed. "Repression was the panacea which [Austrian Prince] Metternich recommended to all governments of Germany, large and small. No doubt the system [put into place after the Napoleonic Wars] secured to Germany and to Europe at large a thirty years' peace, but it could not prevent the accumulation of inflammable material which, after several threatenings, burst forth at last in the conflagration of 1848" (1901c: 117). Tragically, "Germany lost some of her best sons in those miserable years; and if my father [a political liberal] escaped this political persecution, it was probably due to the influence of the reigning Duke and the Duchess, a Princess of Prussia, who knew that he was not a dangerous man, and not likely to blow up the German Diet" (1901c: 118).

Müller proved to be a brilliant student. Though having already mastered Greek and Latin as a youth, and continuing his studies in classics at Leipzig, he soon added Arabic, Sanskrit, and, later on, Persian to his course load. According to Müller's Kollegien-Buch, during his two and a half years at the University of Leipzig, he enrolled in some fifty lecture courses, or roughly ten courses per term, divided among classics, philosophy, and Sanskrit grammar and Indian literature (1901c: 123–125). "Not being satisfied with what seemed to me a mere chewing of the cud in Greek and Latin," Müller recounted, "I betook myself to systematic philosophy, and even during the first terms read more of that than Plato and Aristotle. I belonged to the philosophical societies of [Christian] Weisse, of [M. W.] Drobisch, and of [Hermann] Lotze, a membership in each of which societies entailed a considerable amount of reading and writing" (1901c: 129). Through Weisse and Drobisch, Müller was introduced to the Idealist philosophy of Hegel and to its remedy in the Realist philosophy of one of Hegel's lesser-known rivals, Johann Herbart (1774–1841). Hegel had died in 1831, but his dialectic idealism, which traced the logical and irresistible evolutionary development of the eternal "Idea," as manifested through natural and historical processes, had captured the fancy of German scholars. During Müller's school days, Hegelianism had become something of an intellectual craze. "[T]o be a Hegelian," Müller recalled, "was considered a *sine quâ non,* not only among philosophers, but quite as much among theologians, men of science, lawyers, artists, in fact, in every branch of human knowledge, at least in Prussia. If Christianity in its Protestant form was the state-religion of the kingdom, Hegelianism was its state philosophy" (1901c: 130).

The Hegelian dialectic method that Max Müller learned from Weisse, while true to Hegel, was less sweeping in its scope. Weisse criticized Hegel for his "idealization of history" and for his presumption that facts would bear him out even when historical support for his argument was either inaccurate or unavailable. It was Weisse's "protest of the historical conscience against the demands of the Idea" that intrigued Müller most of all (1901c: 135–136; see also Kitagawa and Strong 1985: 190). For Müller, this disregard for the facts of history, especially with regard to the "growth of religious thought," told against the truth of Hegelian dialectic idealism and, in his own words, "shook my belief in the correctness of Hegel's fundamental principles more than anything else" (1901c: 142). When historical facts and Hegelian idealism collided, Müller noted, the Hegelians would side with their theories, declaring "tant pis pour les faits," too bad for the facts (1901c: 136; see also 1901b: 18). For Müller, this would never do. For, as Müller admitted, "I could not bring myself to admit that the history of religion, nor even the history of philosophy as we know it from Thales to Kant, was really running side by side with [Hegel's] Logic, showing how the leading concepts of the human mind, as elaborated in the Logic, had found successive expression in the history and development of the schools of philosophy as known to us" (1901c: 141).

While the basis of Hegel's concept of Nature is the development of the "Idea" in "space," what is fundamental to Hegel's philosophy of history is the development of "Spirit" in or through "time" (Idea becoming the thesis, Nature the antithesis, and Spirit the synthesis of Hegel's grand dialectic). This "Weltgeist" (world spirit) is behind and parallel to the dialectic of historical development, a development that proceeds from one stage to another until it reaches its highest point. For Hegel, this highest point was nineteenth-century European thought and culture. On this point, Wilhelm Halbfass observes that "Hegel's scheme of the history of philosophy has been designed mainly with regard to the history of Occidental thought from Thales to Kant and to himself. The assumption of a plurality of independent or parallel streams of development was, however, out of the question because of his concept of the 'Weltgeist' and the unity of the world-historical process" (1973: 111). For Hegel, the East and Oriental thought, like the rising sun, stood at the dawn of history; accordingly, the West and Occidental thought stands at the end-point of the sun's triumphant journey across the heavens.

Of the elements of Hegel's philosophy that Weisse did stress, the *logos* ("word" or "reason"), the neo-Platonic creative principle that stands between finite and the Infinite, which for Weisse was "the historical key to Hegel's Idea" (1901c: 133–134), would become in Müller's hands a key component in the evolutionary development of human religion as well as of language and thought (Müller 1893: 361–423; see also Trompf 1978: 58–63). Though Müller did in fact admit Hegelian ideas of progress into his thinking on religious development, seeing Christianity, for instance, as the highest expression of divine *logos* or "reason" (Müller 1867 and 1893: 424–458), in his evolutionary approach to understanding human knowledge, Müller never regarded himself as a Hegelian, let alone a full-blown social evolutionist or Darwinian. Indeed, Müller took great care not to embrace Hegel's "Weltgeist" too tightly.[6] In addition, in his writings against Hegel and against Darwin, Müller often derided both masters, as well as their followers, for forcing facts into their grand theories (1901c: 136; see also 1901b: 1–35, especially pp. 18–19), a charge some of Müller's modern critics have likewise leveled at him.[7]

While Weisse's influence on Müller was notable, it was especially through Drobisch, and his stress on the philosophy of Herbart, that the bearing of Müller's linguistic research took a decidedly historical-critical tack. Until the 1840s, Hegel's Idealism had reigned supreme in the German universities. Then a counterattack came from the so-called Realists, such as Herbart and his disciples, who had set upon Hegel's philosophy of religion and largely succeeded in cutting up and discrediting it (see Trompf 1978: 7–8). For Müller, who had been raised on Kant and who would later translate Kant's *Critique of Pure Reason* into English, Herbart's careful method of analyzing and clarifying concepts—concepts whose original meanings are sometimes lost because of long historical development—acted as "a most useful antidote" to and a helpful defense against Hegel (1901c: 142; see Kitagawa and Strong 1985: 190–191). "This is exactly what I wanted," Müller recounted in his *Autobiography,* "only that occupied as I was with the problem of language, I at once translated the object of [Herbart's] philosophy into a definition of words" (1901c: 142). Müller then continued, "Henceforth the object of my own philosophical occupation was the accurate definition of every word," definitions in which "[a]ll words, such as reason, pure reason, mind, thought, were carefully taken to pieces and traced back, if possible, to their first birth, and then through their further developments. My interest in this analytical process soon took an historical, that is etymological, character in so far as I tried to find out why any words should now mean exactly what, according to our definition, they ought to mean" (1901c: 142–143). In this way, Max Müller had found his method, though not as yet the specific subject matter of his researches.

In 1841, Hermann Brockhaus had become the first occupant of the newly founded chair in Sanskrit at Leipzig, and very early Müller came under the spell of both Brockhaus and the study of the Sanskrit language. Here, for Müller, was a little-known area of study that, despite the disapproval of his classics professors, he believed he could make his own. In Sanskrit, the ancient world of the Vedas and other texts opened before his eyes. "There was a charm in the unknown," he confessed in his *Autobiography,* "a charm also in studying something which my friends and fellow students did not know" (1901c: 147). Under the tutelage of Brockhaus, Müller likewise excelled. By September 1843, at age nineteen, Max Müller completed his doctoral dissertation, an examination of Book Three (*Re Affectibus*) of Spinoza's *Ethica,* a surprising research topic given Müller's remarkable aptitude for classical and Oriental languages. Spinoza's quasi-mystical "pantheistic" philosophy of Nature had been introduced into the German universities by Hegel and Schelling and seems likewise to have influenced Müller's highly romanticized views on the origins of mythology and religion, perhaps even contributing to Müller's perception of God as "the Infinite" as implied by and perceived beyond the field of the finite.[8]

One can see traces of Müller's philosophical eclecticism—combining Spinoza, Kant, Hegel, Herbart, Schopenhauer, and even Darwin—in three "pregnant" comments from his *Autobiography* that summarize his views as they had emerged from his blending of studies in classical literature, Sanskrit, and philosophy at Leipzig. "All knowledge," Müller declared, "whether individual or possessed by mankind at large, must have begun with what the senses can perceive, before it could rise to signify something unperceived by the senses. Only after the blue aether had been perceived and named, was it possible to conceive and speak of the sky as active, as an agent, as a god" (1901c: 149). "All this and much," he continued, "became perfectly intelligible, the step from the visible to the invisible, from the perceived to the conceived, from nature to nature's gods, and from nature's god to a more sublime unseen and spiritual power. All this seemed to pass before our very eyes in the Veda, and then to be reflected in Homer and Pindar" (1901c: 150). And lastly, while "[s]ome details of this restored picture of the world of gods and men in early times, nay, in the very spring of time, [as seen in Homer and the Vedas] may have to be altered," Müller expressed confidence that "nothing could curb the adventurous spirit and keep it from pushing forward and trying to do what seemed to others almost impossible [until now], namely, to watch the growth of the human mind as reflected in the petrifications of language. Language itself spoke to us with a different voice, and a formerly unsuspected meaning" (1901c: 150; cf., Müller 1868: 8–9).

Coinciding Müller's studies at Leipzig, and later at Berlin and Paris, were striking developments in linguistic and anthropological theories that came to exert a direct and defining influence over European opinions regarding the origins of race, national identity, and the evolution of human culture. These developments, which were in some ways nourished by Hegel's philosophy of history and the social application of Darwin's theory of natural selection, fed the emerging concept of "Volksgeist" (Nation Spirit), or the "Weltgeist" (World Spirit) as embodied in a people. This "fittest" of peoples was early on identified with the Aryans ("noble race"), a theoretical construction that explained for many Europeans not only the common roots of their various languages, as traced to pre-Vedic India (hence Indo-European), which they regarded as the "cradle of mankind" (Halbfass 1973: 108), but also proved to them their own cultural superiority over peoples who appeared to have developed very little beyond their primal origins.[9] We will return to this point below.

After completing his doctorate, Max Müller decided to remain at Leipzig to continue his studies in philosophy and Sanskrit. He stayed in Leipzig until early 1844, when he decided to travel to Berlin to hear Franz Bopp (1791–1867), the reputed founder of comparative philology, and to meet the "Nature" philosopher Friedrich von Schelling (1775–1854). Müller found his encounter with the famed but prematurely aged Professor Bopp somewhat disappointing. "In his lectures," Müller recalled, "he simply read his *Comparative Grammar* with a magnifying glass, and added very little that was new" (1901c: 157). On the other hand, his time with Schelling, proved to be much more stimulating. In late 1843, Schelling had announced a new set of lectures on the philosophy of mythology to be given the next year. Müller, of course, could not resist this rare opportunity to hear Schelling expound his views on myth.

In 1841, Schelling had been appointed professor of philosophy at the University of Berlin "with the mission of combating the influence of Hegelianism" (Copleston 1965: 124). Hegel and Schelling had been friendly rivals until Hegel's star began to shine more brightly than Schelling's. Now, a decade after Hegel's passing, Schelling commanded the field, or so he believed, and had come to the Prussian capital to announce the "advent of a new era" in philosophy. At Schelling's inaugural lectures, his listeners included Marx, Engels, Kierkegaard, and Bakunin. Unfortunately, this first series of lectures proved far less successful than Schelling had hoped. Audiences soon lost interest and attendance began to taper off. By 1846, Schelling had discontinued lecturing altogether, devoting himself instead to preparing his few remaining manuscripts for publication, among them his *Philosophy of Mythology* and *Philosophy of Revelation,* both published after his death (Copleston 1965: 124–125; see also Thomas 1985: 70, 77 and Hannay 1993: 2–5).

Despite the disappointing numbers, Schelling's lectures of 1844 were of great interest to Müller, who studiously attended them. Müller was at once intrigued by Schelling's views on the natural history of religion and its possible analogies with the history of language. He spent the next several months visiting Schelling to discuss his ideas further (Trompf 1978: 13–19).[10] Despite this diversion into the "positive" philosophy of Schelling, Müller continued his Sanskrit studies, borrowing textual materials from Bopp when he could. Müller's nine-month stay in Berlin saw the publication of his first book, a German translation of ancient Indian fables, known collectively as the *Hitopadesa* (1844).

The Birth of Comparative Philology

As mentioned above, Max Müller's decision to concentrate his studies in Sanskrit rather than in philosophy was aided in part by his desire to do groundbreaking work in a relatively new field. But perhaps more important, Müller was also influenced by the emerging "scientific" methods of his day, according to which a number of scholars had presented comparative evidence showing, they believed, both racial and linguistic links between Europe and India. While racial and linguistic questions were similarly cast, especially regarding the assumptions that were made in keeping with the Biblical creation and flood stories, they were not identical issues. Müller, for his part, did not dispute the traditional division of races into the postdiluvian Hamite, Shemite, and Japhethic branches (named after Noah's three sons), but when it came to language, he followed Leibniz who had doubted the reigning Adamic theory that held Hebrew as the first human language and, by implication, monotheistic belief as inherent to Adam's Semitic descendants (Poliakov 1974: 188 and Aarsleff 1982: 24–31; cf., Stone 1993 and Eco 1995). Instead, Müller accepted the tripartite division of languages into three different linguistic types, namely Aryan, Semitic, and Turanian (see Müller 1855). Though his own usage of these terms sometimes strayed from language into race—but not "race" as later defined by modern scholars— Müller focused most of his scholarly attention on examining the linguistic character and the religious and mythological elements common to the Aryan family of languages, of which Sanskrit for him became the key link.

The discovery of this key link came as an indirect consequence of eighteenth-century French and British imperialistic designs on India and Southeast Asia. Among the colonial rulers, there were both civil and military advisors, some of whom had scholarly as well as administrative interests in the peoples they had subdued.

These individuals included Sir William Jones (1746–1794), an accomplished student of languages who had pursued a career in law and had come to Calcutta in March 1783 to serve as a supreme court magistrate. This appointment afforded Jones an opportunity to study Sanskrit to "satisfy his literary interests" if not to make it impossible for "'Mohammendan or Hindu lawyers to impose upon us with erroneous opinions'" about traditional Indian law (Aarsleff 1967: 119, 121).

In January 1784, Jones founded the Asiatic Society of Bengal, and in February 1786, he presented a landmark paper to the society, entitled "On the Hindus." Though still a novice in his studies of Sanskrit, Jones's favorable comparison of Sanskrit to Greek and Latin almost immediately created a sensation in Europe. "The Sanscrit language," Jones declared, "whatever be its antiquity, is of a wonderful structure; more perfect than the Greek, more copious than the Latin, and more exquisitely refined than either, yet bearing to both of them a stronger affinity, both in the roots of verbs and in the forms of grammar, than could possibly have been produced by accident; so strong indeed, that no philologer could examine them all three, without believing them to have sprung from some common source, which, perhaps, no longer exists" (quoted in Aarsleff 1967: 133). Scholars of the period read Jones's statement as establishing strong kinship ties among these languages. And thus, as a direct consequence of his researches, the science of comparative philology was born.[11]

Among the philologists who examined this link further was Franz Bopp, perhaps the first scholar to present scientific evidence confirming Jones's assumptions. As Foucault pointed out, "Bopp's analyses were to be of major importance, not only in breaking down the internal composition of a language, but also in defining what language may be in its essence. It is no longer a system of representations which has the power to pattern and recompose other representations; it designates in its *roots* the most constant of actions, states, and wishes. ... " Further, "Language is '*rooted*' not in the things perceived, but in the active subject. And perhaps, in that case, it is a product of will and energy" (1973: 289–290, emphasis added). The "rootedness" of discreet languages in their original forms allowed linguistic analysis to trace kinship ties among language families as well as to record the earliest forms of roots and grammatical structure. The aim was to recover the original language (*die Ursprache*), if not the origins of language itself. Bopp's analyses also subtly implied racial ties not unlike those already assumed in Adamic (or postdiluvian) theories of origins. Though Bopp did not tread far down that path, it was Friedrich Schlegel (1772–1829), who, according to Léon Poliakov, gave Jones's discovery "an anthropological twist by deducing from the relationship of language a relationship of race," one that brought Romantic and nationalistic sentiments to the aid of the fledgling myth of an Aryan *race* of peoples (Poliakov 1974: 190–191; see also Rothermund 1986: 32–45).[12]

The specific work to which Poliakov is referring is Schlegel's 1808 book, *The Language and Wisdom of India* (*Über die Sprache und Weisheit der Indier*), in which Schlegel provided a careful and detailed comparative "grammar" of Sanskrit and its related European languages.[13] The science of comparative philology was furthered by brothers Jakob and Wilhelm Grimm, who showed how language and its dialects can be studied apart from its literature (see Aarsleff 1982: 84–85 and Poliakov 1974: 198). An additional thrust in the development of Jones's discovery came with the publication of Wilhelm von Humboldt's lengthy essay "The Heterogeneity of Language and Its Influence upon the Intellectual Development of Mankind," which appeared in 1836, the year after his death. In it, Humboldt advanced the notion that "the structure of each separate language grew out of and reflected the peculiar mental life of the people who made it, and that each language had therefore its own individuality which separated it from other languages" (Aarsleff 1982: 86; see Wilson 1948: 84–87 and Jespersen 1964: 40–62). It is important to point out that a number of the assumptions behind each of these various linguistic theories mentioned above found their way into the four "sciences" Max Müller attempted to further.[14] How Müller applied these theories depended largely upon his own creative insight and the textual materials that found their way into his linguistic workshop at Oxford.

Müller's Life's Work: The Vedas and Beyond

Max Müller remained in Berlin until March 1845, when he traveled to Paris to meet the famed French Sanskritist, Eugène Burnouf (1801–1852), who invited Müller to attend his lectures on the Rig Veda, the oldest and most revered of the Hindu sacred texts. By June of that year, Müller had been convinced by Burnouf of the need for a critical edition of the Sanskrit text of the Rig Veda, a long and tedious project that would span the next twenty-four years of Müller's life.[15] Because he lacked the necessary manuscripts in Paris to begin the project, in 1846 it was decided that Müller would travel to England to work with a complete set of manuscript copies. These copies were in the possession of the East India Company and were being held at its London archives. Until this time, Müller had struggled to make ends meet. But once it was decided that he would edit the Rig Veda, as well as Sayana's commentary, and that its multiple volumes would be published by the Oxford University Press, Max Müller found himself on a well-defined career path that would guarantee him the time and financial security to pursue his research further.

In London, the twenty-two-year-old Müller met with the Prussian ambassador, Baron Christian von Bunsen (1791–1860), to discuss publication of the Rig Veda and, from there, begin the next leg of his journey as a scholar. It was Bunsen who had arranged matters with the East India Company and the Oxford University Press, and it was Bunsen who would help the young Müller secure his various teaching and research positions at Oxford, at least until Müller had learned English and had established his own reputation in the field. Of that momentous occasion, Müller later recalled:

> More than twenty years have passed since my revered friend Bunsen called me one day into his library at Carlton House Terrace, and announced to me with beaming eyes that the publication of the Rig-veda was secure. ... At last his efforts had been successful ... and Bunsen was the first to announce to me the happy results of his literary diplomacy. 'Now,' he said, 'you have got a work for life—a large block that will take years to plane and polish. But mind,' he added, 'let us have from time to time some chips from your workshop.' I have tried to follow the advice of my departed friend, and I have published almost every year a few articles on such subjects as had engaged my attention, while prosecuting at the same time ... my edition of the Rig-veda, and of other Sanskrit works connected with it. (1867: vii)

In June 1847, before returning to Paris, Max Müller accompanied Bunsen on a visit to Oxford, where Müller, at Bunsen's request, presented a paper on the Bengali language. By the next year, after hurriedly returning to London from Paris in the midst of the revolutions of 1848, Müller came to settle in Oxford, mainly to see to the publication of the Rig Veda—to be completed in 1873—and to continue his pursuit of related research questions. Except for a few brief excursions to the Continent, Müller stayed at Oxford for the remainder of his life.

In 1856, Müller gained immediate public recognition with the publication of his book-length essay, "Comparative Mythology." In this, perhaps his best-known work to scholars outside the study of religion, Müller struck out in a new direction, applying linguistic analysis to the study of mythology "to account in a more intelligible manner for the creation of myths" (1909: 17). For Müller, the sun in its various phenomenal modes was the chief source of ancient myths. But he saw in myths not simply the personification of the sun, the dawn, the twilight, and so on, but a metaphysical correspondence that human thought and, by articulation, human language draws between the perception of nature and the analogues that were used to communicate what was perceived and experienced by the ancients. As Müller explained it, "There is much suffering in nature to those who have eyes for silent grief, and it is this tragedy—the tragedy of nature—which is the lifespring of all the tragedies of the ancient world" (1909: 136–137). Hence, the distinction Müller drew was the one between perception and the naming of perception, between the experience of the world and the articulation of that experience through language.

In 1858, Müller was elected a lifetime fellow of All Souls College, which, along with his stipend as deputy Taylorian professor of modern European languages, provided a sufficient income for him to marry in August 1859. But the heavy demands of family life—Müller and his wife, Georgina, produced a son and three daughters—did not slake Müller's passion for research or slow his pace. He continued to write, publish, lecture, and debate as much as his time would allow. In 1861 and 1863, after recovering from the blow of having lost the Boden Chair of Sanskrit to a less gifted colleague, Müller delivered two well-attended series of lectures on the Science of Language, both of which raised his reputation far above that of his contemporaries in the field. In 1868, Müller was appointed professor of comparative philology, a chair that Oxford University had founded on his behalf.

Though Max Müller had come to be recognized chiefly for his work in comparative philology and mythology, it was his lectures in the "science" of religion that would prove to be his most influential and provocative work. The first series, which he titled "Lectures in the Science of Religion," were given in 1870. In this series, Müller endeavored to widen the study of religion beyond its narrow theological and traditionally Christian province by arguing for a comparative study of religions. Müller's second series of lectures, "On the Origin and Growth of Religion, as Illustrated by the Religions of India," was presented in 1878 as the inaugural Hibbert Lectures. A decade later, he once more expanded his views on the evolutionary, or natural, development of religion in four sets of Gifford Lectures presented in Glasgow between 1888 and 1892. He published these lectures under the titles *Natural Religion* (1889), *Physical Religion* (1891), *Anthropological Religion* (1892), and *Theosophy or Psychological Religion* (1893) respectively.

Both at the beginning and at the conclusion of his second Gifford Lectures, Müller explained the logic behind his threefold scheme. Natural religion—not "nature religion," as his critics, such as Durkheim and Freud, mistakenly read— develops through an innate human capacity to perceive the Infinite. It "manifests itself," he maintained, "under three different aspects, according as its object, what I call the Infinite or Divine, is discovered either in *nature,* or in *man,* or in the *self*" (1891: 1). "It was in these very phenomena of nature," Müller carefully explained, "that [ancient] man perceived for the first time something that startled him out of his animal torpor, that made him ask, What is it? What does it all mean? Whence does it all come?—that forced him ... to look behind the drama of nature for actors and agents ... whom in his language he called superhuman, and, in the end, divine" (1891: 335). The Infinite exists, Müller might say, but was not perceived until human consciousness arose above that of a brute animal. This awareness therefore comes not through a special revelation from the Infinite, or God, but through ancient man's reflection upon his perceptions of the Infinite in physical nature, human nature, and in himself or herself.

Müller then carried forward his examination of ancient man's perception of the Infinite in and beyond the finite in his next three sets of lectures. Accordingly, "Physical" religion, he held, corresponds to the discovery of the Infinite through reflection on nature and to the beliefs about "God" that can be derived from human experience rather than divine revelation, which again Müller rejected as inconsistent with the evolutionary development of human cognition. Similarly, whereas "Anthropological" religion corresponds to the discovery of the Infinite through reflection on human nature, "Psychological" religion emerges from ancient man's reflection on the psyche or soul of man himself. As Peter Byrne puts it: "Each branch [of natural religion] is associated with distinct notions and therefore with different types of religious beliefs. The notion most obviously arising out of physical religion is the concept of the deity. From anthropological religion comes the concept of the soul in relation to the body and thus the belief in personal immortality. Psychological religion gives rise to the notions associated with mysticism and particularly to the concept of a union between human soul and the infinite" (1989: 188; see Müller 1893: 88–90).

Müller argued that the religions of the world—but not exclusively religions with sacred texts (1901b: 11)—can be shown historically to follow this pattern of development. But, while Müller considered Christianity to be at the highest level of development, he did not, at the same time, believe in the uniqueness of the Christian religion. In his essay "Forgotten Bibles," Müller stressed this point, perhaps making his readers a little uneasy. He wrote: "The mere lesson that we are not the only people who have a Bible, that our theologians are not the only theologians who claim for their Bible a divine inspiration, that our Church is not the only Church which has declared that those who do not hold certain doctrines cannot be saved, may have its advantages, if rightly understood"; namely, "how the comparative anatomy of those foreign religions throws light on the questions of the day, on the problems nearest to our own hearts, on our philosophy, and on our own faith" (1901b: 33, 35).

Not surprisingly, controversy over Müller's liberal religious views dogged him. Opposition to his views was indeed strong, with one critic calling his "blasphemous"

Gifford Lectures "nothing less than a crusade against Divine revelation, against Jesus Christ, and against Christianity" (1902, v. 2: 276; see Chaudhuri 1974: 362). By the 1890s, German scholarship had become stereotyped throughout Britain and America as anti-religious, or, more specifically, as anti-Christian. Such uncharitable comments as those above deeply stung Müller's liberal Christian sensibilities. Though in his second Gifford Lectures Müller sounded a more cautious and conciliatory tone, he nevertheless refused to retreat from his position but pressed further his comparative examination of the evolutionary development of religion. And, while in other settings Müller would continue to affirm his broadly Christian sentiments, at the same time he would reassert his firm belief that a comparative study of religion would not undermine the truth of the Christian gospel or a person's faith in it. All the same, as Nirad Chaudhuri points out in his biography of Müller, "It is hardly to be wondered that in the light of such opinions incessantly repeated by Max Müller, orthodox Christians would consider him both un-Christian and anti-Christian, for certainly these views disregarded, if they did not reject, the whole central kerygma of Christianity as a special dispensation of God for man" (1974: 376).

Despite the deep suspicions of English canons and divines, Max Müller was not an anti-religionist but simply styled himself a scientist examining and comparing facts in order to reconstruct a concrete and objective past. In one of his later essays, which he titled "Why I am not an Agnostic," Müller answered these charges. However, he would not relent. Convinced of the facts before him, Müller once more refused to back away from his comparative approach but used the occasion to advance a type of method-ological agnosticism. It is here that Müller, in 1894, a decade before Husserl, intro-duced his readers to the Greek term *epochê* (suspension of judgment) to define his "agnostic" approach to the study of religion, that is, "the proper attitude of the mind towards transcendental questions" that seeks to know by first admitting that it does not know (1901b: 349). "In one sense," he concluded, "I hope I am, and always have been an Agnostic, that is, in relying on nothing but historical facts and in following reason as far as it will take us in matters of the intellect, and in never pretending that conclu-sions are certain which are not demonstrated or demonstrable. This attitude of the mind has always been recognized as the *conditio sine quâ non* of all philosophy" (1901b: 355–356).[16] It is thus, with this liberal attitude of mind toward religions familiar and foreign, that Friedrich Max Müller, in his very prominent life as a scholar, enchanted, enlightened, and, at times, scandalized the literate public of the Victorian era.

Notes

1. See, for example, Lang 1885; Brown 1898; Jordan 1905; Baille 1928; Dorson 1965 and 1968; Munz 1973; de Vries 1977; Littleton 1982; Detienne 1986; Tull 1991; and Durkheim 1995.

2. Eric Sharpe offers two reasons that might account for Müller's loss of popularity among Western scholars: "Max Müller's star had begun to wane long before the end of his career. There were various reasons for this, but the most important were that, on the one hand, he was subjected to a series of severe criticisms from the Darwinian anthropologists, and from Andrew Lang in particular; and on the other, that the science of comparative philol-ogy was moving rapidly into a new phase, which left many of his linguistic assumptions behind" (1986: 46).

3. In his later works, Müller sometimes uses the word "Aryas" to distinguish the family of languages from "Aryans," whose racial referent Müller rejected, though sometimes ambiguously. As Müller himself explained in his book, *Biographies of Words and the Home of the Aryas*: "I have declared again and again that if I say Aryas, I mean neither blood nor bones, nor hair nor skull; I mean simply those who speak an Aryan language. ... To me an ethnologist who speaks of Aryan race, Aryan blood, Aryan eyes and hair, is as great a sinner as a linguist who speaks of a dolichocephalic dictionary or a brachycephalic grammar. It is worse than a Babylonian confusion of tongues—it is downright theft. We have made our terminology for the classification of languages; let ethnologists make their own for the classification of skulls, and hair, and blood" (1888: 120–121). Müller stressed this same point, though not unequivocally (Rothermund 1986: 55–56), in several of his works, including his *Autobiography* (1901c: 31–33; see also Voigt 1967: 5–8, Chaudhuri 1974: 313–315; Leopold 1974; Poliakov 1974: 213–214; Strenski 1996b: 61–63; Clarke 1997: 190–194; Alter 1999: 39–40; Bryant 2001: 22–26, 33–34; and van der Veer 2001: 135, 137–138, and 145). Müller was hailed among Indian nationalists precisely because his views were read as raising the dignity of Indian culture in Western eyes, not lowering it, as social Darwinist theories had done. As Peter van der Veer points out, "Max Müller's insistence on the faculty of language as an essential difference between humans and animals was thus an extremely significant anti-imperialist and antiracist intervention. He had already given up the idea that language was an indicator of race in the light of the Darwinian theory of biological evolution, but he continued to insist on the importance of language and linguistics" (2001: 145).
4. See also Kitagawa and Strong 1985: 198 and an unnecessarily harsh critique in Tull 1991: 42.
5. For a fuller discussion of this last point, see, for example, Kitagawa and Strong 1985: 191, 199–200; Kuper 1988: 51–54; and Byrne 1989: 183–184.
6. In his 1867 Preface to volume one of *Chips from a German Workshop,* Müller also made use of Kantian à priori categories—albeit modified to fit his use—to account for "religion" as a universal phenomenon (see Kitagawa and Strong 1985: 192–196 and van der Veer 2001: 116).
7. See, for instance, Wilson 1948: 89; Detienne 1986: 14–15, 17 and Tull 1991: 42–45.
8. See Voigt 1967: 3; for a brief discussion of the possible influence of German popular romantic culture on Müller's thought, see Strenski 1996a: 294–295 and 1996b: 60–61; I think it is a mistake to place Müller within one or another school of thought. Not only has this led to a number of mischaracterizations of his approaches to language, mythology, and religion—even among his modern supporters, such as Voigt (1967); Chaudhuri (1974); Trompf (1978); and Neufeldt (1980)—but it has also led interpreters to suppose contradictions within his writings, whether in part or as a whole. Müller himself argued that he was not committed to theoretical positions that come and go like so many intellectual *fads,* but to uncovering *facts* through which he might discover truth. Müller was much more versatile and eclectic in his method of and approach to research than most of his critics seemed to have noticed. This is not to say, however, that Müller did not approach his work without certain assumptions about the texts or traditions he studied.
9. For a detailed discussion of Aryan theories and their implications, see, for example, Ergang 1966 [1931]; Leopold 1974; Poliakov 1974; Mosse 1981; Massin 1996; Trautmann 1997; Bryant 2001; and van der Veer 2001; cf., Müller 1855, 1859, 1868, 1883, and 1888, as well as Childe 1970 [1926].
10. According to G. W. Trompf, "If Schelling's more consistent historical approach to religion 'opened up many new views' to Mueller, Schelling's treatment of *ethnos* was doubly stimulating. Combined with Bopp's lectures on the Aryan family of languages, Schelling's conclusions forced Mueller to discern a striking parallel between the histories of language and religion. If all languages had basic elements [i.e., roots or radicals], it was possible that all religions possessed them also. There was, however, a crucial point at which the

analogy broke down. If it were permissible to speak of a family of languages (the Aryan family), how feasible was it to speak of a family of religions?" (1978: 16; see also Chaudhuri 1974: 41–42).

Of Schelling's influence during this brief period, Max Müller himself rather modestly recalled: "I had talked with Schelling and Schopenhauer, and little as I appreciated or understood all their teachings, there were certain aspirations left in my mind which led me far away beyond the historical foundations of Christianity. ... I claimed for each man the liberty of believing in his own Christ ... " (1901c: 289–290). For greater detail concerning this period, see Müller's posthumously published correspondence (1902). It is possible that Schelling's distinction between "negative" and "positive" (or historical) philosophy might have inspired the later distinction Müller himself drew between what he called "theoretical" (or "theoretic") and "historical" schools of thought. Müller's "historical" approach, as distinct from what he identified in his 1884 essay, "Forgotten Bibles," as the Darwinian "theoretical" approach, is reminiscent of Schelling's criticism of Fichte's (and perhaps Hegel's) "negative" philosophy of religion as opposed, again, to his own "positive" (or historical) philosophy (see Copleston 1965: 167–170).

11. Müller himself noted in his *Autobiography:* "It is true, no doubt, that in their first enthusiasm the students of Sanskrit had uttered many exaggerated opinions. Sanskrit was represented as the mother of all languages, instead of being the elder sister of the Aryan family. The beginning of all language, of all thought, of all religion was traced back to India, and when Greek scholars were told that Zeus existed in the Veda under the name Dyaus, there was a great flutter in the dovecots [*sic*] of classical scholarship. Many of these enthusiastic utterances had afterwards to be toned down" (1901c: 148).

12. According to Léon Poliakov, Bopp preferred the term "Indo-European" family of languages to "Indo-Germanic" or, presumably, "Aryan" (1974: 193). As an additional note, Wilhelm Halbfass provides this helpful explanation for German interest in ancient Indian language and literature: "From the very beginning, the motivation of this interest in India was different from that of the earlier [European] interest in China. For the philosophers and literati of the Enlightenment, China was relevant mainly from the viewpoint of religious, political, and ethical alternatives to the Christian Occident, as an example of a more worldly, more 'humanistic' orientation. For the German Romanticists, on the other hand, India was nothing foreign or extraneous; it was a symbol of their own spiritual origin and homeland, their own forgotten depth: The 'Eternal Orient' is the profundity of their own being; India is the 'cradle of mankind' " (Halbfass 1973: 108). This last point, as well as issues related to British interest in Aryan race and language theories, is also addressed in Leopold 1974; Poliakov 1974; Rothermund 1986; Bryant 2001; and van der Veer 2001.

13. See Wilson 1948: 80–84 and Jespersen 1964: 32–40. As Müller himself recalled, "I had read Frederick Schlegel's explanatory book *Über die Sprache und Weisheit der Indier* (1808), and looked into Windischmann's *Die Philosophie im Fortgange der Weltgeschichte* (1827–1834). These books are hardly opened now—they are antiquated, and more than antiquated; they are full of mistakes as to facts, and mistakes as to the conclusions drawn from them. But they had ushered new ideas into the world of thought, and they left on many, as they did on me, that feeling which the digger who prospects for minerals is said to have, that there must be gold beneath the surface, if people would only dig" (1901c: 146). In several of his works, Müller appears to have borrowed Schlegel's comparative style (see, for instance, Müller 1909: 22–29, 41, 54–55, and 67).

14. For instance, in his famous 1878 *Lectures on the Origin and Growth of Religion*, Müller affirmed the connection between the development of religious thought and its expression in religious language, and especially the importance of India in tracing this link. As he had emphasized on many occasions, "what we can watch and study in India better than anywhere else [regarding the 'genesis and growth of religion'] is, how religious thoughts and religious language arise, how they gain force, how they spread, changing their forms

as they pass from mouth to mouth, from mind to mind, yet always retaining some faint contiguity with the spring from which they rose at first" (1879: 127).

15. According to Kitagawa and Strong, it was Burnouf "who first suggested the theory of the origins of deities in a disease or confusion of language: what was a *nomen* [name] became a *numen* [spirit], a view Müller was to argue for at great length and for which he was to become famous" (1985: 183), or rather, infamous.

16. The late anthropologist E. E. Evans-Pritchard made a similar observation: Müller, he wrote, "was not prepared to go as far as some of his more extreme German colleagues, not just because at Oxford in those days it was dangerous to be an agnostic, but from conviction ... but he got fairly near their position, and by tacking and veering in his many books to avoid it, he rendered his thought sometimes ambiguous and opaque" (1965: 21).

References and Sources

Primary Sources

Müller, F. Max. 1855. *The Languages of the Seat of War in the East* (2nd edition). London: Williams and Norgate.

——. 1859. *A History of Ancient Sanskrit Literature*. London: Williams and Norgate.

——. 1867. "Preface." Pp. iiv–xxxiii in *Chips from a German Workshop* (vol. 1: Essays on the Science of Religion). London: Longmans, Green, & Co.

——. 1868. *On the Stratification of Language*. London: Longmans, Green, Reader, and Dyer.

——. 1872. *Lectures on the Science of Religion* (with essay "Buddhist Nihilism" and translation of "The Dhammapada"). New York: Scribner, Armstrong & Co.

——. 1873a. *Introduction to the Science of Religion*. London: Longmans, Green, & Co.

——. 1873b. *Lectures on Mr. Darwin's Philosophy of Language*. London: Longmans, Green, & Co.

——. 1873c. *Lectures on the Science of Language* (2 vols.; 7th ed.; originally delivered in 1861 and 1863). London: Longmans, Green, & Co.

——. 1878. "On the Origin of Reason." *The Contemporary Review* 31 (February): 465–493.

——. 1879. *Lectures on the Origin and Growth of Religion as Illustrated by the Religions of India* (first edition, 1878; new edition, 1882). London: Longmans, Green, & Co.

——. 1881. *Selected Essays on Language, Mythology, and Religion* (2 vols.). London: Longmans, Green, & Co.

——. 1883. *India: What Can It Teach Us?* New York: John W. Lovell Co.

——. 1887a. *The Science of Thought*. London: Longmans, Green, & Co.

——. 1887b. *Three Introductory Lectures on the Science of Thought*. Chicago: The Open Court Publishing Co.

——. 1888. *Biographies of Words and the Home of the Aryas*. London: Longmans, Green, & Co.

——. 1889. *Natural Religion*. London: Longmans, Green, & Co.

——. 1891. *Physical Religion*. London: Longmans, Green, & Co.

——. 1892. *Anthropological Religion*. London: Longmans, Green, & Co.

——. 1893. *Theosophy, or Psychological Religion*. London: Longmans, Green, & Co.

——. 1894. *Three Lectures on the Vedânta Philosophy*. London: Longmans, Green, & Co.

——. 1895. *Chips from a German Workshop* (2 vols, in 1867; 5 vols. in 1881; subsequently revised and reprinted). New York: Charles Scribner's Sons.

——. 1897. *Contributions to the Science of Mythology* (2 vols.). London: Longmans, Green, & Co.

——. 1899a. *The Six Systems of Indian Philosophy*. London: Longmans, Green, & Co.

——. 1899b. *Three Lectures on the Science of Language, etc., with a Supplement, My Predecessors* (3rd ed.). Chicago: The Open Court Publishing Co.

———. 1901a. *Last Essays: First Series, Essays on the Science of Religion*. London: Longmans, Green, & Co.

———. 1901b. *Last Essays: Second Series, Essays on the Science of Religion*. London: Longmans, Green, & Co.

———. 1901c. *My Autobiography: A Fragment*. New York: Charles Scribner's Sons.

———. 1902. *The Life and Letters of the Right Honorable Friedrich Max Müller, edited by his wife* (2 vols.). London: Longmans, Green, & Co.

———. 1909. *Comparative Mythology: An Essay* (A. Smythe Palmer, ed.). London: G. Routledge and Sons; New York: E. P. Dutton and Co.

Secondary Sources

Aarsleff, Hans. 1967. *The Study of Language in England, 1780–1860*. Princeton, NJ: Princeton University Press.

———. 1982. *From Locke to Saussure: Essays on the Study of Language and Intellectual History*. Minneapolis: University of Minnesota Press.

Alter, Stephen G. 1999. *Darwinism and the Linguistic Image: Language, Race, and Natural Theology in the Nineteenth Century*. Baltimore: The Johns Hopkins University Press.

Baille, John. 1928. *The Interpretation of Religion*. New York: Charles Scribner's Sons.

Brown, Robert, Jr. 1898. *Semitic Influence in Hellenic Mythology*. London: Williams and Norgate.

Bryant, Edwin. 2001. *The Quest for the Origins of Vedic Culture: The Indo-Aryan Migration Debate*. New York: Oxford University Press.

Byrne, Peter. 1989. *Natural Religion and the Nature of Religion: The Legacy of Deism*. London and New York: Routledge.

Cassirer, Ernst. 1953. *Language and Myth*. New York: Dover Publications, Inc.

Chaudhuri, Nirad C. 1974. *Scholar Extraordinary: The Life of Professor the Rt Hon. Friedrich Max Müller*. New York: Oxford University Press.

Childe, V. Gordon. 1970 [1926]. *The Aryans: A Study of Indo-European Origins*. Port Washington, NY: Kennikat Press.

Clarke, J. J. 1997. *Oriental Enlightenment: The Encounter Between Asian and Western Thought*. New York: Routledge.

Copleston, Frederick. 1965. *A History of Philosophy* (vol. 7, pt. 1). Garden City, NY: Image Books/Doubleday.

Davidson, Donald. 1990. *Inquiries into Truth and Interpretation*. Oxford: Clarendon Press.

Detienne, Marcel. 1981. "The Myth of 'Honeyed Orpheus.'" Pp. 95–109 in *Myth, Religion, and Society*, edited by R. L. Gordon. Cambridge: Cambridge University Press.

———. 1986. *The Creation of Mythology* (Margaret Cook, trans.). Chicago: The University of Chicago Press.

de Vries, Jan. 1977. *Perspectives in the History of Religions* (Kees W. Bolle, trans.). Berkeley: University of California Press.

Dorson, Richard M. 1965. "The Eclipse of Solar Mythology." Pp. 25–63 in *Myth: A Symposium*, edited by Thomas A. Sebeok. Bloomington: Indiana University Press.

———. 1968. *The British Folklorists: A History*. London: Routledge and Kegan Paul.

Durkheim, Emile. 1995 [1912]. *The Elementary Forms of the Religious Life* (Karen E. Fields, trans.) New York: The Free Press.

Eco, Umberto. 1995. *The Search for the Perfect Language* (James Fentress, trans.). Oxford: Basil Blackwell.

Ergang, Robert R. 1966 [1931]. *Herder and the Foundations of German Nationalism*. New York: Octagon Books.

Evans-Pritchard, E. E. 1965. *Theories of Primitive Religion*. Oxford: Clarendon Press.

Feldman, Burton, and Robert D. Richardson. 1972. *The Rise of Modern Mythology, 1680–1860*. Bloomington: Indiana University Press.

Foucault, Michel. 1973. *The Order of Things: An Archaeology of Human Sciences* (a translation of *Les Mots et Les Choses*). New York: Vantage Press.

Girardot, N. J. 2002. "Max Müller's *Sacred Books* and the Nineteenth-Century Production of the Comparative Science of Religion." *History of Religions* 41 (February): 213–250.

Halbfass, Wilhelm. 1973. "Hegel on the Philosophy of the Hindus." Pp. 107–122 in *German Scholars on India: Contributions to Indian Studies* (vol. 1), edited by the Cultural Department of the Embassy of the Federal Republic of Germany. Varanasi: The Chowkhamba Sanskrit Series Office.

Hannay, Alastair. 1993. *Kierkegaard*. London: Routledge.

Jespersen, Otto. 1964 [1922]. *Language: Its Nature, Development, and Origin*. London: George Allen & Unwin.

Jordan, Louis Henry. 1905. *Comparative Religion: Its Genesis and Growth*. Edinburgh: T. & T. Clark [reprinted by Scholars Press, Atlanta, 1986].

Kitagawa, Joseph M., and John S. Strong. 1985. "Friedrich Max Müller and the Comparative Study of Religion." Pp. 179–213 in *Nineteenth Century Religious Thought in the West* (vol. 3), edited by Ninian Smart, John Clayton, Stephen Katz, and Patrick Sherry. Cambridge: Cambridge University Press.

Kissane, James. 1962. "Victorian Mythology." *Victorian Studies* 6: 5–27.

Kuper, Adam. 1988. *The Invention of Primitive Society*. London & New York: Routledge.

Lang, Andrew. 1885. *Custom and Myth*. New York: Harper & Brothers.

Leopold, Joan. 1974. "British Applications of the Aryan Theory of Race to India, 1850–1870." *The English Historical Review* 89 (July): 578–603.

Littleton, C. Scott. 1982. *The New Comparative Mythology* (3rd edition). Berkeley: University of California Press.

Massin, Benoit. 1996. "From Virchow to Fischer: Physical Anthropology and 'Modern Race Theories' in Wilhelmine Germany." Pp. 79–154 in *"Volkgeist" as Method and Ethic* (History of Anthropology, vol. 8), edited by George W. Stocking, Jr. Madison: University of Wisconsin Press.

Masuzawa, Tomoko. 1993. *In Search of Dreamtime: The Quest for the Origin of Religion*. Chicago: The University of Chicago Press.

Morris, Brian. 1987. *Anthropological Studies of Religion: An Introductory Text*. Cambridge: Cambridge University Press.

Mosse, George L. 1981. *The Crisis of German Ideology: Intellectual Origins of the Third Reich*. New York: Howard Fertig.

Munz, Peter. 1973. *When the Golden Bough Breaks*. London: Routledge & Kegan Paul.

Noire, Ludwig. 1879. *Max Müller and the Philosophy of Language*. London: Longmans, Green, & Co.

Poliakov, Léon. 1974. *The Aryan Myth: A History of Racist and Nationalist Ideas in Europe* (Edmund Howard, trans.). New York: Basic Books.

Preus, J. Samuel. 1987. *Explaining Religion: Criticism and Theory from Bodin to Freud*. New Haven: Yale University Press.

Rothermund, Dietmar. 1986. *The German Intellectual Quest for India*. New Delhi: Manohar.

Stephen, Leslie and Sidney Lee (eds.). 1917. *The Dictionary of National Biography* (vol. 22, Supplement). Oxford: Oxford University Press.

Stocking, George W., Jr. 1987. *Victorian Anthropology*. New York: The Free Press.

Stone, Jon R. 1993. "The Medieval Mappaemundi: Toward an Archaeology of Sacred Cartography." *Religion* 23 (July): 197–216.

—— (ed.). 1998. *The Craft of Religious Studies*. New York: St. Martin's Press.

Strenski, Ivan. 1987. *Four Theories of Myth in Twentieth-Century History*. Iowa City: University of Iowa Press.

——. 1996a. "Misreading Max Müller." *Method & Theory in the Study of Religion* 8(3): 291–296.

——. 1996b. "The Rise of Ritual and the Hegemony of Myth: Sylvain Lévi, the Durkheimians, and Max Müller." Pp. 52–81 in *Myth and Method,* edited by Laurie L. Patton and Wendy Doniger. Charlottesville: University of Virginia Press.

Thomas, J. Heywood. 1985. "Fichte and Schelling." Pp. 41–79 in *Nineteenth Century Religious Thought in the West* (vol. 1), edited by Ninian Smart, John Clayton, Stephen Katz, and Patrick Sherry. Cambridge: Cambridge University Press.

Trautmann, Thomas R. 1997. *Aryans and British India*. Berkeley: University of California Press.

Trompf, G. W. 1978. *Friedrich Max Müller: As a Theorist of Comparative Religion*. Bombay: Shakuntala Publishing House.

Tull, Herman. 1991. "F. Max Müller and A. B. Keith: 'Twaddle', the 'Stupid' Myth, and the Disease of Indology." *Numen* 38 (June): 27–58.

van der Veer, Peter. 2001. *Imperial Encounters: Religion and Modernity in India and Britain*. Princeton: Princeton University Press.

Voigt, Johannes H. 1967. *Max Mueller: The Man and His Ideas*. Calcutta: Firma K. L. Mukhopadhyay.

Waardenburg, J. J. (ed.). 1973. *Classical Approaches to the Study of Religion* (vol. 1). The Hague: Mouton.

Wiebe, Donald. 1999. *The Politics of Religious Studies*. New York: St. Martin's Press.

Wilson, Robert A. 1948. *The Miraculous Birth of Language*. New York: Philosophical Library.

Chapter 1
Semitic Monotheism[1] (1860)

A work such as M. Renan's *Histoire Générale et Système Comparé des Langues Sémitiques* can only be reviewed chapter by chapter. It contains a survey not only, as its title would lead us to suppose, of the Semitic languages, but of the Semitic languages and nations; and considering that the whole history of the civilised world has hitherto been acted by two races only, the Semitic and the Aryan, with occasional interruptions produced by the inroads of the Turanian race, M. Renan's work comprehends in reality half of the history of the ancient world. We have received as yet the first volume only of this important work, and before the author had time to finish the second, he was called upon to publish a second edition of the first, which appeared in 1858, with important additions and alterations.

In writing the history of the Semitic race it is necessary to lay down certain general characteristics common to all the members of that race, before we can speak of nations so widely separated from each other as the Jews, the Babylonians, Phœnicians, Carthaginians, and Arabs, as one race or family. The most important bond that binds these scattered tribes together into one ideal whole is to be found in their language. There can be as little doubt that the dialects of all the Semitic nations are derived from one common type as there is about the derivation of French, Spanish, and Italian from Latin, or of Latin, Greek, German, Celtic, Slavonic, and Sanskrit from the primitive idiom of the ancestors of the Aryan race. The evidence of language would by itself be quite sufficient to establish the fact that the Semitic nations descended from common ancestors, and constitute what, in the science of language, may be called a distinct race. But M. Renan was not satisfied with the single criterion of the relationship of the Semitic tribes, and he has endeavoured to draw, partly from his own observations, partly from the suggestions of other scholars, such as Ewald and Lassen, a more complete portrait of the Semitic man. This was no easy task. It was like drawing the portrait of a whole family, omitting all that is peculiar to each individual member, and yet preserving the features that constitute the general family likeness. The result has been what might be expected. Critics most familiar with one or the other branch of the Semitic family have each and all protested that they can see no likeness in the portrait. It seems to some to contain features which it ought not to contain, whereas others miss the very expression which appears to them most striking.

The following is a short abstract of what M. Renan considers the salient points in the Semitic character: "Their character," he says,

> is religious rather than political, and the mainspring of their religion is the conception of the unity of God. Their religious phraseology is simple, and free from mythological

elements. Their religious feelings are strong, exclusive, intolerant, and sustained by a fervour which finds its peculiar expression in prophetic visions. Compared to the Aryan nations, they are found deficient in scientific and philosophical originality. Their poetry is chiefly subjective or lyrical, and we look in vain among their poets for excellence in epic and dramatic compositions. Painting and the plastic arts have never arrived at a higher than the decorative stage. Their political life has remained patriarchal and despotic, and their inability to organise on a large scale has deprived them of the means of military success. Perhaps the most general feature of their character is a negative one—their inability to perceive the general and the abstract, whether in thought, language, religion, poetry, or politics; and, on the other hand, a strong attraction towards the individual and personal, which makes them monotheistic in religion, lyrical in poetry, monarchical in politics, abrupt in style, and useless for speculation.

One cannot look at this bold and rapid outline of the Semitic character without perceiving how many points it contains which are open to doubt and discussion. We shall confine our remarks to one point, which, in our mind, and, as far as we can see, in M. Renan's mind likewise, is the most important of all, namely, the supposed monotheistic tendency of the Semitic race. M. Renan asserts that this tendency belongs to the race by instinct—that it forms the rule, not the exception; and he seems to imply that without it the human race would never have arrived at the knowledge or worship of the One God.

If such a remark had been made fifty years ago, it would have roused little or no opposition. "Semitic" was then used in a more restricted sense, and hardly comprehended more than the Jews and Arabs. Of this small group of people it might well have been said, with such limitations as are tacitly implied in every general proposition on the character of individuals or nations, that the work set apart for them by a Divine providence in the history of the world was the preaching of a belief in one God. Three religions have been founded by members of that more circumscribed Semitic family—the Jewish, the Christian, the Mohammedan; and all three proclaim, with the strongest accent, the doctrine that there is but one God.

Of late, however, not only have the limits of the Semitic family been considerably extended, so as to embrace several nations notorious for their idolatrous worship, but the history of the Jewish and Arab tribes has been explored so much more fully, that even there traces of a wide-spreading tendency to polytheism have come to light.

The Semitic family is divided by M. Renan into two great branches, differing from each other in the form of their monotheistic belief, yet both, according to their historian, imbued from the beginning with the instinctive faith in one God:

1. The nomad branch, consisting of Arabs, Hebrews, and the neighbouring tribes of Palestine, commonly called the descendants of Terah; and
2. The political branch, including the nations of Phœnicia, of Syria, Mesopotamia, and Yemen.

Can it be said that all these nations, comprising the worshippers of Elohim, Jehovah, Sabaoth, Moloch, Nisroch, Rimmon, Nebo, Dagon, Ashtaroth, Baal or Bel, Baal-peor, Baal-zebub, Chemosh, Milcom, Adrammelech, Annamelech, Nibhaz and Tartak, Ashima, Nergal, Succoth-benoth, the Sun, Moon, planets, and all the host of heaven, were endowed with a monotheistic instinct? M. Renan admits that monotheism has always had its principal bulwark in the nomadic branch, but he maintains that it has by no means been so unknown among the members of the political branch as is commonly supposed. But where are the criteria by which, in the same manner as their dialects, the religions of the Semitic races could be distinguished from the religions of the Aryan and Turanian races? We can recognise any Semitic dialect by the triliteral character of its roots. Is it possible to discover similar radical elements in all the forms of faith, primary or secondary, primitive or derivative, of the Semitic tribes? M. Renan thinks that it is. He imagines that he hears the key-note of a pure monotheism through all the wild shoutings of the priests of Baal and other Semitic idols, and he denies the presence of that key-note in any of the religious systems of the Aryan nations, whether Greeks or Romans, Germans or Celts, Hindus or Persians. Such an assertion could not but rouse considerable opposition, and so strong seems to have been the remonstrances addressed to M. Renan by several of his colleagues in the French Institute that without awaiting the publication of the second volume of his great work, he has thought it right to publish part of it as a separate pamphlet. In his *Nouvelles Considérations sur le Caractère Général des Peuples Sémitiques, et en particulier sur leur Tendance au Monothéisme,* he endeavours to silence the objections raised against the leading idea of his history of the Semitic race. It is an essay which exhibits not only the comprehensive knowledge of the scholar, but the warmth and alacrity of the advocate. With M. Renan the monotheistic character of the descendants of Shem is not only a scientific tenet, but a moral conviction. He wishes that his whole work should stand or fall with this thesis, and it becomes, therefore, all the more the duty of the critic, to inquire whether the arguments which he brings forward in support of his favourite idea are valid or not.

It is but fair to M. Renan that, in examining his statements, we should pay particular attention to any slight modifications which he may himself have adopted in his last memoir. In his history he asserts with great confidence, and somewhat broadly, that "le monothéisme résume et explique tous les caractères de la race Sémitique." In his later pamphlet he is more cautious. As an experienced pleader he is ready to make many concessions in order to gain all the more readily our assent to his general proposition. He points out himself with great candour the weaker points of his argument, though, of course, only in order to return with unabated courage to his first position: that of all the races of mankind the Semitic race alone was endowed with the instinct of monotheism. As it is impossible to deny the fact that the Semitic nations, in spite of this supposed monotheistic instinct, were frequently addicted to the most degraded forms of a polytheistic idolatry, and that even the Jews, the most monotheistic of all, frequently provoked the anger of the Lord by burning incense to other gods, M. Renan remarks that when he speaks of a nation in general he only speaks of the intellectual aristocracy of that nation. He appeals in self-defense to the manner in which historians lay down the character of modern nations. "The French," he says, "are repeatedly called '*une nation spirituelle,*' and yet no one would wish to assert either that every Frenchman is *spirituel,* or that no one could be *spirituel* who is not a Frenchman." Now, here we may grant to M. Renan that if we speak of "*esprit*" we naturally think of the intellectual minority only, and not of the whole bulk of a nation; but if we speak of religion, the case is different. If we say that the

French believe in one God only, or that they are Christians, we speak not only of the intellectual aristocracy of France but of every man, woman, and child born and bred in France. Even if we say that the French are Roman Catholics, we do so only because we know that there is a decided majority in France in favour of the unreformed system of Christianity. But if, because some of the most distinguished writers of France have paraded their contempt for all religious dogmas, we were to say broadly that the French are a nation without religion, we should justly be called to order for abusing the legitimate privileges of generalisation. The fact that Abraham, Moses, Elijah, and Jeremiah were firm believers in one God could not be considered sufficient to support the general proposition that the Jewish nation was monotheistic by instinct. And if we remember that among the other Semitic races we should look in vain for even four such names, the case would seem to be desperate to any one but M. Renan.

We cannot believe that M. Renan would be satisfied with the admission that there had been among the Jews a few leading men who believed in one God, or that the existence of but one God was an article of faith not quite unknown among the other Semitic races; yet he has hardly proved more. He has collected, with great learning and ingenuity, all traces of monotheism in the annals of the Semitic nations; but he has taken no pains to discover the traces of polytheism, whether faint or distant, which are disclosed in the same annals. In acting the part of an advocate he has for a time divested himself of the nobler character of the historian.

If M. Renan had looked with equal zeal for the scattered vestiges both of a monotheistic and of a polytheistic worship, he would have drawn, perhaps, a less striking, but we believe a more faithful portrait of the Semitic man. We may accept all the facts of M. Renan, for his facts are almost always to be trusted; but we cannot accept his conclusions, because they would be in contradiction to other facts which M. Renan places too much in the background, or ignores altogether. Besides there is something in the very conclusions to which he is driven by his too partial evidence which jars on our ears, and betrays a want of harmony in the premises on which he builds. Taking his stand on the fact that the Jewish race was the first of all the nations of the world to arrive at the knowledge of one God, M. Renan proceeds to argue that, if their monotheism had been the result of a persevering mental effort—if it had been a discovery like the philosophical or scientific discoveries of the Greeks—it would be necessary to admit that the Jews surpassed all other nations of the world in intellect and vigour of speculation. This, he admits, is contrary to fact:

Apart la supériorité de son culte, le peuple juif n'en a aucune autre; c'est un des peuples les moins doués pour la science et la philosophie parmi les peuples de l'antiquité; il n'a une grande position ni politique ni militaire. Ses institutions sont purement conservatrices; les prophètes, qui représentent excellemment son génie, sont des hommes essentiellement réactionnaires, se reportant toujours vers un idéal antérieur. Comment expliquer, au sein d'une société aussi étroite et aussi peu développée, une révolution d'idées qu'Athènes et Alexandrie n'ont pas réussi à accomplir?[2]

M. Renan then defines the monotheism of the Jews, and of the Semitic nations in general, as the result of a low, rather than of a high state of intellectual cultivation:

"Il s'en faut," he writes (p. 40), "que le monothéisme soit le produit d'une race qui a des idées exaltées en fait de religion; c'est en réalité le fruit d'une race qui a peu de besoins religieux. C'est comme *minimum* de religion, en fait de dogmes et en fait de pratiques extérieures, que le monothéisme est surtout accommodé aux besoins des populations nomades."[3]

But even this *minimum* of religious reflection, which is required, according to M. Renan, for the perception of the unity of God, he grudges to the Semitic nations, and he is driven in the end (p. 73) to explain the Semitic Monotheism as the result of a *religious instinct*, analogous to the instinct that led each race to the formation of its own language.

Here we miss the clearness and precision which distinguish most of M. Renan's works. It is always dangerous to transfer expressions from one branch of knowledge to another. The word "instinct" has its legitimate application in natural history, where it is used of the unconscious acts of unconscious beings. We say that birds build their nests by instinct, that fishes swim by instinct, that cats catch mice by instinct; and, though no natural philosopher has yet explained what instinct is, yet we accept the term as a conventional expression for an unknown power working in the animal world.

If we transfer this word to the unconscious acts of conscious beings, we must necessarily alter its definition. We may speak of an instinctive motion of the arm, but we only mean a motion which has become so habitual as to require no longer any special effort of the will.

If, however, we transfer the word to the conscious thoughts of conscious beings, we strain the word beyond its natural capacities, we use it in order to avoid other terms which would commit us to the admission either of innate ideas or inspired truths. We use a word in order to avoid a definition. It may sound more scientific to speak of a monotheistic instinct rather than of the inborn image or the revealed truth of the One living God; but is instinct less mysterious than revelation? Can there be an instinct without an instigation or an instigator? And whose hand was it that instigated the Semitic mind to the worship of one God? Could the same hand have instigated the Aryan mind to the worship of many gods? Could the monotheistic instinct of the Semitic race, if an instinct, have been so frequently obscured, or the polytheistic instinct of the Aryan race, if an instinct, so completely annihilated, as to allow the Jews to worship on all the high places round Jerusalem, and the Greeks and Romans to become believers in Christ? Fishes never fly, and cats never catch frogs. These are the difficulties into which we are led; and they arise simply and solely from our using words for their sound rather than for their meaning. We begin by playing with words, but in the end the words will play with us.

There are, in fact, various kinds of monotheism, and it becomes our duty to examine more carefully what they mean and how they arise. There is one kind of monotheism, though it would more properly be called theism, or henotheism, which forms the birthright of every human being. What distinguishes man from all other creatures, and not only raises him above the animal world, but removes him altogether from the confines of a merely natural existence, is the feeling of sonship inherent in and inseparable from human nature. That feeling may find expression in a thousand ways, but there breathes through all of them the inextinguishable conviction, "It is He that hath made us, and not we ourselves." That feeling of sonship may with some races manifest itself in fear and trembling, and it may drive whole generations into religious madness and devil worship. In other countries it may tempt the creature into a fatal familiarity with the Creator, and end in an apotheosis of man, or a headlong plunging of the human into the divine. It may take, as with the Jews, the form of a simple assertion that "Adam was the son of God," or it may be clothed in the mythological phraseology of the Hindus, that Manu, or man, was the descendant of Svayambhu, the Self-existing. But, in some form or other, the feeling of dependence on a higher Power breaks through in all the religions of the world, and explains to us the meaning of St. Paul, "that God, though in times past He suffered all nations to walk in their own ways, nevertheless He left not Himself without witness, in that He did good and gave us rain from heaven, and fruitful seasons filling our hearts with food and gladness."

This primitive intuition of God and the ineradicable feeling of dependence on God could only have been the result of a primitive revelation, in the truest sense of that word. Man, who owed his existence to God, and whose being centred and rested in God, saw and felt God as the only source of his own and of all other existence. By the very act of the creation, God had revealed Himself. There He was, manifested in His works, in all His majesty and power, before the face of those to whom He had given eyes to see and ears to hear, and into whose nostrils He had breathed the breath of life, even the Spirit of God.

This primitive intuition of God, however, was in itself neither monotheistic nor polytheistic, though it might become either, according to the expression which it took in the languages of man. It was this primitive intuition that supplied either the subject or the predicate in all the religions of the world, and without it no religion, whether true or false, whether revealed or natural, could have had even its first beginning. It is too often forgotten by those who believe that a polytheistic worship was the most natural unfolding of religious life, that polytheism must everywhere have been preceded by a more or less conscious theism. In no language does the plural exist before the singular. No human mind could have conceived the idea of gods without having previously conceived the idea of a god. It would be, however, quite as great a mistake to imagine, because the idea of a god must exist previously to that of gods, that therefore a belief in One God preceded everywhere the belief in many gods. A belief in God as exclusively One involves a distinct negation of more than one God, and that negation is possible only after the conception, whether real or imaginary, of many gods.

The primitive intuition of the Godhead is neither monotheistic nor polytheistic, and it finds its most natural expression in the simplest and yet the most important article of faith—that God is God. This must have been the faith of the ancestors of mankind previously to any division of race or confusion of tongues. It might seem, indeed, as if in such a faith the oneness of God, though not expressly asserted, was implied, and that it existed, though latent, in the first revelation of God. History, however, proves that the question of oneness was yet undecided in that primitive faith, and that the intuition of God was not yet secured against the illusions of a double vision. There are, in reality, two kinds of oneness which, when we enter into metaphysical discussions, must be carefully distinguished, and which for practical purposes are well kept separate by the definite and indefinite articles. There is one kind of oneness which does not exclude the idea of plurality; there is another which does. When we say that Cromwell was a Protector of England, we do not assert that he was the only protector. But if we say that he was the Protector of England, it is understood that he was the only man who enjoyed that title. If, therefore, an expression had been given to that primitive intuition of the Deity, which is the mainspring of all later religion, it would have been—"There is a God," but not yet "There is but 'One God.'" The latter form of faith, the belief in One God, is properly called monotheism, whereas the term of henotheism would best express the faith in a single God.

We must bear in mind that we are here speaking of a period in the history of mankind when, together with the awakening of ideas, the first attempts only were being made at expressing the simplest conceptions by means of a language most simple, most sensuous, and most unwieldy. There was as yet no word sufficiently reduced by the wear and tear of thought to serve as an adequate expression for the abstract idea of an immaterial and supernatural Being. There were words for walking and shouting, for cutting and burning, for dog and cow, for house and wall, for sun and moon, for day and night. Every object was called by some quality which had struck the eye as most peculiar and characteristic. But what quality should be predicated of that Being of which man knew as yet nothing but its existence? Language possessed as yet no auxiliary verbs. The very idea of being, without the attributes of quality or action, had never entered into the human mind. How then was that Being to be called which had revealed its existence, and continued to make itself felt by everything that most powerfully impressed the awakening mind, but which as was yet known only like a subterraneous spring by the waters which it poured forth with inexhaustible strength? When storm and lightning drove a father with his helpless family to seek refuge in the forests, and the fall of mighty trees crushed at his side those who were most dear to him, there were, no doubt, feelings of terror and awe, of helplessness and dependence, in the human heart which burst forth in a shriek for pity or help from the only Being that could command the storm. But there was no name by which He could be called. There might be names for the storm-wind and the thunderbolt, but these were not the names applicable to Him that rideth upon the heaven of heavens, which were of old. Again, when after a wild and tearful night the sun dawned in the morning, smiling on man—when after a dreary and death-like winter spring came again with its sunshine and flowers, there were feelings of joy and gratitude, of love and adoration in the heart of every human being, but though there were names for the sun and the spring, for the bright sky and the brilliant dawn, there was no word by which to call the source of all this gladness, the giver of light and life.

At the time when we may suppose that the first attempts at finding a name for God were made, the divergence of the languages of mankind had commenced. We cannot dwell here on the causes which led to the multiplicity of human speech; but whether we look on the confusion of tongues as a natural or supernatural event, it was an event which the science of language has proved to have been inevitable. The ancestors of the Semitic and the Aryan nations had long become unintelligible to each other in their conversations on the most ordinary topics, when they each in their own way began to look for a proper name for God. Now one of the most striking differences between the Aryan and the Semitic forms of speech was this: In the

Semitic languages the roots expressive of the predicates that were to serve as the proper names of any subjects, remained so distinct within the body of a word, that those who used the word were unable to forget its predicative meaning, and retained in most cases a distinct consciousness of its appellative power. In the Aryan languages, on the contrary, the significative element, or the root of a word, was apt to become so completely absorbed by the derivative elements, whether prefixes or suffixes, that most substantives ceased almost immediately to be appellative, and were changed into mere names or proper names. What we mean can best be illustrated by the fact that the dictionaries of Semitic languages are mostly arranged according to their roots. When we wish to find the meaning of a word in Hebrew or Arabic we first look for its root, whether triliteral or biliteral, and then look in the dictionary for that root and its derivatives. In the Aryan languages, on the contrary, such an arrangement would be extremely inconvenient. In many words it is impossible to detect the radical element. In others after the root is discovered, we find that it has not given birth to any other derivatives which would throw their converging rays of light on its radical meaning. In other cases, again, such seems to have been the boldness of the original name-giver that we can hardly enter into the idiosyncrasy which assigned such a name to such an object.

This peculiarity of the Semitic and Aryan languages must have had the greatest influence on the formation of their religious phraseology. The Semitic man would call on God in adjectives only, or in words which always conveyed a predicative meaning. Every one of his words was more or less predicative, and he was therefore restricted in his choice to such words as expressed some one or other of the abstract qualities of the Deity. The Aryan man was less fettered in his choice. Let us take an instance. Being startled by the sound of thunder, he would at first express his impression by the single phrase, "It thunders,"—βροντᾷ. Here the idea of God is understood rather than expressed, very much in the same manner as the Semitic proper names "Zabd" (present), "Abd" (servant), "Aus" (present) are habitually used for "Abd-allah," "Zabd-allah," "Aus-allah,"—the servant of God, the gift of God. It would be more in accordance with the feelings and thoughts of those who first used these so-called impersonal verbs to translate them by "He thunders," "He rains," "He snows." Afterward, instead of the simple impersonal verb "he thunders," another expression naturally suggested itself. The thunder came from the sky, the sky was frequently called "Dyaus" (the bright one), in Greek Ζεὺς; and though it was not the bright sky which thundered, but the dark, yet "Dyaus" had already ceased to be an expressive predicate; it had become a traditional name, and hence there was nothing to prevent an Aryan man from saying "Dyaus," or "the sky thunders," in Greek Ζεὺς βροντᾷ. Let us here mark the almost irresistible influence of language on the mind. The word "Dyaus," which at first meant "bright," had lost its radical meaning, and now meant simply "sky." It then entered into a new stage. The idea which had first been expressed by the pronoun or the termination of the third person, "He thunders," was taken up into the word "Dyaus," or "sky." "He thunders," and "Dyaus thunders," became synonymous expressions; and by the mere habit of speech "He" became "Dyaus," and "Dyaus" became "He." Henceforth "Dyaus" remained as an apellative of that unseen though ever present Power, which had revealed its existence to man from the beginning, but which remained without a name long after every beast of the field and every fowl of the air had been named by Adam.

Now, what happened in this instance with the name of "Dyaus" happened again and again with other names. When men felt the presence of God in the great and strong wind, in the earthquake, or the fire, they said at first, "He storms," "He shakes," "He burns." But they likewise said, the storm ("Marut") blows, the fire ("Agni") burns, the subterraneous fire ("Vulcanus") upheaves the earth. And after a time the result was the same as before, and the words meaning originally wind or fire were used, under certain restrictions, as names of the unknown God. As long as all these names were remembered as mere names or attributes of one and the same Divine Power, there was as yet no polytheism, though, no doubt, every new name threatened to obscure more and more the primitive intuition of God. At first, the names of God, like fetishes or statues, were honest attempts at expressing or representing an idea which could never find an adequate expression or representation. But the *eidôlon*, or likeness, became an *idol;* the *nomen*, or name, lapsed into a *numen*, or demon, as soon as they were drawn away from their original intention. If the Greeks had remembered that Zeus was but a name or symbol of the Deity, there would have been no more harm in calling God by that name than by any other. If they had remembered that Kronos, and Uranos, and Apollon were all but so many attempts at naming the various sides, or manifestations, or aspects, or persons of the Deity, they might have used these names in the hours of their various needs, just as the Jews called on Jehovah, Elohim, and Sabaoth, or as Roman Catholics implore the help of Nunziata, Dolores, and Notre-Dame-de-Grace.

What, then, is the difference between the Aryan and Semitic nomenclature for the Deity? Why are we told that the pious invocations of the Aryan world turned into a blasphemous mocking of the Deity, whereas the Semitic nations are supposed to have found from the first the true name of God? Before we look anywhere else for an answer to the question, we must look to language itself, and here we see that the Semitic dialects could never, by any possibility, have produced such names as the Sanskrit "Dyaus" (Zeus), "Varuna" (Uranos), "Marut" (Storm, Mars), or "Ushas" (Eos). They had no doubt names for the bright sky, for the tent of heaven, and for the dawn. But these names were so distinctly felt as appellatives, that they could never be thought of as proper names, whether as names of the Deity, or as names of deities. This peculiarity has been illustrated with great skill by M. Renan. We differ from him when he tries to explain the difference between the mythological phraseology of the Aryan and the theological phraseology of the Semitic races, by assigning to each a peculiar theological instinct. We cannot, in fact, see how the admission of such an instinct, i.e., of an unknown and incomprehensible power, helps us in any way whatsoever to comprehend this curious mental process. His problem, however, is exactly the same as ours, and it would be impossible to state that problem in a more telling manner than he has done.

"The rain," he says (p. 79), "is represented, in all the primitive mythologies of the Aryan race, as the fruit of the embraces of Heaven and Earth." "The bright sky," says Æschylus, in a passage which one might suppose was taken from the Vedas, "loves to penetrate the earth; the earth on her part aspires to the heavenly marriage.

Rain falling from the loving sky impregnates the earth, and she produces for mortals pastures of the flocks and the gifts of Ceres." In the Book of Job,[4] on the contrary, it is God who tears open the waterskins of heaven (38.37) who opens the courses for the floods (38.25), who engenders the drops of dew (38.28):

> He draws towards Him the mists from the waters,
> Which pour down as rain, and form their vapours.
> Afterwards the clouds spread them out,
> They fall as drops on the crowds of men (Job 36.27, 28)

> He charges the night with damp vapours,
> He drives before Him the thunder-bearing cloud.
> It is driven to one side or the other by His command.
> To execute all that He ordains
> On the face of the universe,
> Whether it be to punish His creatures
> Or to make thereof a proof of his mercy (Job 37.11–13)

Or, again, Proverbs 30.4:

> Who hath gathered the wind in His fists? Who hath bound the waters in a garment? Who hath established all the ends of the earth? What is His name, and what is His Son's name, if thou canst tell?

It has been shown by ample evidence from the Rig-veda how many myths were suggested to the Aryan world by various names of the dawn, the day-spring of life. The language of the ancient Aryans of India had thrown out many names for that heavenly apparition, and every name, as it ceased to be understood, became, like a decaying seed, the germ of an abundant growth of myth and legend. Why should not the same have happened to the Semitic names for the dawn? Simply and solely because the Semitic words had no tendency to phonetic corruption; simply and solely because they continued to be felt as appellatives, and would inevitably have defeated every attempt at mythological phraseology such as we find in India and Greece. When the dawn is mentioned in the Book of Job (9.7), it is God "who commandeth the sun and it riseth not, and sealeth up the stars." It is His power which causeth the day-spring to know its place, that it might take hold of the ends of the earth, that the wicked might be shaken out of it (Job 38.12,13; Renan, *Livre de Job*, pref. 71). "Shahar," the dawn, never becomes an independent agent; she is never spoken of as Eos rising from the bed of her husband Tithonos (the setting sun), solely and simply because the word retained its power as an appellative, and thus could not enter into any mythological metamorphosis.

Even in Greece there are certain words which have remained so pellucid as to prove unfit for mythological refraction. "Selene" in Greek is so clearly the moon that her name would pierce through the darkest clouds of myth and fable. Call her "Hecate," and she will bear any disguise, however fanciful. It is the same with the Latin "Luna." She is too clearly the moon to be mistaken for anything else; but call her "Lucina," and she will readily enter into various mythological phases. If, then, the names of sun and moon, of thunder and lightning, of light and day, of night and dawn, could not yield to the Semitic races fit appellatives for the Deity, where were they to be found? If the names of Heaven or Earth jarred on their ears as names unfit for the Creator, where could they find more appropriate terms? They would not have objected to real names such as "Jupiter Optimus Maximus," or Ζεὺς κύδιστος μέγιστος, if such words could have been framed in their dialects, and the names of Jupiter and Zeus could have been so ground down as to become synonymous with the general term for "God." Not even the Jews could have given a more exalted definition of the Deity than that of *Optimus Maximus*—the Best and the Greatest; and their very name of God, "Jehovah," is generally supposed to mean no more than what the Peleiades of Dodona said of Zeus: Ζεὺς ἦν, Ζεὺς ἐστίν, Ζεὺς ἔσσεται, ὦ μεγάλε Ζεῦ, "He was, He is, He will be, O great Zeus!" Not being able to form such substantives as Dyaus or Varuna or Indra, the descendants of Shem fixed on the predicates which in the Aryan prayers follow the name of the Deity, and called Him the Best and the Greatest, the Lord and King. If we examine the numerous names of the Deity in the Semitic dialects we find that they are all adjectives, expressive of moral qualities. There is "El," strong; "Bel" or "Baal," Lord; "Beel-samin," Lord of Heaven; "Adonis" (in Phœnicia), Lord; "Marnas" (at Gaza), our Lord; "Shet," Master, afterwards a demon; "Moloch," "Milcom," "Malika," King; "Eliun," the Highest (the God of Melchisedek); "Ram" and "Rimmon," the Exalted; and many more names, all originally adjectives and expressive of certain general qualities of the Deity, but all raised by one or the other of the Semitic tribes to be the names of God or of that idea which the first breath of life, the first sight of this world, the first consciousness of existence, had forever impressed and implanted in the human mind.

But do these names prove that the people who invented them had a clear and settled idea of the unity of the Deity? Do we not find among the Aryan nations that the same superlatives, the same names of Lord and King, of Master and Father, are used when the human mind is brought face to face with the Divine, and the human heart pours out in prayer and thanksgiving the feelings inspired by the presence of God? "Brahman," in Sanskrit, meant originally Power, the same as El. It resisted for a long time the mythological contagion, but at last it yielded like all other names of God, and became the name of one God. By the first man who formed or fixed these names, "Brahman," like El, and like every name of God, was meant, no doubt, as the best expression that could be found for the image reflected from the Creator upon the mind of the creature. But in none of these words can we see any decided proof that those who framed them had arrived at the clear perception of One God, and were thus secured against the danger of polytheism. Like Dyaus, like Indra, like Brahman, Baal and El and Moloch were names of God, but not yet of the One God.

And we have only to follow the history of these Semitic names in order to see that, in spite of their superlative meaning, they proved no stronger bulwarks against polytheism than the Latin *Optimus Maximus*. The very names which we saw explained before as meaning the Highest, the Lord, the Master, are represented in the Phœnician mythology as standing to each other in the relation of Father and Son (Renan, p. 60). There is hardly one single Semitic tribe which did not at times forget the original meaning of the names by which they called on God. If the Jews had remembered the meaning of El, the Omnipotent, they could not have worshipped Baal, the Lord, as different from El. But as the Aryan tribes bartered the names of their gods, and were glad to add the worship of Zeus to that of Uranos, the worship of Apollon to that of Zeus, the worship of Hermes to that of Apollon, the Semitic nations likewise were ready to try the gods of their neighbours. If there had been in the Semitic race a truly monotheistic instinct, the history of those nations would become perfectly unintelligible. Nothing is more difficult to overcome than an instinct; "naturam expellas furcâ, tamen usque recurret."[5]

But the history even of the Jews is made up of an almost uninterrupted series of relapses into polytheism. Let us admit, on the contrary, that God had in the beginning revealed Himself as the same to the ancestors of the whole human race. Let us then observe the natural divergence of the languages of man, and consider the peculiar difficulties that had to be overcome in framing names for God, and the peculiar manner in which they were overcome in the Semitic and Aryan languages, and everything that follows will be intelligible. If we consider the abundance of synonymes [sic] into which all ancient languages burst out at their first starting—if we remember that there were hundreds of names for the earth and the sky, the sun and the moon—we shall not be surprised at meeting with more than one name for God both among the Semitic and the Aryan nations. If we consider how easily the radical or significative elements of words were absorbed and obscured in the Aryan, and how they stood out in bold relief in the Semitic languages, we shall appreciate the difficulty which the Shemites experienced in framing any name that should not seem to take too one-sided a view of the Deity by predicating but one quality, whether strength, dominion, or majesty; and we shall equally perceive the snare which their very language laid for the Aryan nations, by supplying them with a number of words which, though they seemed harmless as meaning nothing except what by tradition or definition they were made to mean, yet were full of mischief owing to the recollections which, at any time, they might revive.

"Dyaus" in itself was as good a name as any for God, and in some respects more appropriate than its derivative "deva," the Latin *deus,* which the Romance nations still use without meaning any harm. But "Dyaus" had meant "sky" for too long a time to become entirely divested of all the old myths or sayings which were true of Dyaus, the sky, but could only be retained as fables, if transferred to Dyaus, God. Dyaus, the Bright, might be called the husband of the earth; but when the same myth was repeated of Zeus, the god, then Zeus became the husband of Demeter, Demeter became a goddess, a daughter sprang from their union, and all the sluices of mythological madness were opened. There were a few men, no doubt, at all times, who saw through this mythological phraseology, who called on God, though they called him "Zeus," or "Dyaus," or "Jupiter." Xenophanes, one of the earliest Greek heretics, boldly maintained that there was but "one God, and that he was not like unto men, either in body or mind."[6] A poet in the Veda asserts distinctly, "They call him 'Indra,' 'Mitra,' 'Varuna,' 'Agni'; then he is 'the well-winged heavenly Garutmat;' that which is One the wise call it many ways,—they call it 'Agni,' 'Yama,' 'Mâtarisvan.' "[7]

But, on the whole, the charm of mythology prevailed among the Aryan nations, and a return to the primitive intuition of God and a total negation of all gods were rendered more difficult to the Aryan than to the Semitic man. The Semitic man had hardly ever to resist the allurements of mythology. The names with which he invoked the Deity did not trick him by their equivocal character. Nevertheless, these Semitic names, too, though predicative in the beginning, became subjective, and from being the various names of One Being, lapsed into names of various beings. Hence arose a danger which threatened well-nigh to bar to the Semitic race the approach to the conception and worship of the One God.

Nowhere can we see this danger more clearly than in the history of the Jews. The Jews had, no doubt, preserved from the beginning the idea of God, and their names of God contained nothing but what might by right be ascribed to Him. They worshipped a single God, and, whenever they fell into idolatry, they felt that they had fallen away from God. But that God, under whatever name they invoked Him, was especially their God, their own national God, and His existence did not exclude the existence of other gods or demons. Of the ancestors of Abraham and Nachor, even of their father Terah, we know that in old time, when they dwelt on the other side of the flood, they served other gods (Joshua 24.2). At the time of Joshua these gods were not yet forgotten, and instead of denying their existence altogether, Joshua only exhorts the people to put away the gods which their fathers served on the other side of the flood and in Egypt, and to serve the Lord: "Choose ye this day," he says, "whom you will serve; whether the gods which your fathers served that were on the other side of the flood, or the gods of the Amorites, in whose land ye dwell; but as for me and my house, we will serve the Lord."

Such a speech, exhorting the people to make their choice between various gods, would have been unmeaning if addressed to a nation which had once conceived the unity of the Godhead. Even images of the gods were not unknown to the family of Abraham, for, though we know nothing of the exact form of the *teraphim,* or images which Rachel stole from her father, certain it is that Laban calls them his gods (Genesis 31.19, 30). But what is much more significant than these traces of polytheism and idolatry is the hesitating tone in which some of the early patriarchs speak of their God. When Jacob flees before Esau into Padan-Aram and awakes from his vision at Bethel, he does not profess his faith in the One God, but he bargains, and says, "If God will be with me, and will keep me in this way that I go, and will give me bread to eat, and raiment to put on, so that I come again to my father's house in peace, then shall the Lord be my God: and this stone, which I have set for a pillar, shall be God's house: and of all that thou shalt give me, I will surely give the tenth unto thee" (Genesis 28.20–22). Language of this kind evinces not only a temporary want of faith in God, but it shows that the conception of God had not yet acquired that complete universality which alone deserves to be called monotheism, or belief in the One God. To him who has seen God face to face there is no longer any escape or doubt as to who is to be his god; God is his god, whatever befall. But this Jacob learnt not until he had struggled and wrestled with God, and committed himself to His care at the very time when no one else could have saved him. In that struggle Jacob asked for the true name of God, and he learnt from God that His name was secret (Genesis 32.29). After that, his God was no longer one of many gods. His faith was not like the faith of Jethro (Exodus 27.11), the priest of Midian, the father-in-law of Moses, who when he heard of all that God had done for Moses acknowledged that God (Jehovah) was greater than all gods (Elohim). This is not yet faith in the One God. It is a faith hardly above the faith of the people who were halting between Jehovah and Baal, and who only when they saw what the Lord did for Elijah, fell on their faces and said, "The Lord He is the God."

And yet this limited faith in Jehovah as the God of the Jews, as a God more powerful than the gods of the heathen, as a God above all gods, betrays itself again and again in the history of the Jews. The idea of many gods is there, and wherever that idea exists, wherever the plural of god is used in earnest, there is polytheism. It is not so much the names of Zeus, Hermes, etc., which constitute the polytheism of the Greeks; it is the plural θεοί, gods, which contains the fatal spell. We do not know what M. Renan means when he says that Jehovah with the Jews "n'est pas le plus grand entre plusieurs dieux; c'est le Dieu unique." It was so with Abraham, it was so after Jacob had been changed into Israel, it was so with Moses, Elijah, and Jeremiah. But what is the meaning of the very first commandment, "Thou shalt have no other gods before me?" Could this command have been addressed to a nation to whom the plural of God was a nonentity? It might be answered that the plural of God was to the Jews as revolting as it is to us; that it was revolting to their faith, if not to their reason. But how was it that their language tolerated the plural of a word which excludes plurality as much as the word for the centre of a sphere? No man who had clearly perceived the unity of God, could say with the Psalmist (86.8), "Among the gods there is none like unto Thee, O Lord, neither are there any works like unto Thy works." Though the same poet says, "Thou art God alone," he could not have compared God with other gods, if his idea of God had really reached that all-embracing character which it had with Abraham, Moses, Elijah, and Jeremiah. Nor would God have been praised as the "great king above all gods" by a poet in whose eyes the gods of the heathen had been recognised as what they were—mighty shadows, thrown by the mighty works of God, and intercepting for a time the pure light of the Godhead.

We thus arrive at a different conviction from that which M. Renan has made the basis of the history of the Semitic race. We can see nothing that would justify the admission of a monotheistic instinct, granted to the Semitic, and withheld from the Aryan race. They both share in the primitive intuition of God, they are both exposed to dangers in framing names for God, and they both fall into polytheism. What is peculiar to the Aryan race is their mythological phraseology, superadded to their polytheism; what is peculiar to the Semitic race is their belief in a national god—in a god chosen by his people as his people had been chosen by him.

No doubt M. Renan might say that we ignored his problem, and that we have not removed the difficulties which drove him to the admission of a monotheistic instinct. How is the fact to be explained, he might ask, that the three great religions of the world in which the unity of the Deity forms the key-note, are of Semitic origin, and that the Aryan nations, wherever they have been brought to a worship of the One God, invoke Him with names borrowed from the Semitic languages?

But let us look more closely at the facts before we venture on theories. Mohammedanism, no doubt, is a Semitic religion, and its very core is monotheism. But did Mohammed invent monotheism? Did he invent even a new name of God? (Renan, p. 23). Not at all. His object was to destroy the idolatry of the Semitic tribes of Arabia, to dethrone the angels, the Jin, the sons and daughters who had been assigned to Allah, and to restore the faith of Abraham in one God (Renan, p. 37).

And how is it with Christianity? Did Christ come to preach a faith in a new God? Did He or His disciples invent a new name of God? No, Christ came not to destroy, but to fulfill; and the God whom He preached was the God of Abraham.

And who is the God of Jeremiah, of Elijah, and of Moses? We answer again, the God of Abraham.

Thus the faith in the One living God, which seemed to require the admission of a monotheistic instinct, grafted in every member of the Semitic family, is traced back to one man, to him "in whom all families of the earth shall be blessed" (Genesis 12.3; Acts 3.25; Galatians 3.8). If from our earliest childhood we have looked upon Abraham, the friend of God, with love and veneration; if our first impressions of a truly god-fearing life were taken from him, who left the land of his fathers to live a stranger in the land whither God had called him, who always listened to the voice of God, whether it conveyed to him the promise of a son in his old age, or the command to sacrifice that son, his only son Isaac, his venerable figure will assume still more majestic proportions when we see in him the life-spring of that faith which was to unite all the nations of the earth, and the author of that blessing which was to come on the Gentiles through Jesus Christ.

And if we are asked how this one Abraham possessed not only the primitive intuition of God as He had revealed Himself to all mankind, but passed through the denial of all other gods to the knowledge of the one God, we are content to answer that it was by a special Divine Revelation. We do not indulge in theological phraseology, but we mean every word to its fullest extent. The Father of Truth chooses His own prophets, and He speaks to them in a voice stronger than the voice of thunder. It is the same inner voice through which God speaks to all of us. That voice may dwindle away, and become hardly audible; it may lose its Divine accent, and sink into the language of worldly prudence; but it may also, from time to time, assume its real nature with the chosen of God, and sound into their ears as a voice from Heaven. A "divine instinct" may sound more scientific, and less theological; but in truth it would neither be an appropriate name for what is a gift or grace accorded to but few, nor would it be a more scientific, i.e., a more intelligible word than "special revelation."

The important point, however, is not whether the faith of Abraham should be called a divine instinct or a revelation; what we wish here to insist on is that that instinct, or that revelation, was special, granted to one man, and handed down from him to Jews, Christians, and Mohammedans—to all who believe in the God of Abraham. Nor was it granted to Abraham entirely as a free gift. Abraham was tried and tempted before he was trusted by God. He had to break with the faith of his fathers; he had to deny the gods who were worshipped by his friends and neighbours. Like all the friends of God, he had to hear himself called an infidel and atheist, and in our own days he would have been looked upon as a madman for attempting to slay his son. It was through special faith that Abraham received his special revelation, not through instinct, not through abstract meditation, not through ecstatic visions. We want to know more of that man than we do; but, even with the little we know of him, he stands before us as a figure second only to one in the whole history of the world. We see his zeal for God, but we never see him contentious. Though Melchisedek worshipped God under a different name, invoking Him as Eliun, the Most High, Abraham at once acknowledged in Melchisedek a worshipper and priest of the true God, or Elohim, and paid him tithes. In the very name of Elohim we seem to trace the conciliatory spirit of Abraham. Elohim is a plural, though it is followed by the verb in the singular. It is generally said that the genius of the Semitic languages countenances the use of plurals for abstract conceptions, and that when Jehovah is called Elohim, the plural should be translated by "the Deity." We do not deny the fact, but we wish for an explanation, and an explanation is suggested by the various phases through which, as we saw, the conception of God passed in the ancient history of the Semitic mind. Eloah was at first the name for God, and as it is found in all the dialects of the Semitic family except the Phœnician (Renan, p. 61), it may probably be considered as the most ancient name of the Deity, sanctioned at a time when the original Semitic speech had not yet branched off into national dialects. When this name was first used in the plural, it could only have signified, like every plural, many Eloahs; and such a plural could only have been formed after the various names of God had become the names of independent deities, i.e., during a polytheistic stage.

The transition from this into the monotheistic stage could be effected in two ways: either by denying altogether the existence of the Elohim, and changing them into devils, as the Zoroastrians did with the Devas of their Brahmanic ancestors; or by taking a higher view, and looking upon the Elohim as so many names invented with the honest purpose of expressing the various aspects of the Deity, though in time diverted from their original purpose. This is the view taken by St. Paul of the religion of the Greeks when he came to declare unto them "Him whom they *ignorantly* worshipped," and the same view was taken by Abraham. Whatever the names of the Elohim, worshipped by the numerous clans of his race, Abraham saw that all the Elohim were meant for God, and thus Elohim, comprehending by one name everything that ever had been or could be called divine, became the name with which the monotheistic age was rightly inaugurated—a plural, conceived and constructed as a singular. Jehovah was all the Elohim, and therefore there could be no other God. From this point of view the Semitic name of the Deity, Elohim, which seemed at first not only ungrammatical but irrational, becomes perfectly clear and intelligible, and it proves better than anything else that the true monotheism could not have risen except on the ruins of a polytheistic faith. It is easy to scoff at the gods of the heathen, but a cold-hearted philosophical negation of the gods of the ancient world is more likely to lead to Deism or Atheism than to a belief in the One living God, the Father of all mankind, "who hath made of one blood all nations of men, for to dwell on all the face of the earth; and hath determined the times before appointed, and the bounds of their habitation; that they should seek the Lord, if haply they might feel after Him, and find Him, though He be not far from every one of us; for in Him we live, and move, and have our being; as certain also of your own poets have said, 'For we are also His offspring' " (Acts 17.26–28).

Taking this view of the historical growth of the idea of God, many of the difficulties which M. Renan has to overcome by most elaborate, and sometimes hairsplitting arguments, disappear at once. M. Renan, for instance, dwells much on Semitic proper names in which the names of the Deity occur; and he thinks that, like the Greek names "Theodoros" or "Theodotos," instead of "Zenodotos," they prove the existence of a faith in one God. We should say they may or may not. As "Devadatta," in Sanskrit, may mean either "given by God," or "given by the gods," so every proper name which M. Renan quotes, whether of Jews, or Edomites,

Ishmaelites, Ammonites, Moabites, and Themanites, whether from the Bible, or from Arab historians, from Greek authors, Greek inscriptions, the Egyptian papyri, the Himyaritic and Sinaitic inscriptions and ancient coins, are all open to two interpretations. "The servant of Baal" may mean the servant of the Lord, but it may also mean the servant of Baal, as one of many lords, or even the servant of the Baalim or the Lords. The same applies to all other names. "The gift of El" may mean "the gift of the only strong God"; but it may likewise mean "the gift of the El," as one of many gods, or even "the gift of the Els," in the sense of the strong gods. Nor do we see why M. Renan should take such pains to prove that the name of "Orotal" or "Orotulat," mentioned by Herodotus (III.8), may be interpreted as the name of a supreme deity; and that "Alilat," mentioned by the same traveller, should be taken not as the name of a goddess, but as a feminine noun expressive of the abstract sense of the deity. Herodotus says distinctly that Orotal was a deity like Bacchus; and Alilat, as he translates her name by Οὐρανίη, must have appeared to him as a goddess, and not as the Supreme Deity. One verse of the Koran is sufficient to show that the Semitic inhabitants of Arabia worshipped not only gods, but goddesses also. "What think ye of Allat, al Uzza, and Manah, that other third goddess?" (Sura 53, al-Najm)

If our view of the development of the idea of God be correct, we can perfectly understand how, in spite of this polytheistic phraseology, the primitive intuition of God should make itself felt from time to time, long before Mohammed restored the belief of Abraham in one God. The old Arabic prayer mentioned by Abulfarag may be perfectly genuine: "I dedicate myself to thy service, O God! Thou hast no companion, except thy companion, of whom thou art absolute master, and of whatever is his." The verse pointed out to M. Renan by M. Caussin de Perceval from the Moallaka of Zoheyr, was certainly anterior to Mohammed: "Try not to hide your secret feelings from the sight of Allah; Allah knows all that is hidden." But these quotations serve no more to establish the universality of the monotheistic instinct in the Semitic race than similar quotations from the Veda would prove the existence of a conscious monotheism among the ancestors of the Aryan race. There too we read, "Agni knows what is secret among mortals" (Rig-veda VIII.39, 6); and again: "He, the upholder of order, Varuna, sits down among his people; he, the wise, sits there to govern. From thence perceiving all wondrous things, he sees what has been and what will be done."[8]

But in these very hymns, better than anywhere else, we learn that the idea of supremacy and omnipotence ascribed to one god did by no means exclude the admission of other gods, or names of God. All the other gods disappear from the vision of the poet while he addresses his own God, and he only who is to fulfill his desires stands in full light before the eyes of the worshipper as the supreme and only God.

The Science of Religion is only just beginning, and we must take care how we impede its progress by preconceived notions or too hasty generalisations. During the last fifty years the authentic documents of the most important religions of the world have been recovered in a most unexpected and almost miraculous manner. We have now before us the canonical books of Buddhism; the Zend-Avesta of Zoroaster is no longer a sealed book; and the hymns of the Rig-veda have revealed a state of religion anterior to the first beginnings of that mythology which in Homer and Hesiod stands before us as a mouldering ruin. The soil of Mesopotamia has given back the very images once worshipped by the most powerful of the Semitic tribes, and the cuneiform inscriptions of Babylon and Nineveh have disclosed the very prayers addressed to Baal or Nisroch. With the discovery of these documents a new era begins in the study of religion. We begin to see more clearly every day what St. Paul meant in his sermon at Athens. But as the excavator at Babylon or Nineveh, before he ventures to reconstruct the palaces of these ancient kingdoms, sinks his shafts into the ground slowly and circumspectly lest he should injure the walls of the ancient palaces which he is disinterring; as he watches every corner-stone lest he mistake their dark passages and galleries, and as he removes with awe and trembling the dust and clay from the brittle monuments lest he destroy their outlines, and obliterate their inscriptions, so it behooves the student of the history of religion to set to work carefully, lest he should miss the track, and lose himself in an inextricable maze. The relics which he handles are more precious than the ruins of Babylon; the problems he has to solve are more important than the questions of ancient chronology; and the substructions which he hopes one day to lay bare are the world-wide foundations of the eternal kingdom of God.

We look forward with the highest expectations to the completion of M. Renan's work, and though English readers will differ from many of the author's views, and feel offended now and then at his blunt and unguarded language, we doubt not that they will find his volumes both instructive and suggestive. They are written in that clear and brilliant style which has secured to M. Renan the rank of one of the best writers of French, and which throws its charm even over the dry and abstruse inquiries into the grammatical forms and radical elements of the Semitic languages.

Notes

1. *Histoire Générale et Système Comparé des Langues Sémitiques.* Par Ernest Renan, Membre de l'Institut. Seconde édition. Paris, 1858; and *Nouvelles Considérations sur le Caractère Général des Peuples Sémitiques et en particulier sur leur Tendance au Monothéisme.* Par Ernest Renan, Paris, 1859.
2. Translation: "The Jewish people have no superiority other than the superiority of its religion; of the peoples of antiquity, they are the least scientifically and philosophically gifted; they have no high position either in politics, or in the military. Their institutions are purely conservative; the prophets who represent their spirit so well are men who are essentially reactionary, always referring back to earlier ideals. How else do we explain, at the heart of a society so narrow and so undeveloped, a revolution of ideas that even Athens and Alexandria were unable to accomplish?"—Sondra Bacharach.
3. Translation: "It is implausible that monotheism be the product of a people who have exalted ideas by way of religion; in reality, it's the fruits of a race who has few religious needs. As if it were the *minimum* of religion, in terms of dogmas and of exterior practices, monotheism has above all adapted to the needs of the nomadic populations."—Sondra Bacharach
4. We give the extracts according to M. Renan's translation of the Book of Job (Paris, 1859, Michel Lévy).
5. "You may drive out Nature with a pitchfork, but she will always find a way back" (Horace)—The Editor.
6. Xenophanes, about contemporary with Cyrus, as quoted by Clemens Alex., *Strom.* v, p. 601.
7. Max Müller, *History of Ancient Sanskrit Literature,* p. 567.
8. Ibid., p. 536.

CHAPTER 2

LECTURE ON THE VEDAS, OR THE SACRED BOOKS
OF THE BRAHMANS[1] (1865)

I have brought with me one volume of my edition of the Veda, and I should not
wonder if it were the first copy of the work which has ever reached this busy town
of Leeds. Nay, I confess I have some misgivings that I may have undertaken a hopeless task, and I begin to doubt whether I shall succeed in explaining to you the interest which I feel for this ancient collection of sacred hymns—an interest which has
never failed me while devoting to the publication of this voluminous work the best
twenty years of my life. Many times have I been asked, But what is the Veda? Why
should it be published? What are we likely to learn from a book composed nearly
four thousand years ago, and intended from the beginning for an uncultivated race
of mere heathens and savages—a book which the natives of India have never
published themselves, although, to the present day, they profess to regard it as the
highest authority for their religion, morals, and philosophy? Are we, the people of
England, or of Europe, in the nineteenth century, likely to gain any new light on religious, moral, or philosophical questions from the old songs of the Brahmans? And is
it so very certain that the whole book is not a modern forgery, without any substantial claims to that high antiquity which is ascribed to it by the Hindus, so that all the
labour bestowed upon it would not only be labour lost, but [would] throw discredit
on our powers of discrimination, and make us a laughing-stock among the shrewd
natives of India? These and similar questions I have had to answer many times when
asked by others, and some of them when asked by myself, before embarking on so
hazardous an undertaking as the publication of the Rig-veda and its ancient
commentary. And I believe I am not mistaken in supposing that many of those who
to-night have honoured me with their presence may have entertained similar doubts
and misgivings when invited to listen to a lecture "On the Vedas, or the Sacred Books
of the Brahmans."

I shall endeavour, therefore, as far as this is possible within the limits of one lecture,
to answer some of these questions, and to remove some of these doubts, by explaining to you, first, what the Veda really is; and, secondly, what importance it possesses,
not only to the people of India, but to ourselves in Europe—and here again, not only
to the student of Oriental languages, but to every student of history, religion, or
philosophy; to every man who has once felt the charm of tracing that mighty stream
of human thought on which we ourselves are floating onward, back to its distant
mountain-sources; to every one who has a heart for whatever has once filled the hearts
of millions of human beings with their noblest hopes, and fears, and aspirations; to

every student of mankind in the fullest sense of that full and weighty word. Whoever claims that noble title must not forget, whether he examines the highest achievements of mankind in our own age, or the miserable failures of former ages, what man is, and in whose image and after whose likeness man was made. Whether listening to the shrieks of the Shaman sorcerers of Tartary, or to the odes of Pindar, or to the sacred songs of Paul Gerhard; whether looking at the pagodas of China, or the Parthenon of Athens, or the Cathedral of Cologne; whether reading the sacred books of the Buddhists, of the Jews, or of those who worship God in spirit and in truth, we ought to be able to say, like the Emperor Maximilian, "Homo sum, humani nihil a me alienum puto"; or, translating his words somewhat freely, "I am a man; nothing pertaining to man I deem foreign to myself." Yes, we must learn to read in the history of the whole human race something of our own history; and as in looking back on the story of our own life, we all dwell with a peculiar delight on the earliest chapters of our childhood, and try to find there the key to many of the riddles of our later life, it is but natural that the historian, too, should ponder with most intense interest over the few relics that have been preserved to him of the childhood of the human race. These relics are few indeed, and therefore very precious; and this I may venture to say, at the outset and without fear of contradiction, that there exists no literary relic that carries us back to a more primitive, or, if you like, more childlike state in the history of man than the Veda.[2]

As the language of the Veda, the Sanskrit, is the most ancient type of the English of the present day (Sanskrit and English are but varieties of one and the same language), so its thoughts and feelings contain in reality the first roots and germs of that intellectual growth which by an unbroken chain connects our own generation with the ancestors of the Aryan race—with those very people who at the rising and setting of the sun listened with trembling hearts to the songs of the Veda, that told them of bright powers above, and of a life to come after the sun of their own lives had set in the clouds of the evening. Those men were the true ancestors of our race; and the Veda is the oldest book we have in which to study the first beginnings of our language, and of all that is embodied in language. We are by nature Aryan, Indo-European, not Semitic: our spiritual kith and kin are to be found in India, Persia, Greece, Italy, Germany; not in Mesopotamia, Egypt, or Palestine. This is a fact that ought to be clearly perceived, and constantly kept in view, in order to understand the importance which the Veda has for us, after the lapse of more than three thousand years, and after ever so many changes in our language, thought, and religion.

Whatever the intrinsic value of the Veda, if it simply contained the names of kings, the description of battles, the dates of famines, it would still be, by its age alone, the most venerable of books. Do we ever find much beyond such matters in Egyptian hieroglyphics, or in Cuneiform inscriptions? In fact, what does the ancient history of the world before Cyrus, before 500 B.C., consist of, but meagre lists of Egyptian, Babylonian, Assyrian dynasties? What do the tablets of Karnak, the palaces of Nineveh, and the cylinders of Babylon tell us about the thoughts of men? All is dead and barren, nowhere a sigh, nowhere a jest, nowhere a glimpse of humanity. There has been but one oasis in that vast desert of ancient Asiatic history, the history of the Jews. Another such oasis is the Veda. Here, too, we come to a stratum of ancient thought, of ancient feelings, hopes, joys, and fears—of ancient religion. There is perhaps too little of kings and battles in the Veda, and scarcely anything of the chronological framework of history. But poets, surely, are better than kings; hymns and prayers are more worth listening to than the agonies of butchered armies; and guesses at truth more valuable than unmeaning titles of Egyptian or Babylonian despots. It will be difficult to settle whether the Veda is "the oldest of books," and whether some of the portions of the Old Testament may not be traced back to the same or even an earlier date than the oldest hymns of the Veda. But in the Aryan world, the Veda is certainly the oldest book, and its preservation amounts almost to a marvel.

It is nearly twenty years ago since my attention was first drawn to the Veda, while attending, in the years 1846 and 1847, the lectures of Eugène Burnouf at the Collège de France. I was then looking out, like most young men at that time of life, for some great work, and without weighing long the difficulties which had hitherto prevented the publication of the Veda, I determined to devote all my time to the collection of the materials necessary for such an undertaking. I had read the principal works of the later Sanskrit literature, but had found little there that seemed to be more than curious. But to publish the Veda, a work that had never before been published in India or in Europe, that occupied in the history of Sanskrit literature the same position which the Old Testament occupies in the history of the Jews, the New Testament in the history of modern Europe, the Koran in the history of Mohammedanism—a work which fills a gap in the history of the human mind, and promises to bring us nearer than any other work to the first beginnings of Aryan language and Aryan thought—this seemed to me an undertaking not altogether unworthy a man's life. What added to the charm of it was that it had once before been undertaken by Frederick Rosen, a young German scholar, who died in England before he had finished the first book, and that after his death no one seemed willing to carry on his work. What I had to do, first of all, was to copy not only the text, but the commentary of the Rig-veda, a work which when finished will fill six of these large volumes. The author, or rather the compiler of this commentary, Sâyana Âkârya, lived about fourteen hundred [years] after Christ, that is to say, about as many centuries after [the beginning of our era] as the poets of the Veda lived before [it]. Yet through the three thousand years which separate the original poetry of the Veda from the latest commentary, there runs an almost continuous stream of tradition, and it is from it, rather than from his own brain, that Sâyana draws his explanations of the sacred texts. Numerous manuscripts, more or less complete, more or less inaccurate, of Sâyana's classical work, existed in the then Royal Library at Paris, in the Library of the East India House, then in Leadenhall Street, and in the Bodleian Library at Oxford.

But to copy and collate these manuscripts was by no means all. A number of other works were constantly quoted in Sâyana's commentary, and these quotations had all to be verified. It was necessary first to copy these works, and to make indexes to all of them, in order to be able to find any passage that might be referred to in the larger commentary. Many of these works have since been published in Germany and France, but they were not to be procured twenty years ago. The work, of course, proceeded but slowly, and many times I doubted whether I should be able to carry it through. Lastly came the difficulty—and by no means the smallest—[of] who was

Table 1.1

Sanskrit	Greek	Gothic	Anglo-Saxon	German
véda	οἶδα	vait	wât	ich weiss
véttha	οἶσθα	vaist	wâst	du weisst
véda	οἶδε	vait	wât	er weiss
vidvá	—	vitu	—	—
vidáthuh	ἴστον	vituts	—	—
vidátuh	ἴστον	—	—	—
vidmá	ἴσμεν	vitum	witon	wir wissen
vidá	ἴστε	vituth	wite	ihr wisset
vidúh	ἴσασι	vitun	witan	sie wissen

to publish a work that would occupy about six thousand pages in quarto, all in Sanskrit, and of which probably not a hundred copies would ever be sold. Well, I came to England in order to collect more materials at the East India House and at the Bodleian Library, and thanks to the exertions of my generous friend Baron Bunsen, and of the late Professor Wilson, the Board of Directors of the East India Company decided to defray the expenses of a work which, as they stated in their letter, "is in a peculiar manner deserving of the patronage of the East India Company, connected as it is with the early religion, history, and language of the great body of their Indian subjects." It thus became necessary for me to take up my abode in England, which has since become my second home. The first volume was published in 1849, the second in 1853, the third in 1856, the fourth in 1862. The materials for the remaining volumes are ready, so that, if I can but make leisure, there is little doubt that before long the whole work will be complete.

Now, first, as to the name. Veda means originally knowing or knowledge, and this name is given by the Brahmans not to one work, but to the whole body of their most ancient sacred literature. Veda is the same word which appears in the Greek οἶδα, I know, and in the English, wise, wisdom, to wit [See Table 1.1]. The name of Veda is commonly given to four collections of hymns, which are respectively known by the names of "Rig-veda," "Yagur-veda," "Sâma-veda," and "Atharva-veda"; but for our own purposes, namely for tracing the earliest growth of religious ideas in India, the only important, the only real Veda, is the Rig-veda.

The other so-called Vedas, which deserve the name of Veda no more than the Talmud deserves the name of Bible, contain chiefly extracts from the Rig-veda, together with sacrificial formulas, charms, and incantations, many of them, no doubt, extremely curious, but never likely to interest any one except the Sanskrit scholar by profession.

The Yagur-veda and Sâma-veda may be described as prayer-books, arranged according to the order of certain sacrifices, and intended to be used by certain classes of priests. Four classes of priests were required in India at the most solemn sacrifices:

1. The officiating priests, manual labourers, and acolytes, who have chiefly to prepare the sacrificial ground, to dress the altar, slay the victims, and pour out the libations.

2. The choristers, who chant the sacred hymns.
3. The reciters or readers, who repeat certain hymns.
4. The overseers or bishops, who watch and superintend the proceedings of the other priests, and ought to be familiar with all the Vedas.

The formulas and verses to be muttered by the first class are contained in the Yagur-veda-sanhitâ. The hymns to be sung by the second class are in the Sâma-veda-sanhitâ.

The Atharva-veda is said to be intended for the Brahman or overseer, who is to watch the proceedings of the sacrifice, and to remedy any mistake that may occur.[3]

Fortunately the hymns to be recited by the third class were not arranged in a sacrificial prayer-book, but were preserved in an old collection of hymns, containing all that had been saved of ancient, sacred, and popular poetry, more like the Psalms than like a ritual; a collection made for its own sake, and not for the sake of any sacrificial performances.

I shall, therefore, confine my remarks to the Rig-veda, which in the eyes of the historical student is the Veda *par excellence*. Now Rig-veda means the Veda of hymns of praise, for *Rich*, which before the initial soft letter of Veda is changed to *Rig*, is derived from a root which in Sanskrit means to celebrate.

In the Rig-veda we must distinguish again between the original collection of the hymns or Mantras, called the "Sanhitâ" or the collection, being entirely metrical and poetical, and a number of prose works, called "Brâhmanas" and "Sûtras," written in prose, and giving information on the proper use of the hymns at sacrifices, on their sacred meaning, on their supposed authors, and similar topics. These works, too, go by the name of "Rig-veda": but though very curious in themselves, they are evidently of a much later period and of little help to us in tracing the beginnings of religious life in India. For that purpose we must depend entirely on the hymns, such as we find them in the Sanhitâ or the collection of the Rig-veda.

Now this collection consists of ten books, and contains altogether 1,028 hymns. As early as about 600 B.C., we find that in the theological schools of India every verse, every word, every syllable of the Veda had been carefully counted. The number of verses as computed in treatises of that date varies from 10,402 to 10,622; that of the words is 153,826, that of the syllables 432,000.[4] With these numbers, and with the description given in these early treatises of each hymn, of its metre, its deity, its number of verses, our modern manuscripts of the Veda correspond as closely as could be expected.

I say our modern manuscripts, for all our manuscripts are modern, and very modern. Few Sanskrit manuscripts are more than four or five hundred years old, the fact being that in the damp climate of India no paper will last for more than a few centuries. How, then, you will naturally ask, can it be proved that the original hymns were composed between 1200 and 1500 before the Christian era, if our manuscripts only carry us back to about the same date after the Christian era? It is not very easy to bridge over this gulf of nearly three thousand years, but all I can say is that, after carefully examining every possible objection that can be made against the date of the Vedic hymns, their claim to that high antiquity which is ascribed to them has not, as far as I can judge, been shaken. I shall try to explain on what kind of evidence these claims rest.

You know that we possess no manuscript of the Old Testament in Hebrew older than about the tenth century after the Christian era; yet the Septuagint translation by itself would be sufficient to prove that the Old Testament, such as we now read it, existed in manuscript previous, at least, to the third century before our era. By a similar train of argument, the works to which I referred before, in which we find every hymn, every verse, every word and syllable of the Veda accurately counted by native scholars about five or six hundred years before Christ, guarantee the existence of the Veda, such as we now read it, as far back at least as five or six hundred years before Christ. Now in the works of that period, the Veda is already considered not only as an ancient, but as a sacred book; and, more than this, its language had ceased to be generally intelligible. The language of India had changed since the Veda was composed, and learned commentaries were necessary in order to explain to the people, then living, the true purport, nay, the proper pronunciation, of their sacred hymns.

But more than this. In certain exegetical compositions, which are generally comprised under the name of "Sûtras," and which are contemporary with, or even anterior to, the treatises on the theological statistics just mentioned, not only are the ancient hymns represented as invested with sacred authority, but that other class of writings, the Brâhmanas, standing half-way between the hymns and the Sûtras, have likewise been raised to the dignity of a revealed literature. These Brâhmanas, you will remember, are prose treatises, written in illustration of the ancient sacrifices and of the hymns employed at them. Such treatises would only spring up when some kind of explanation began to be wanted both for the ceremonial and for the hymns to be recited at certain sacrifices; and we find, in consequence, that in many cases the authors of the Brâhmanas had already lost the power of understanding the text of the ancient hymns in its natural and grammatical meaning, and that they suggested the most absurd explanations of the various sacrificial acts, most of which, we may charitably suppose, had originally some rational purpose. Thus it becomes evident that the period during which the hymns were composed must have been separated by some centuries, at least, from the period that gave birth to the Brâhmanas, in order to allow time for the hymns growing unintelligible and becoming invested with a sacred character. Secondly, the period during which the Brâhmanas were composed must be separated by some centuries from the authors of the Sûtras, in order to allow time for further changes in the language and more particularly for the growth of a new theology, which ascribed to the Brâhmanas the same exceptional and revealed character which the Brâhmanas themselves ascribed to the hymns. So that we want previously to 600 B.C., when every syllable of the Veda was counted, at least two strata of intellectual and literary growth, of two or three centuries each; and are thus brought to 1100 or 1200 B.C. as the earliest time when we may suppose the collection of the Vedic hymns to have been finished. This collection of hymns again contains, by its own showing, ancient and modern hymns—the hymns of the sons together with the hymns of their fathers and earlier ancestors; so that we cannot well assign a date more recent than 1200 to 1500 before our era, for the original composition of those simple hymns, which up to the present day are regarded by the Brahmans with the same feelings with which a Mohammedan regards the Koran, a Jew the Old Testament, a Christian his Gospel.

That the Veda is not quite a modern forgery can be proved, however, by more tangible evidence. Hiouen-thsang, a Buddhist pilgrim, who travelled from China to India in the years 629 to 645, and who, in his diary, translated from Chinese into French by M. Stanislas Julien, gives the names of the four Vedas, mentions some grammatical forms peculiar to the Vedic Sanskrit, and states that at his time young Brahmans spent all their time, from the seventh to the thirtieth year of their age, in learning these sacred texts. At the time when Hiouen-thsang was travelling in India, Buddhism was clearly on the decline. But Buddhism was originally a reaction against Brahmanism, and chiefly against the exclusive privileges which the Brahmans claimed, and which from the beginning were represented by them as based on their revealed writings, the Vedas, and hence beyond the reach of human attacks. Buddhism, whatever the date of its founder, became the state religion of India under Asoka, the Constantine of India, in the middle of the third century B.C. This Asoka was the third king of a new dynasty founded by Kandragupta, the well-known contemporary of Alexander and Seleucus, about 315 B.C. The preceding dynasty was that of the Nandas, and it is under this dynasty that the traditions of the Brahmans place a number of distinguished scholars, whose treatises on the Veda we still possess, such as Saunaka, Kâtyâyana, Âsvalâyana, and others. Their works, and others written with a similar object and in the same style, carry us back to about 600 B.C. This period of literature, which is called the Sûtra period, was preceded, as we saw, by another class of writings, the Brâhmanas, composed in a very prolix and tedious style, and containing lengthy lucubrations on the sacrifices and on the duties of the different classes of priests. Each of the three or four Vedas, or each of the three or four classes of priests, has its own Brâhmanas and its own Sûtras; and as the Brâhmanas are presupposed by the Sûtras, while no Sûtra is ever quoted by the Brâhmanas, it is clear that the period of the Brâhmana literature must have preceded the period of the Sûtra literature. There are, however, old and new Brâhmanas; and there are in the Brâhmanas themselves long lists of teachers who handed down old Brâhmanas or composed new ones; so that it seems impossible to accommodate the whole of that literature in less than two centuries, from about 800 to 600 B.C.

Before, however, a single Brâhmana could have been composed, it was not only necessary that there should have been one collection of ancient hymns, like that contained in the ten books of the Rig-veda, but the three or four classes of priests must have been established; the officiating priests and the choristers must have had their special prayer-books; nay, these prayer-books must have undergone certain changes, because the Brâhmanas presuppose different texts, called "sâkhâs," of each of these prayer-books, which are called the "Yagur-veda-sanhitâ," the "Sâma-veda-sanhitâ," and the "Atharva-veda-sanhitâ." The work of collecting the prayers for the different classes of priests, and of adding new hymns and formulas for purely sacrificial purposes, belonged probably to the tenth century B.C.; and three generations more would, at least, be required to account for the various readings adopted in the prayer-books by different sects, and invested with a kind of sacred authority, long before the composition of even the earliest among the Brâhmanas. If, therefore, the years from about 1000 to 800 B.C. are assigned to this collecting age, the time before 1000 B.C. must be set apart for the free and natural growth of what was then national and religious, but not yet sacred and sacrificial poetry. How far back this period extends it is impossible to tell; it is enough if the hymns of the Rig-veda can be traced to a period anterior to 1000 B.C.

Much in the chronological arrangement of the three periods of Vedic literature that are supposed to have followed the period of the original growth of the hymns must of necessity be hypothetical, and has been put forward rather to invite than to silence criticism. In order to discover truth, we must be truthful ourselves, and must welcome those who point out our errors as heartily as those who approve and confirm our discoveries. What seems, however, to speak strongly in favour of the historical character of the three periods of Vedic literature is the uniformity of style which marks the productions of each. In modern literature we find, at one and the same time, different styles of prose and poetry cultivated by one and the same author. A Goethe writes tragedy, comedy, satire, lyrical poetry, and scientific prose; but we find nothing like this in primitive literature. The individual is there much less prominent, and the poet's character disappears in the general character of the layer of literature to which he belongs. It is the discovery of such large layers of literature following each other in regular succession which inspires the critical historian with confidence in the truly historical character of the successive literary productions of ancient India.

As in Greece there is an epic age of literature, where we should look in vain for prose or dramatic poetry; as in that country we never meet with real elegiac poetry before the end of the eighth century, nor with iambics before the same date; as even in more modern times rhymed heroic poetry appears in England with the Norman Conquest, and in Germany the Minnesänger rise and set with the Swabian dynasty—so, only in a much more decided manner, we see in the ancient and spontaneous literature of India an age of poets followed by an age of collectors and imitators, that age to be succeeded by an age of theological prose writers, and this last by an age of writers of scientific manuals. New wants produced new supplies, and nothing sprang up or was allowed to live, in prose or poetry, except what was really wanted. If the works of poets, collectors, imitators, theologians, and teachers were all mixed up together—if the Brâhmanas quoted the Sûtras, and the hymns alluded to the Brâhmanas—an historical restoration of the Vedic literature of India would be almost an impossibility. We should suspect artificial influences, and look with small confidence on the historical character of such a literary agglomerate. But he who would question the antiquity of the Veda must explain how the layers of literature were formed that are super-imposed over the original stratum of the poetry of the Rishis; he who would suspect a literary forgery must show how, when, and for what purpose, the 1,000 hymns of the Rig-veda could have been forged, and have become the religious, moral, political, and literary life of the ancient inhabitants of India.

The idea of revelation, and I mean more particularly book-revelation, is not a modern idea, nor is it an idea peculiar to Christianity. Though we look for it in vain in the literature of Greece and Rome, we find the literature of India saturated with this idea from beginning to end. In no country, I believe, has the theory of revelation been so minutely elaborated as in India. The name for revelation in Sanskrit is "Sruti," which means hearing; and this title distinguishes the Vedic hymns, and, at a later time, the Brâhmanas also, from all other works, which, however sacred and authoritative to the Hindu mind, are admitted to have been composed by human authors. The Laws of Manu, for instance, according to the Brahmanic theology, are not revelation; they are not Sruti, but only Smriti, which means recollection or tradition. If these laws, or any other work of authority, can be proved on any point to be at variance with a single passage of the Veda, their authority is at once overruled. According to the orthodox views of Indian theologians, not a single line of the Veda was the work of human authors. The whole Veda is in some way or other the work of the Deity; and even those who received the revelation, or, as they express it, those who saw it, were not supposed to be ordinary mortals, but beings raised above the level of common humanity, and less liable, therefore, to error in the reception of revealed truth. The views entertained of revelation by the orthodox theologians of India are far more minute and elaborate than those of the most extreme advocates of verbal inspiration in Europe. The human element, called "paurusheyatva" in Sanskrit, is driven out of every corner or hiding-place; and as the Veda is held to have existed in the mind of the Deity before the beginning of time, every allusion to historical events, of which there are not a few, is explained away with a zeal and ingenuity worthy of a better cause.

But let me state at once that there is nothing in the hymns themselves to warrant such extravagant theories. In many a hymn, the author says plainly that he or his friends made it to please the gods; that he made it as a carpenter makes a chariot (Rv. I. 130, 6; V. 2, 11), or like a beautiful vesture (Rv. V. 29, 15); that he fashioned it in his heart and kept it in his mind (Rv. I. 171, 2); that he expects, as his reward, the favour of the god whom he celebrates (Rv. IV. 6, 21). But though the poets of the Veda knew nothing of the artificial theories of verbal inspiration, they were not altogether unconscious of higher influences: nay, they speak of their hymns as "god-given" ("devattam," Rv. III. 37, 4). One poet says:

O god (Indra) have mercy, give me my daily bread!
Sharpen my mind, like the edge of iron.
Whatever I now may utter, longing for thee, do thou accept it;
Make me possessed of God! (Rv. VI. 47, 10).

Another utters for the first time the famous hymn, the "Gâyatrî," which now for more than three thousand years has been the daily prayer of every Brahman, and is still repeated every morning by millions of pious worshippers: "Let us meditate on the adorable light of the divine Creator: may he rouse our minds."[5] This consciousness of higher influences, or of divine help in those who uttered for the first time the simple words of prayer, praise, and thanksgiving, is very different, however, from the artificial theories of verbal inspiration which we find in the later theological writings; it is indeed but another expression of that deep-felt dependence on the Deity, of that surrender and denial of all that seems to be self, which was felt more or less by every nation, but by none, I believe, more strongly, more constantly, than by the Indian. "It is He that has made it"—namely, the prayer in which the soul of the poet has thrown off her burden—is but a variation of "It is He that has made us," which is the key-note of all religion, whether ancient or modern, whether natural or revealed.

I must say no more to-night of what the Veda is, for I am very anxious to explain to you, as far as it is possible, what I consider to be the real importance of the Veda to the student of history, to the student of religion, to the student of mankind.

In the study of mankind there can hardly be a subject more deeply interesting than the study of the different forms of religion; and much as I value the Science of Language for the aid which it lends us in unraveling some of the most complicated tissues of the human intellect, I confess that to my mind there is no study more absorbing than that of the Religions of the World—the study, if I may so call it, of the various languages in which man has spoken to his Maker, and of that language in which his Maker "at sundry times and in divers manners" spake to man.

To my mind the great epochs in the world's history are marked not by the foundation or the destruction of empires, by the migrations of races, or by French revolutions. All this is outward history, made up of events that seem gigantic and overpowering to those only who cannot see beyond and beneath. The real history of man is the history of religion—the wonderful ways by which the different families of the human race advanced towards a truer knowledge and a deeper love of God. This is the foundation that underlies all profane history: it is the light, the soul, and life of history, and without it all history would indeed be profane.

On this subject there are some excellent works in English, such as Mr. Maurice's *Lectures on the Religions of the World*, or Mr. Hardwick's *Christ and Other Masters*; in German, I need only mention Hegel's *Philosophy of Religion*, out of many other learned treatises on the different systems of religion in the East and the West. But in all these works religions are treated very much as languages were treated during the last century. They are rudely classed, either according to the different localities in which they prevailed, just as in Adelung's *Mithridates* you find the languages of the world classified as European, African, American, Asiatic, etc.; or according to their age, as formerly languages used to be divided into ancient and modern; or according to their respective dignity, as languages used to be treated as sacred or profane, as classical or illiterate. Now, you know that the Science of Language has sanctioned a totally different system of classification; and that the Comparative Philologist ignores altogether the division of languages according to their locality, or according to their age, or according to their classical or illiterate character. Languages are now classified genealogically, i.e., according to their real relationship; and the most important languages of Asia, Europe, and Africa—that is to say, of that part of the world on which what we call the history of man has been acted—have been grouped together into three great divisions, the Aryan or Indo-European Family, the Semitic Family, and the Turanian Class. According to that division you are aware that English, together with all the Teutonic languages of the Continent, Celtic, Slavonic, Greek, Latin with its modem offshoots, such as French and Italian, Persian, and Sanskrit, are so many varieties of one common type of speech: that Sanskrit, the ancient language of the Veda, is no more distinct from the Greek of Homer, or from the Gothic of Ulfilas, or from the Anglo-Saxon of Alfred, than French is from Italian. All these languages together form one family, one whole, in which every member shares certain features in common with all the rest, and is at the same time distinguished from the rest by certain features peculiarly its own.

The same applies to the Semitic family, which comprises, as its most important members, the Hebrew of the Old Testament, the Arabic of the Koran, and the ancient languages on the monuments of Phœnicia and Carthage, of Babylon and Assyria. These languages, again, form a compact family, and differ entirely from the other family, which we called Aryan or Indo-European. The third group of languages, for we can hardly call it a family, comprises most of the remaining languages of Asia, and counts among its principal members the Tungusic, Mongolic, Turkic, Samoyedic, and Finnic, together with the languages of Siam, the Malay Islands, Thibet, and Southern India. Lastly, the Chinese language stands by itself, as monosyllabic, the only remnant of the earliest formation of human speech.

Now I believe that the same division which has introduced a new and natural order into the history of languages, and has enabled us to understand the growth of human speech in a manner never dreamt of in former days, will be found applicable to a scientific study of religions. I shall say nothing to-night of the Semitic or Turanian or Chinese religions, but confine my remarks to the religions of the Aryan family. These religions, though more important in the ancient history of the world, as the religions of the Greeks and Romans, of our own Teutonic ancestors and of the Celtic and Slavonic races, are nevertheless of great importance even at the present day. For although there are no longer any worshippers of Zeus, or Jupiter, of Wodan, Esus,[6] or Perkunas,[7] the two religions of Aryan origin which still survive, Brahmanism and Buddhism, claim together a decided majority among the inhabitants of the globe. Out of the whole population of the world:

31.2 per cent. are Buddhists
13.4 per cent. are Brahmanists
———
44.6

which together gives us 44 per cent. for what may be called living Aryan religions. Of the remaining 56 per cent., 15.7 are Mohammedans, 8.7 per cent. nondescript Heathens, 30.7 per cent. Christians, and only 0.3 per cent. Jews.

Now, as a scientific study of the Aryan languages became possible only after the discovery of Sanskrit, a scientific study of the Aryan religion dates really from the discovery of the Veda. The study of Sanskrit brought to light the original documents of three religions, the Sacred Books of the Brahmans, the Sacred Books of the Magians, the followers of Zoroaster, and the Sacred Books of the Buddhists. Fifty years ago, these three collections of sacred writings were all but unknown, their very existence was doubted, and there was not a single scholar who could have translated a line of the Veda, a line of the Zend-Avesta, or a line of the Buddhist Tripitaka. At present large portions of these, the canonical writings of the most ancient and most important religions of the Aryan race, are published and deciphered, and we begin to see a natural progress, and almost a logical necessity, in the growth of these three systems of worship. The oldest, most primitive, most simple form of Aryan faith finds its expression in the Veda. The Zend-Avesta represents in its language, as well as in its thoughts, a branching off from that more primitive stem; a more or less conscious opposition to the worship of the gods of nature, as adored in the Veda, and a striving after a more spiritual, supreme, moral deity, such as Zoroaster proclaimed under the name of "Ahura-mazda," or Ormuzd. Buddhism, lastly, marks a decided schism; a decided antagonism against the established religion of the Brahmans; a denial of the true divinity of the Vedic gods; and a proclamation of new philosophical and social doctrines.

Without the Veda, therefore, neither the reforms of Zoroaster nor the new teaching of Buddha would have been intelligible: we should not know what was behind them, or what forces impelled Zoroaster and Buddha to the founding of new religions; how much they received, how much they destroyed, how much they created. Take but one word in the religious phraseology of these three systems. In the Veda the gods are called Deva. This word in Sanskrit means bright—brightness or light being one of the most general attributes shared by the various manifestations of the Deity, invoked in the Veda, as Sun, or Sky, or Fire, or Dawn, or Storm. We can see, in fact, how in the minds of the poets of the Veda, deva, from meaning bright, came gradually to mean divine. In the Zend-Avesta the same word "daêva" means evil spirit. Many of the Vedic gods, with Indra at their head, have been degraded to the position of daêvas, in order to make room for Ahura-mazda, the Wise Spirit, as the supreme deity of the Zoroastrians. In his confession of faith the follower of Zoroaster declares: "I cease to be a worshipper of the daêvas." In Buddhism, again, we find these ancient Devas, Indra and the rest, as merely legendary beings, carried about at shows, as servants of Buddha, as goblins or fabulous heroes; but no longer either worshipped or even feared by those with whom the name of Deva had lost every trace of its original meaning. Thus this one word deva marks the mutual relations of these three religions. But more than this. The same word deva is the Latin *deus*, thus pointing to that common source of language and religion, far beyond the heights of the Vedic Olympus, from which the Romans, as well as the Hindus, draw the names of their deities, and the elements of their language as well as of their religion.

The Veda, by its language and its thoughts, supplies that distant background in the history of all the religions of the Aryan race, which was missed indeed by every careful observer, but which formerly could be supplied by guess-work only. How the Persians came to worship Ormuzd; how the Buddhists came to protest against temples and sacrifices; how Zeus and the Olympian gods came to be what they are in the mind of Homer; or how such beings as Jupiter and Mars came to be worshipped by the Italian peasant—all these questions, which used to yield material for endless and baseless speculations, can now be answered by a simple reference to the hymns of the Veda. The religion of the Veda is not the source of all the other religions of the Aryan world, nor is Sanskrit the mother of all the Aryan languages. Sanskrit, as compared to Greek and Latin, is an elder sister, not a parent: Sanskrit is the earliest deposit of Aryan speech, as the Veda is the earliest deposit of Aryan faith. But the religion and incipient mythology of the Veda possess the same simplicity and transparency which distinguish the grammar of Sanskrit from Greek, Latin, or German grammar. We can watch in the Veda ideas and their names growing, which in Persia, Greece, and Rome we meet with only as full-grown or as fast decaying. We get one step nearer to that distant source of religious thought and language which has fed the different national streams of Persia, Greece, Rome, and Germany; and we begin to see clearly what ought never to have been doubted, that there is no religion without God, or, as St. Augustine expressed it, that "there is no false religion which does not contain some elements of truth."

I do not wish by what I have said to raise any exaggerated expectations as to the worth of these ancient hymns of the Veda, and the character of that religion which they indicate rather than fully describe. The historical importance of the Veda can hardly be exaggerated; but its intrinsic merit, and particularly the beauty or elevation of its sentiments, have by many been rated far too high. Large numbers of the Vedic hymns are childish in the extreme: tedious, low, commonplace. The gods are constantly invoked to protect their worshippers, to grant them food, large flocks, large families, and a long life; for all which benefits they are to be rewarded by the praises and sacrifices offered day after day, or at certain seasons of the year. But hidden in this rubbish there are precious stones. Only in order to appreciate them justly, we must try to divest ourselves of the common notions about Polytheism, so repugnant not only to our feelings, but to our understanding.

No doubt, if we must employ technical terms, the religion of the Veda is Polytheism, not Monotheism. Deities are invoked by different names, some clear and intelligible, such as "Agni," fire; "Sûrya," the sun; "Ushas," dawn; "Maruts," the storms; "Prithivî," the earth; "Âp," the waters; "Nadî," the rivers; others such as "Varuna," "Mitra," "Indra," which have become proper names, and disclose but dimly their original application to the great aspects of nature, the sky, the sun, the day. But whenever one of these individual gods is invoked, they are not conceived as limited by the powers of others, as superior or inferior in rank. Each god is to the mind of the supplicant as good as all gods. He is felt, at the time, as a real divinity—as supreme and absolute—without a suspicion of those limitations which, to our mind, a plurality of gods *must* entail on every single god. All the rest disappear for a moment from the vision of the poet, and he only who is to fulfill their desires stands in full light before the eyes of the worshippers. In one hymn, ascribed to Manu, the poet says: "Among you, O gods, there is none that is small, none that is young; you are all great indeed." And this is indeed the key-note of the ancient Aryan worship. Yet it would be easy to find in the numerous hymns of the Veda, passages in which almost ever important deity is represented as supreme and absolute. Thus in one hymn, Agni (fire) is called "the ruler of the universe," "the lord of men," "the wise king, the father, the brother, the son, the friend of man"; nay, all the powers and names of the other gods are distinctly ascribed to Agni. But though Agni is thus highly exalted, nothing is said to disparage the divine character of the other gods. In another hymn another god, Indra, is said to be greater than all: "The gods," it is said, "do not reach thee, Indra, nor men; thou overcomest all creatures in strength." Another god, Soma, is called the king of the world, the king of heaven and earth, the conqueror of all. And what more could human language achieve, in trying to express the idea of a divine and supreme power, than what another poet says of another god, Varuna: "Thou art lord of all, of heaven and earth; thou art the king of all, of those who are gods, and of those who are men!"

This surely is not what is commonly understood by Polytheism. Yet it would be equally wrong to call it Monotheism. If we must have a name for it, I should call it *Kathenotheism*. The consciousness that all the deities are but different names of one and the same godhead, breaks forth indeed here and there in the Veda. But it is far from being general. One poet, for instance, says:

> They call him Indra, Mitra, Varuna, Agni;
> Then he is the beautiful-winged heavenly Garutmat:
> That which is One the wise call it in divers manners:
> They call it Agni, Yama, Mâtarisvan (Rv. I. 164, 46).

And again: "Wise poets make the beautiful-winged, though he is one, manifold by words" (Rv. X. 114, 5).

I shall read you a few Vedic verses, in which the religious sentiment predominates, and in which we perceive a yearning after truth, and after the true God, untrammeled as yet by any names or any traditions:[8]

1. In the beginning there arose the golden Child—He was the one born lord of all that is. He stablished [*sic*] the earth, and this sky;—Who is the God to whom we shall offer our sacrifice?

2. He who gives life, He who gives strength; whose command all the bright gods revere; whose shadow is immortality, whose shadow is death;—Who is the God to whom we shall offer our sacrifice?

3. He who through His power is the one king of the breathing and awakening world—He who governs all, man and beast;—Who is the God to whom we shall offer our sacrifice?

4. He whose greatness these snowy mountains, whose greatness the sea proclaims, with the distant river—He whose these regions are, as it were His two arms;—Who is the God to whom we shall offer our sacrifice?

5. He through whom the sky is bright and the earth firm—He through whom the heaven was stablished [*sic*]—nay, the highest heaven—He who measured out the light in the air;—Who is the God to whom we shall offer our sacrifice?

6. He to whom heaven and earth, standing firm by His will, look up, trembling inwardly—He over whom the rising sun shines forth;—Who is the God to whom we shall offer our sacrifice?

7. Wherever the mighty water-clouds went, where they placed the seed and lit the fire, thence arose He who is the sole life of the bright gods;—Who is the God to whom we shall offer our sacrifice?

8. He who by His might looked even over the water-clouds, the clouds which gave strength and lit the sacrifice; He who alone is God above all gods;—Who is the God to whom we shall offer our sacrifice?

9. May He not destroy us—He the creator of the earth; or He, the righteous, who created the heaven; He also created the bright and mighty waters;—Who is the God to whom we shall offer our sacrifice? (Rv. X. 121).[9]

The following may serve as specimens of hymns addressed to individual deities whose names have become the centres of religious thought and legendary traditions; deities, in fact, like Jupiter, Apollo, Mars, or Minerva, no longer mere germs, but fully developed forms of early thought and language:

Hymn to Indra (Rv. I. 53)[10]

1. Keep silence well![11] We offer praises to the great Indra in the house of the sacrificer. Does he find treasure for those who are like sleepers? Mean praise is not valued among the munificent.

2. Thou art the giver of horses, Indra, thou art the giver of cows, the giver of corn, the strong lord of wealth; the old guide of man, disappointing no desires, a friend to friends: to him we address this song.

3. O powerful Indra, achiever of many works, most brilliant god—all this wealth around here is known to be thine alone: take from it, conqueror, bring it hither!; do not stint the desire of the worshipper who longs for thee!

4. On these days thou art gracious, and on these nights,[12] keeping off the enemy from our cows and from our stud. Tearing[13] the fiend night after night with the help of Indra, let us rejoice in food, freed from haters.

5. Let us rejoice, Indra, in treasure and food, in wealth of manifold delight and splendour. Let us rejoice in the blessing of the gods, which gives us the strength of offspring, gives us cows first and horses.

6. These draughts inspired thee, O lord of the brave!; these were vigour, these libations, in battles, when for the sake of the poet, the sacrificer, thou struckest down irresistibly ten thousands of enemies.

7. From battle to battle[14] thou advancest bravely, from town to town thou destroyest all this with might, when thou, Indra, with Nâmî as thy friend, struckest down from afar the deceiver Namuki.

8. Thou hast slain Karnaga and Parnaya with the brightest spear of Atithigva. Without a helper thou didst demolish the hundred cities of Vangrida, which were beseiged by Rigisvan.

9. Thou hast felled down with the chariot-wheel these twenty kings of men, who had attacked the friendless Susravas,[15] and gloriously the sixty thousand and ninety-nine forts.

10. Thou, Indra, hast succoured Susravas with thy succors, Tûrvayâna with thy protections. Thou hast made Kutsa, Atithigva, and Âyu subject to this mighty youthful king.

11. We who in future, protected by the gods, wish to be thy most blessed friends, we shall praise thee, blessed by thee with offspring, and enjoying henceforth a longer life.

The next hymn is one of many addressed to Agni as the god of fire, not only the fire as a powerful element, but likewise the fire of the hearth and the altar, the guardian of the house, the minister of the sacrifice, the messenger between gods and men:

Hymn to Agni (Rv. II. 6)

1. Agni, accept this log which I offer to thee, accept this my service; listen well to these my songs.

2. With this log, O Agni, may we worship thee, thou son of strength, conqueror of horses!; and with this hymn, thou high-born!

3. May we thy servants serve thee with songs, O granter of riches, thou who lovest songs and delightest in riches.

4. Thou lord of wealth and giver of wealth, be thou wise and powerful; drive away from us the enemies!

5. He gives us rain from heaven, he gives us inviolable strength, he gives us food a thousand-fold.

6. Youngest of the gods, their messenger, their invoker, most deserving of worship, come, at our praise, to him who worships thee and longs for thy help.

7. For thou, O sage, goest wisely between these two creations (heaven and earth, gods and men), like a friendly messenger between two hamlets.
8. Thou art wise, and thou hast been pleased; perform thou, intelligent Agni, the sacrifice without interruption, sit down on this sacred grass!

The following hymn, partly laudatory, partly deprecatory, is addressed to the Maruts or Rudras, the Storm-gods:

Hymn to the Maruts (Rv. I. 39)[16]

1. When you thus from afar cast forward your measure, like a blast of fire, through whose wisdom is it, through whose design? To whom do you go, to whom, ye shakers (of the earth)?
2. May your weapons be firm to attack, strong also to withstand! May yours be the more glorious strength, not that of the deceitful mortal!
3. When you overthrow what is firm, O ye men, and whirl about what is heavy, ye pass through the trees of the earth, through the clefts of the rocks.
4. No real foe of yours is known in heaven, nor in earth, ye devourers of enemies! May strength be yours, together with your race, O Rudras, to defy even now.
5. They make the rocks to tremble, they tear asunder the kings of the forest. Come on, Maruts, like madmen, ye gods, with your whole tribe.
6. You have harnessed the spotted deer to your chariots, a red deer draws as leader. Even the earth listened at your approach, and men were frightened.
7. O Rudras, we quickly desire your help for our race. Come now to us with help, as of yore, thus for the sake of the frightened Kanva.
8. Whatever fiend, roused by you or roused by mortals, attacks us, tear him from us by your power, by your strength, by your aid.
9. For you, worshipful and wise, have wholly protected Kanva. Come to us, Maruts, with your whole help, as quickly as lightnings come after the rain.
10. Bounteous givers, ye possess whole strength, whole power, ye shakers (of the earth). Send, O Maruts, against the proud enemy of the poets, an enemy, like an arrow.

The following is a simple prayer addressed to the Dawn:

Hymn to Ushas (Rv. VII. 77)

1. She shines upon us, like a young wife, rousing every living being to go to his work. When the fire had to be kindled by men, she made the light by striking down darkness.
2. She rose up, spreading far and wide, and moving everywhere. She grew in brightness, wearing her brilliant garment. The mother of the cows (the mornings), the leader of the days, she shone gold-coloured, lovely to behold.
3. She, the fortunate, who brings the eye of the gods, who leads the white and lovely steed (of the sun), the Dawn was seen revealed by her rays, with brilliant treasures, following every one.

4. Thou art a blessing where thou art near, drive far away the unfriendly; make the pasture wide, give us safety! Scatter the enemy, bring riches! Raise up wealth to the worshipper, thou mighty Dawn.

5. Shine for us with thy best rays, thou bright Dawn, thou who lengthenest our life, thou the love of all, who givest us food, who givest us wealth in cows, horses, and chariots.

6. Thou daughter of the sky, thou high-born Dawn, whom the Vasishthas magnify with songs, give us riches high and wide: all ye gods protect us always with your blessings.

I must confine myself to shorter extracts, in order to be able to show to you that all the principal elements of real religion are present in the Veda. I remind you again that the Veda contains a great deal of what is childish and foolish, though very little of what is bad and objectionable. Some of its poets ascribe to the gods sentiments and passions unworthy of the Deity, such as anger, revenge, delight in material sacrifices; they likewise represent human nature on a low level of selfishness and worldliness. Many hymns are utterly unmeaning and insipid, and we must search patiently before we meet, here and there, with sentiments that come from the depth of the soul, and with prayers in which we could join ourselves. Yet there are such passages, and they are the really important passages, as marking the highest points to which the religious life of the ancient poets of India had reached: and it is to these that I shall now call your attention.

First of all, the religion of the Veda knows of no idols. The worship of idols in India is a secondary formation, a later degradation of the more primitive worship of ideal gods.

The gods of the Veda are conceived as immortal: passages in which the birth of certain gods is mentioned have a physical meaning: they refer to the birth of the day, the rising of the sun, the return of the year.

The gods are supposed to dwell in heaven, though several of them, as, for instance, Agni, the god of fire, are represented as living among men, or as approaching the sacrifice, and listening to the praises of their worshippers.

Heaven and earth are believed to have been made or to have been established by certain gods. Elaborate theories of creation, which abound in the later works, the Brâhmanas, are not to be found in the hymns. What we find are such passages as: "Agni held the earth, he stablished [sic] the heaven by truthful words" (Rv. I. 67, 3); "Varuna stemmed asunder the wide firmaments; he lifted on high the bright and glorious heaven; he stretched out apart the starry sky and the earth" (Rv. VII. 86, 1).

More frequently, however, the poets confess their ignorance of the beginning of all things, and one of them exclaims: "Who has seen the first-born? Where was the life, the blood, the soul of the world? Who went to ask this from any that knew it?" (Rv. I. 164, 4).[17]

Or again: "What was the forest, what was the tree out of which they shaped heaven and earth? Wise men, ask this indeed in your mind, on what he stood when he held the worlds?" (Rv. X. 81, 4).

I now come to a more important subject. We find in the Veda what few would have expected to find there, the two ideas so contradictory to the human understanding and yet so easily reconciled in every human heart: God has established the eternal laws of right and wrong, he punishes sin and rewards virtue, and yet the same God is willing to forgive; just, yet merciful; a judge, and yet a father. Consider, for instance, the following lines: "His path is easy and without thorns, who does what is right." (Rv. I. 41, 4).

And again: "Let man fear Him who holds the four (dice), before he throws them down (i.e., God who holds the destinies of men in his hand); let no man delight in evil words!" (Rv. I. 41, 9).

And then consider the following hymns, and imagine the feelings which alone could have prompted them:

Hymn to Varuna (Rv. VII. 89)

1. Let me not yet, O Varuna, enter into the house of clay; have mercy, almighty, have mercy!
2. If I go along trembling, like a cloud driven by the wind; have mercy, almighty, have mercy!
3. Through want of strength, thou strong and bright god, have I gone wrong; have mercy, almighty, have mercy!
4. Thirst came upon the worshipper, though he stood in the midst of the waters; have mercy, almighty, have mercy!
5. Whenever we men, O Varuna, commit an offense before the heavenly host, whenever we break the law through thoughtlessness; punish us not, O god, for that offense.

And again (Rv. VII. 86):

1. Wise and mighty are the works of him who stemmed asunder the wide firmaments (heaven and earth). He lifted on high the bright and glorious heaven; he stretched out apart the starry sky and the earth.
2. Do I say this to my own self? How can I get unto Varuna? Will he accept my offering without displeasure? When shall I, with a quiet mind, see him propitiated?
3. I ask, O Varuna, wishing to know this my sin. I go to ask the wise. The sages all tell me the same Varuna it is who is angry with thee.
4. Was it an old sin, O Varuna, that thou wishest to destroy thy friend, who always praises thee? Tell me, thou unconquerable lord, and I will quickly turn to thee with praise, freed from sin.
5. Absolve us from the sins of our fathers, and from those which we committed with our own bodies. Release Vasishtha, O king, like a thief who has feasted on stolen oxen; release him like a calf from the rope.
6. It was not our own doing, O Varuna, it was necessity (or temptation), an intoxicating draught, passion, dice, thoughtlessness. The old is there to mislead the young; even sleep brings unrighteousness.
7. Let me without sin give satisfaction to the angry god, like a slave to his bounteous lord. The lord god enlightened the foolish; he, the wisest, leads his worshipper to wealth.
8. O lord Varuna, may this song go well to thy heart! May we prosper in keeping and acquiring! Protect us, O gods, always with your blessings!

The consciousness of sin is a prominent feature in the religion of the Veda; so is likewise the belief that the gods are able to take away from man the heavy burden of his sins. And when we read such passages as "Varuna is merciful even to him who has committed sin" (Rv. VII. 87, 7), we should surely not allow the strange name of Varuna to jar on our ears, but should remember that it is but one of the many names which men invented in their helplessness to express their ideas of the Deity, however partial and imperfect.

The next hymn, which is taken from the Atharva-veda (IV. 16), will show how near the language of the ancient poets of India may approach to the language of the Bible:[18]

1. The great lord of these worlds sees as if he were near. If a man thinks he is walking by stealth, the gods know it all.
2. If a man stands or walks or hides, if he goes to lie down or to get up, what two people sitting together whisper, King Varuna knows it, he is there as the third.
3. This earth, too, belongs to Varuna, the king, and this wide sky with its ends far apart. The two seas (the sky and the ocean) are Varuna's loins; he is also contained in this small drop of water.
4. He who should flee far beyond the sky, even he would not be rid of Varuna, the king. His spies proceed from heaven towards this world; with thousand eyes they overlook this earth.
5. King Varuna sees all this, what is between heaven and earth, and what is beyond. He has counted the twinklings of the eyes of men. As a player throws the dice, he settles all things.
6. May all thy fatal nooses, which stand spread out seven by seven and threefold, catch the man who tells a lie, may they pass by him who tells the truth.

Another idea which we find in the Veda is that of *faith*: not only in the sense of trust in the gods, in their power, their protection, their kindness, but in that of belief in their existence. The Latin word *credo*, I believe, is the same as the Sanskrit "sraddhâ," and this sraddhâ occurs in the Veda: Rv. I. 102, 2: "Sun and moon go on in regular succession, that we may see, Indra, and believe"; Rv. I. 104, 6: "Destroy not our future offspring, O Indra, for we have believed in thy great power"; [and] Rv. I. 55, 5: "When Indra hurls again and again his thunderbolt, then they believe in the brilliant god."[19]

A similar sentiment, namely, that men only believe in the gods when they see their signs and wonders in the sky, is expressed by another poet:

> Thou, Indra, never findest a rich man to be thy friend;
> Wine-swillers despise thee.
> But when thou thunderest, when thou gatherest (the clouds),
> Then thou art called, like a father (Rv. VIII. 21, 14).

And with this belief in god, there is also coupled that doubt, that true skepticism, if we may so call it, which is meant to give to faith its real strength. We find passages, even in these early hymns, where the poet asks himself, whether there is really such a god as Indra—a question immediately succeeded by an answer, as if given to the poet by Indra himself. Thus we read:

> If you wish for strength, offer to Indra a hymn of praise:
> A true hymn, if Indra truly exist;
> For some one says, Indra does not exist!
> Who has seen him? Whom shall we praise? (Rv. VIII. 100, 3).

Then Indra answers through the poet: "Here am I, O worshipper, behold me here! In might I surpass all things."

Similar visions occur elsewhere, where the poet, after inviting a god to a sacrifice, or imploring his pardon for his offenses, suddenly exclaims that he has seen the god, and that he feels that his prayer is granted. For instance:

Hymn to Varuna (Rv. I. 25)

1. However we break thy laws from day to day, men as we are, O god, Varuna,
2. Do not deliver us unto death, nor to the blow of the furious, nor to the wrath of the spiteful!
3. To propitiate thee, O Varuna, we unbend thy mind with songs, as the charioteer a weary steed.
4. Away from me they flee dispirited, intent only on gaining wealth; as birds to their nests.
5. When shall we bring hither the man, who is victory to the warriors; when shall we bring Varuna, the wide-seeing, to be propitiated?
[6. They (Mitra and Varuna) take this in common; gracious, they never fail the faithful giver.]
7. He who knows the place of the birds that fly through the sky, who on the waters knows the ships;
8. He, the upholder of order, who knows the twelve months with the offspring of each, and knows the month that is engendered afterwards;
9. He who knows the track of the wind, of the wide, the bright, the mighty; and knows those who reside on high;
10. He, the upholder of order, Varuna, sits down among his people; he, the wise, sits there to govern.
11. From thence perceiving all wondrous things, he sees what has been and what will be done.
12. May he, the wise Âditya, make our paths straight all our days; may he prolong our lives!
13. Varuna, wearing golden mail, has put on his shining cloak; the spies sat down around him.
14. The god whom the scoffers do not provoke, nor the tormentors of men, nor the plotters of mischief;
15. He, who gives to men glory, and not half glory, who gives it even to our own selves;
16. Yearning for him, the far-seeing, my thoughts move onwards, as kine move to their pastures.

17. Let us speak together again, because my honey has been brought: that thou mayest eat what thou likest, like a friend.
18. Did I see the god who is to be seen by all, did I see the chariot above the earth? He must have accepted my prayers.
19. O hear this my calling, Varuna, be gracious now; longing for help, I have called upon thee.
20. Thou, O wise god, art lord of all, of heaven and earth: listen on thy way.
21. That I may live, take from me the upper rope, loose the middle, and remove the lowest!

In conclusion, let me tell you that there is in the Veda no trace of *metempsychosis*, or that transmigration of souls from human to animal bodies, which is generally supposed to be a distinguishing feature of Indian religion. Instead of this, we find what is really the *sine quâ non* of all real religion, a belief in immortality, and in personal immortality. Without a belief in personal immortality, religion surely is like an arch resting on one pillar, like a bridge ending in an abyss. We cannot wonder at the great difficulties felt and expressed by Bishop Warburton, and other eminent divines, with regard to the supposed total absence of the doctrine of immortality or personal immortality in the Old Testament; and it is equally startling that the Sadducces who sat in the same council with the high-priest, openly denied the resurrection.[20] However, though not expressly asserted anywhere, a belief in personal immortality is taken for granted in several passages of the Old Testament, and we can hardly think of Abraham or Moses as without a belief in life and immortality. But while this difficulty, so keenly felt with regard to the Jewish religion, ought to make us careful in the judgments which we form of other religions, and teach us the wisdom of charitable interpretation, it is all the more important to mark that in the Veda, passages occur where immortality of the soul, personal immortality, and personal responsibility after death, are clearly proclaimed. Thus we read: "He who gives alms goes to the highest place in heaven; he goes to the gods" (Rv. I. 125, 56).

Another poet, after rebuking those who are rich and do not communicate, says: "The kind mortal is greater than the great in heaven!"

Even the idea, so frequent in the later literature of the Brahmans, that immortality is secured by a son, seems implied, unless our translation deceives us, in one passage of the Veda: "Asmé (íti) vîrah marutah sushmî astu gánânâm yáh ásurah vi dhartâ, apâh yéna su-kshitáye tárema, ádha svám ókah abhí vah syáma." "O Maruts, may there be to us a strong son, who is a living ruler of men: through whom we may cross the waters on our way to the happy abode; then may we come to your own house!" (VII. 56, 24).

One poet prays that he may see again his father and mother after death (Rv. I. 24, 1); and the fathers (Pitris) are invoked almost like gods, oblations are offered to them, and they are believed to enjoy, in company with the gods, a life of never-ending felicity (Rv. X. 15, 16).

We find this prayer addressed to Soma:

Where there is eternal light, in the world where the sun is placed, in that immortal imperishable world place me, O Soma!

Where king Vaivasvata reigns, where the secret place of heaven is, where these mighty waters are, there make me immortal!

Where life is free, in the third heaven of heavens, where the worlds are radiant, there make me immortal!

Where wishes and desires are, where the bowl of the bright Soma is, where there is food and rejoicing, there make me immortal!

Where there is happiness and delight, where joy and pleasure reside, where the desires of our desire are attained, there make me immortal! (Rv. IX. 113, 7) [21]

Whether the old Rishis believed likewise in a place of punishment for the wicked, is more doubtful, though vague allusions to it occur in the Rig-veda, and more distinct descriptions are found in the Atharva-veda. In one verse it is said that the dead is rewarded for his good deeds, that he leaves or casts off all evil, and, glorified, takes his new body (Rv. X. 14, 8).[22] The dogs of Yama, the king of the departed, present some terrible aspects, and Yama is asked to protect the departed from them (Rv. X. 14, 11). Again, a pit (karta) is mentioned into which the lawless are said to be hurled down (Rv. IX. 73, 8), and into which Indra casts those who offer no sacrifices (Rv. I. 121, 13). One poet prays that the Âdityas may preserve him from the destroying wolf and from falling into the pit (Rv. II. 29, 6). In one passage we read that "those who break the commandments of Varuna and who speak lies are born for that deep place" (Rv. IV. 5, 5).[23]

Surely the discovery of a religion like this, as unexpected as the discovery of the jaw-bone of Abbeville, deserves to arrest our thoughts for a moment, even in the haste and hurry of this busy life. No doubt, for the daily wants of life, the old division of religions into true and false is quite sufficient; as for practical purposes we distinguish only between our own mother-tongue on the one side, and all other foreign languages on the other. But, from a higher point of view, it would not be right to ignore the new evidence that has come to light; and as the study of geology has given us a truer insight into the stratification of the earth, it is but natural to expect that a thoughtful study of the original works of three of the most important religions of the world, Brahmanism, Magism, and Buddhism, will modify our views as to the growth or history of religion, as to the hidden layers of religious thought beneath the soil on which we stand. Such inquiries should be undertaken without prejudice and without fear: the evidence is placed before us; our duty is to sift it critically, to weigh it honestly, and to wait for the results.

Three of these results, to which, I believe, a comparative study of religions is sure to lead, I may state before I conclude this lecture:

First, we shall learn that religions in their most ancient form, or in the minds of their authors, are generally free from many of the blemishes that attach to them in later times;

Second, we shall learn that there is hardly one religion which does not contain some truth, some important truth; truth sufficient to enable those who seek the Lord and feel after Him, to find Him in their hour of need;

Last, we shall learn to appreciate better than ever what we have in our own religion. No one who has not examined patiently and honestly the other religions of the world, can know what Christianity really is, or can join with such truth and sincerity in the words of St. Paul: "I am not ashamed of the Gospel of Christ."

Notes

1. Delivered at the Philosophical Institution, Leeds, in March 1865. Some of the points touched upon in this lecture have been more fully treated in my *History of Ancient Sanskrit Literature*. As the second edition of this work has been out of print for several years, I have here quoted a few passages from it in full.
2. "In the sciences of law and society, old means not old in chronology, but in structure: that is most archaic which lies nearest to the beginning of human progress considered as a development; and that is most modern which is furthest removed from that beginning" (J. F. McLennan, *Primitive Marriage*, p. 8).
3. *History of Ancient Sanskrit Literature*, p. 449.
4. *History of Ancient Sanskrit Literature*, second edition, p. 219 seq.
5. "Tat Savitur varenyam bhargo devasya dhîmahi, dhiyo yo nah prakodayât." Colebrooke, *Miscellaneous Essays*, i, 30. Many passages bearing on this subject have been collected by Dr. Muir in the third volume of his *Sanskrit Texts*, p. 114 seq.
6. Mommsen, *Inscriptiones Helveticæ*, 40. Becker, *Die inschriftlichen Überreste der Keltischen Sprache*, in *Beiträge zur Vergleichenden Sprachforschung*, vol. 3, p. 341. Lucan, *Phars.*, 1, 445, "horrensque feris altaribus Hesus."
7. Cf., G. Bühler, *Über Parjanya*, in Benfey's *Orient und Occident*, vol. 1, p. 214. In the Old Irish, *arg*, a drop, has been pointed out as derived from the same root as paraganya.
8. *History of Ancient Sanskrit Literature*, p. 569.
9. A last verse is added, which entirely spoils the poetical beauty and the whole character of the hymn. Its later origin seems to have struck even native critics, for the author of the Pada text did not receive it. "O Pragâpati, no other than thou hast embraced all these created things; may what we desired when we called on thee, be granted to us, may we be lords of riches."
10. I subjoin for some of the hymns here translated, the translation of the late Professor Wilson, in order to show what kind of difference there is between the traditional rendering of the Vedic hymns, as adopted by him, and their interpretation according to the rules of modern scholarship:

1. We ever offer fitting praise to the mighty Indra, in the dwelling of the worshipper, by which he (the deity) has quickly acquired riches, as (a thief) hastily carries (off the property) of the sleeping. Praise ill expressed is not valued among the munificent.·
2. Thou Indra, art the giver of horses, of cattle, of barley, the master and protector of wealth, the foremost in liberality, (the being) of many days; thou disappointest not desires (addressed to thee); thou art a friend to our friends: such an Indra, we praise.
3. Wise and resplendent Indra, the achiever of great deeds, the riches that are spread around are known to be thine: having collected them, victor (over thy enemies), bring them to us: disappoint not the expectation of the worshipper who trusts in thee.
4. Propitiated by these offerings, by these libations, dispel poverty with cattle and horses: may we, subduing our adversary, and relieved from enemies by Indra (pleased) by our libations, enjoy together abundant food.
5. Indra, may we become possessed of riches, and of food; and with energies agreeable to many, and shining around, may we prosper through thy divine favour, the source of prowess, of cattle, and of horses.
6. Those who were thy allies (the Maruts), brought thee joy: protector of the pious, those libations and oblations (that were offered thee on slaying Vritra), yielded thee delight, when thou, unimpeded by foes, didst destroy the ten thousand obstacles opposed to him who praised thee and offered thee libations.
7. Humiliator (of adversaries), thou goest from battle to battle, and destroyest by thy might city after city: with thy foe-prostrating associate (the thunderbolt), thou, Indra, didst slay afar off the deceiver named Namuki.

8. Thou hast slain Karanga and Parnaya with thy bright gleaming spear, in the cause of Atithigva: unaided, thou didst demolish the hundred cities of Vangrida, when besieged by Rigisvan.

9. Thou, renowned Indra, overthrewest by thy not-to-be-overtaken chariot-wheel, the twenty kings of men, who had come against Susravas, unaided, and their sixty thousand and ninety and nine followers.

10. Thou, Indra, hast preserved Susravas by thy succour, Tûrvayâna, by thy assistance: thou hast made Kutsa, Atithigva, and Âyu subject to the mighty though youthful Susravas.

11. Protected by the gods, we remain, Indra, at the close of the sacrifice, thy most fortunate friends: we praise thee, as enjoying through thee excellent offspring, and a long and prosperous life.

11. Favete linguis. [Favor with thy tongue—The Editor.]

12. Cf., Rv. I. 112, 25, "dyúbhir aktúbhih," by day and by night; also Rv. III. 31, 16; M. M., "Todtenbestattung," p. v.

13. Professor Benfey reads "durayantah," but all manuscripts that I know, without exception, read "darayantah."

14. For a different translation, see Roth, in *Deutsche Monatsschrift*, p. 89.

15. See Spiegel, *Erân*, p. 269, on Khai Khosru=Susravas.

16. Professor Wilson translates as follows:

1. When, Maruts, who make (all things) tremble, you direct your awful (vigour) downwards from afar, as light (descends from heaven), by whose worship, by whose praise (are you attracted)? To what (place of sacrifice), to whom, indeed, do you repair?

2. Strong be your weapons for driving away (your) foes, firm in resisting them: yours be the strength that merits praise, not (the strength) of a treacherous mortal.

3. Directing Maruts, when you demolish what is stable, when you scatter what is ponderous, then you make your way through the forest (trees) of earth and the defiles of the mountains.

4. Destroyers of foes, no adversary of yours is known above the heavens, nor (is any) upon earth: may your collective strength be quickly exerted, sons of Rudra, to humble (your enemies).

5. They make the mountains tremble, they drive apart the forest trees. Go, divine Maruts, whither you will, with all your progeny, like those intoxicated.

6. You have harnessed the spotted deer to your chariot; the red deer yoked between them (aids to) drag the car: the firmament listens for your coming, and men are alarmed.

7. Rudras, we have recourse to your assistance for the sake of our progeny: come quickly to the timid Kanva, as you formerly came, for our protection.

8. Should any adversary, instigated by you, or by man, assail us, withhold from him food and strength and your assistance.

9. Praketasas, who are to be unreservedly worshipped, uphold (the sacrificer) Kanva: come to us, Maruts, with undivided protective assistances, as the lightnings (bring) the rain.

10. Bounteous givers, you enjoy unimpaired vigour: shakers (of the earth), you possess undiminished strength: Maruts, let loose your anger like an arrow, upon the wrathful enemy of the Rishis.

17. *History of Ancient Sanskrit Literature*, p. 26, note.

18. This hymn was first pointed out by Professor Roth in a dissertation on the Atharva-veda (Tübingen, 1856), and it has since been translated and annotated by Dr. Muir, in his article on the *Vedic Theogony and Cosmogony*, p. 31.

19. During violent thunder-storms, the natives of New Holland are so afraid of War-ru-gu-ra, the evil spirit, that they seek shelter even in caves haunted by Ingnas, subordinate demons, which at other times they would enter on no account. There, in silent terror,

they prostrate themselves with their faces to the ground, waiting until the spirit, having expended his fury, shall retire to Uta (hell) without having discovered their hiding-place. *Transactions of Ethnological Society*, vol. 3, p. 229; Oldfield, *The Aborigines of Australia*.

20. Acts of the Apostles 22.30–23.6.

21. Professor Roth, after quoting several passages from the Veda in which a belief in immortality is expressed, remarks with great truth: "We here find, not without astonishment, beautiful conceptions on immortality expressed in unadorned language with childlike conviction. If it were necessary, we might find here the most powerful weapons against the view which has lately been revived, and proclaimed as new, that Persia was the only birthplace of the idea of immortality, and that even the nations of Europe had derived it from that quarter. As if the religious spirit of every gifted race was not able to arrive at it by its own strength." *Journal of the German Oriental Society*, vol. 4, p. 427. See Dr. Muir's article on "Yama," in the *Journal of the Royal Asiatic Society*, p. 10.

22. M. M., "Die Todtenbestattung bei den Brahmanen," *Zeitschrift der Deutschen Morgenländischen Gesellschaft*, vol. 9, p. xii.

23. Dr. Muir, article on "Yama," p. 18.

CHAPTER 3

PREFACE TO *CHIPS FROM A GERMAN WORKSHOP* (1867)

More than twenty years have passed since my revered friend Bunsen called me one day into his library at Carlton House Terrace and announced to me with beaming eyes that the publication of the Rig-veda was secure. He had spent many days in seeing the Directors of the East India Company, and explaining to them the importance of this work, and the necessity of having it published in England. At last his efforts had been successful, the funds for printing my edition of the text and commentary of the Sacred Hymns of the Brahmans had been granted, and Bunsen was the first to announce to me the happy result of his literary diplomacy. "Now," he said, "you have got a work for life—a large block that will take years to plane and polish. But mind," he added, "let us have from time to time some chips from your workshop."

I have tried to follow the advice of my departed friend, and I have published almost every year a few articles on such subjects as had engaged my attention, while prosecuting at the same time, as far as altered circumstances would allow, my edition of the Rig-veda, and of other Sanskrit works connected with it. These articles were chiefly published in the *Edinburgh* and *Quarterly* Reviews, in the *Oxford Essays*, and *Macmillan's* and *Fraser's* Magazines, in the *Saturday Review*, and in *The Times*. In writing them my principal endeavor has been to bring out even in the most abstruse subjects the points of real interest that ought to engage the attention of the public at large, and never to leave a dark nook or corner without attempting to sweep away the cobwebs of false learning, and let in the light of real knowledge. Here, too, I owe much to Bunsen's advice; and when last year I saw in Cornwall the large heaps of copper ore piled up around the mines, like so many heaps of rubbish, while the poor people were asking for coppers to buy bread, I frequently thought of Bunsen's words, "Your work is not finished when you have brought the ore from the mine: it must be sifted, smelted, refined and coined before it can be of real use, and contribute towards the intellectual food of mankind." I can hardly hope that in this my endeavor to be clear and plain, to follow the threads of every thought to the very ends, and to place the web of every argument clearly and fully before my readers, I have always been successful. Several of the subjects treated in these essays are, no doubt, obscure and difficult: but there is no subject, I believe, in the whole realm of human knowledge, that cannot be rendered clear and intelligible, if we ourselves have perfectly mastered it. And now while the two last volumes of my edition of the Rig-veda are passing through the press, I thought the time had come for gathering up a few armfuls of these chips and splinters, throwing away what seemed worthless, and putting the rest into some kind of shape, in order to clear my workshop for other work.

The first and second volumes which I am now publishing contain essays on the early thoughts of mankind, whether religious or mythological, and on early traditions and customs. There is to my mind no subject more absorbing than tracing the origin and first growth of human thought; not theoretically, or in accordance with the Hegelian laws of thought, or the Comtean epochs; but historically, and like an Indian trapper, spying for every footprint, every layer, every broken blade that might tell and testify of the former presence of man in his early wanderings and searchings after light and truth.

In the languages of mankind, in which everything new is old and everything old is new, an inexhaustible mine has been discovered for researches of this kind. Language still bears the impress of the earliest thoughts of man, obliterated, it may be, buried under new thoughts, yet here and there still recoverable in their sharp original outline. The growth of language is continuous, and by continuing our researches backward from the most modern to the most ancient strata, the very elements and roots of human speech have been reached, and with them the elements and roots of human thought. What lies beyond the beginnings of language, however interesting it may be to the physiologist, does not yet belong to the history of man, in the true and original sense of that word. Man means the thinker, and the first manifestation of thought is speech.

But more surprising than the continuity in the growth of language is the continuity in the growth of religion. Of religion, too, as of language, it may be said that in it everything new is old, and everything old is new, and that there has been no entirely new religion since the beginning of the world. The elements and roots of religion were there as far back as we can trace the history of man; and the history of religion, like the history of language, shows us throughout a succession of new combinations of the same radical elements. An intuition of God, a sense of human weakness and dependence, a belief in a Divine government of the world, a distinction between good and evil, and a hope of a better life—these are some of the radical elements of all religions. Though sometimes hidden, they rise again and again to the surface. Though frequently distorted, they tend again and again to their perfect form. Unless they had formed part of the original dowry of the human soul, religion itself would have remained an impossibility, and the tongues of angels would have been to human ears but as sounding brass or a tinkling cymbal. If we once understand this clearly, the words of St. Augustine, which have seemed startling to many of his admirers, become perfectly clear and intelligible, when he says: "What is now called the Christian religion, has existed among the ancients, and was not absent from the beginning of the human race, until Christ came in the flesh: from which time the true religion, which existed already, began to be called Christian."[1] From this point of view the words of Christ too, which startled the Jews, assume their true meaning, when He said to the centurion of Capernaum: "Many shall come from the east and the west, and shall sit down with Abraham, and Isaac, and Jacob, in the kingdom of heaven."

During the last fifty years the accumulation of new and authentic materials for the study of the religions of the world has been most extraordinary; but such are the difficulties in mastering these materials that I doubt whether the time has yet come for attempting to trace, after the model of the Science of Language, the definite outlines of the Science of Religion. By a succession of the most fortunate circumstances, the canonical books of three of the principal religions of the ancient world have lately been recovered: the Veda, the Zend-Avesta, and the Tripitaka. But not only have we thus gained access to the most authentic documents from which to study the ancient religion of the Brahmans, the Zoroastrians, and the Buddhists, but by discovering the real origin of Greek, Roman, and likewise of Teutonic, Slavonic, and Celtic mythology, it has become possible to separate the truly religious elements in the sacred traditions of these nations from the mythological crust by which they are surrounded, and thus to gain a clearer insight into the real faith of the ancient Aryan world.

If we turn to the Semitic world, we find that although but few new materials have been discovered from which to study the ancient religion of the Jews, yet a new spirit of inquiry has brought new life into the study of the sacred records of Abraham, Moses, and the Prophets; and the recent researches of Biblical scholars, though starting from the most opposite points, have all helped to bring out the historical interest of the Old Testament, in a manner not dreamt of by former theologians. The same may be said of another Semitic religion, the religion of Mohammed, since the Koran and the literature connected with it were submitted to the searching criticism of real scholars and historians. Some new materials for the study of the Semitic religions have come from the monuments of Babylon and Nineveh. The very images of Bel and Nisroch now stand before our eyes, and the inscriptions on the tablets may hereafter tell us even more of the thoughts of those who bowed their knees before them. The religious worship of the Phœnicians and Carthaginians has been illustrated by Movers from the ruins of their ancient temples, and from scattered notices in classical writers; nay, even the religious ideas of the nomads of the Arabian peninsula, previous to the rise of Mohammedanism, have been brought to light by the patient researches of Oriental scholars.

There is no lack of idols among the ruined and buried temples of Egypt with which to reconstruct the pantheon of that primeval country: nor need we despair of recovering more and more of the thoughts buried under the hieroglyphics of the inscriptions, or preserved in hieratic and demotic manuscripts, if we watch the brilliant discoveries that have rewarded the patient researches of the disciples of Champollion.

Besides the Aryan and Semitic families of religion, we have in China three recognized forms of public worship, the religion of Confucius, that of Lao-tse, and that of Fo (Buddha); and here, too, recent publications have shed new light, and have rendered an access to the canonical works of these religions, and an understanding of their various purports, more easy, even to those who have not mastered the intricacies of the Chinese language.

Among the Turanian nations, a few only, such as the Finns and the Mongolians, have preserved some remnants of their ancient worship and mythology, and these too have lately been more carefully collected and explained by d'Ohson, Castrèn, and others.

In America, the religions of Mexico and Peru had long attracted the attention of theologians; and of late years the impulse imparted to ethnological researches has induced travellers and missionaries to record any traces of religious life that could be discovered among the savage inhabitants of Africa, America, and the Polynesian islands.

It will be seen from these few indications, that there is no lack of materials for the student of religion; but we shall also perceive how difficult it is to master such vast materials. To gain a full knowledge of the Veda, or the Zend-Avesta, or the Tripitaka, of the Old Testament, the Koran, or the sacred books of China, is the work of a whole life. How then is one man to survey the whole field of religious thought, to classify the religions of the world according to definite and permanent criteria, and to describe their characteristic features with a sure and discriminating hand?

Nothing is more difficult to seize than the salient features, the traits that constitute the permanent expression and real character of a religion. Religion seems to be the common property of a large community, and yet it not only varies in numerous sects, as language does in its dialects, but it escapes our firm grasp till we can trace it to its real habitat, the heart of one true believer. We speak glibly of Buddhism and Brahmanism, forgetting that we are generalizing on the most intimate convictions of millions and millions of human souls, divided by half the world and by thousands of years.

It may be said that at all events where a religion possesses canonical books, or a definite number of articles, the task of the student of religion becomes easier, and this, no doubt, is true to a certain extent. But even then we know that the interpretation of these canonical books varies, so much so that sects appealing to the same revealed authorities—as, for instance, the founders of the Vedânta and the Sânkhya systems—accuse each other of error, if not of willful error or heresy. Articles, too, though drawn up with a view to define the principal doctrines of a religion, lose much of their historical value by the treatment they receive from subsequent schools; and they are frequently silent on the very points which make religion what it is.

A few instances may serve to show what difficulties the student of religion has to contend with, before he can hope firmly to grasp the facts on which his theories are to be based.

Roman Catholic missionaries who had spent their lives in China, who had every opportunity, while staying at the court of Pekin, of studying in the original the canonical works of Confucius and their commentaries, who could consult the greatest theologians then living, and converse with the crowds that thronged the temples of the capital, differed diametrically in their opinions as to the most vital points in the state religion of China. Lecomte, Fouquet, Prémare, and Bouvet thought it undeniable that Confucius, his predecessors and his disciples, had entertained the noblest ideas on the constitution of the universe, and had sacrificed to the true God in the most ancient temple of the earth. According to Maigrot, Navarette, on the contrary, and even according to the Jesuit Longobardi, the adoration of the Chinese was addressed to inanimate tablets, meaningless inscriptions, or, in the best case, to coarse ancestral spirits and beings without intelligence.[2] If we believe the former, the ancient deism of China approached the purity of the Christian religion; if we listen to the latter, the absurd fetichism [sic] of the multitude degenerated amongst the educated, into systematic materialism and atheism. In answer to the peremptory texts quoted by one party, the other adduced the glosses of accredited interpreters, and the dispute of the missionaries who had lived in China, and knew Chinese, had to be settled in the last instance by a decision of the see of Rome.

There is hardly any religion that has been studied in its sacred literature, and watched in its external worship with greater care than the modern religion of the Hindus, and yet it would be extremely hard to give a faithful and intelligible description of it. Most people who have lived in India would maintain that the Indian religion, as believed in and practiced at present by the mass of the people, is idol worship and nothing else. But let us hear one of the mass of the people, a Hindu of Benares, who in a lecture delivered before an English and native audience defends his faith and the faith of his forefathers against such sweeping accusations.

"If by idolatry," he says,

> is meant a system of worship which confines our ideas of the Deity to a mere image of clay or stone; which prevents our hearts from being expanded and elevated with lofty notions of the attributes of God; if this is what is meant by idolatry, we disclaim idolatry, we abhor idolatry, and deplore the ignorance or uncharitableness of those that charge us with this groveling system of worship. ... But if, firmly believing, as we do, in the omnipresence of God, we behold, by the aid of our imagination, in the form of an image any of His glorious manifestations, ought we to be charged with identifying them with the matter of the image, whilst during those moments of sincere and fervent devotion we do not even think of matter? If at the sight of a portrait of a beloved and venerated friend no longer existing in this world, our heart is filled with sentiments of love and reverence; if we fancy him present in the picture, still looking upon us with his wonted tenderness and affection, and then indulge our feelings of love and gratitude, should we be charged with offering the grossest insult to him—that of fancying him to be no other than a piece of painted paper? ... We really lament the ignorance or uncharitableness of those who confound our representative worship with the Phœnician, Grecian, or Roman idolatry as represented by European writers, and then charge us with polytheism in the teeth of thousands of texts in the Purânas, declaring in clear and unmistakable terms that there is but one God who manifests Himself as Brahma, Vishnu, and Rudra (Siva), in His functions of creation, preservation, and destruction.[3]

In support of these statements, this eloquent advocate quotes numerous passages from the sacred literature of the Brahmans, and he sums up his view of the three manifestations of the Deity in the words of their great poet Kalidâsa, as translated by Mr. Griffith:

> In those Three Persons the One God was shown:
> Each First in place, each Last—not one alone;
> Of Siva, Vishnu, Brahma, each may be
> First, second, third, among the Blessed Three.

If such contradictory views can be held and defended with regard to religious systems still prevalent amongst us, where we can cross-examine living witnesses, and appeal to chapter and verse in their sacred writings, what must the difficulty be when we have to deal with the religions of the past? I do not wish to disguise these difficulties which are inherent in a comparative study of the religions of the world. I rather dwell on them strongly, in order to show how much care and caution is required in so difficult a subject, and how much indulgence should be shown in judging of the shortcomings and errors that are unavoidable in so comprehensive a study. It was supposed at one time that a comparative analysis of the languages of mankind must transcend the powers of man; and yet, by the combined and well directed efforts of many scholars, great results have here been obtained, and the principles that must guide the student of the Science of Language are now firmly established. It will be the same with the Science of Religion. By a proper division of labor, the materials that are still wanting will be collected and published and translated, and when that is done, surely man will never rest till he has discovered the purpose that runs through the religions of mankind, and till he has reconstructed the true *Civitas Dei* on foundations as wide as the ends of the world. The Science of Religion may be the last of the sciences which man is destined to elaborate; but when it is elaborated, it will change the aspect of the world, and give a new life to Christianity itself.

The Fathers of the Church, though living in much more dangerous proximity to the ancient religions of the Gentiles, admitted freely that a comparison of Christianity and other religions was useful. "If there is any agreement," Basilius remarked, "between their (the Greeks') doctrines and our own, it may benefit us to know them: if not, then to compare them and to learn how they differ, will help not a little towards confirming that which is the better of the two."[4]

But this is not the only advantage of a comparative study of religions. The Science of Religion will for the first time assign to Christianity its right place among the religions of the world; it will show for the first time fully what was meant by the fullness of time; it will restore to the whole history of the world, in its unconscious progress towards Christianity, its true and sacred character.

Not many years ago great offense was given by an eminent writer who remarked that the time had come when the history of Christianity should be treated in a truly historical spirit—in the same spirit in which we treat the history of other religions, such as Brahmanism, Buddhism, or Mohammedanism. And yet what can be truer? He must be a man of little faith, who would fear to subject his own religion to the same critical tests to which the historian subjects all other religions. We need not surely crave a tender or merciful treatment for that faith which we hold to be the only true one. We should rather challenge for it the severest tests and trials, as the sailor would for the good ship to which he intrusts [*sic*] his own life, and the lives of those who are most dear to him. In the Science of Religion, we can decline no comparisons, nor claim any immunities for Christianity, as little as the missionary can, when wrestling with the subtle Brahman, or the fanatical Mussulman, or the plain speaking Zulu. And if we send out our missionaries to every part of the world to face every kind of religion, to shrink from no contest, to be appalled by no objections, we must not give way at home or within our own hearts to any misgivings, lest a comparative study of the religions of the world could shake the firm foundations on which we must stand or fall.

To the missionary, more particularly, a comparative study of the religions of mankind will be, I believe, of the greatest assistance. Missionaries are apt to look upon all other religions as something totally distinct from their own, as formerly they used to describe the languages of barbarous nations as something more like the twittering of birds than the articulate speech of men. The Science of Language has taught us that there is order and wisdom in all languages, and even the most degraded jargons contain the ruins of former greatness and beauty. The Science of Religion, I hope, will produce a similar change in our views of barbarous forms of faith and worship; and missionaries, instead of looking only for points of difference, will look out more anxiously for any common ground, any spark of the true light that may still be revived, any altar that may be dedicated afresh to the true God.[5]

And even to us at home, a wider view of the religious life of the world may teach many a useful lesson. Immense as is the difference between our own and all other religions of the world—and few can know that difference who have not honestly examined the foundations of their own as well as of other religions—the position which believers and unbelievers occupy with regard to their various forms of faith is very much the same all over the world. The difficulties which trouble us, have troubled the hearts and minds of men as far back as we can trace the beginnings of religious life. The great problems touching the relation of the Finite to the Infinite, of the human mind as the recipient, and of the Divine Spirit as the source of truth, are old problems indeed; and while watching their appearance in different countries, and their treatment under varying circumstances, we shall be able, I believe, to profit ourselves, both by the errors which others committed before us, and by the truth which they discovered. We shall know the rocks that threaten every religion in this changing and shifting world of ours, and having watched many a storm of religious controversy and many a shipwreck in distant seas, we shall face with greater calmness and prudence the troubled waters at home.

If there is one thing which a comparative study of religions places in the clearest light, it is the inevitable decay to which every religion is exposed. It may seem almost like a truism, that no religion can continue to be what it was during the lifetime of its founder and its first apostles. Yet it is but seldom borne in mind that without constant reformation, i.e., without a constant return to its fountain-head, every religion, even the most perfect, nay the most perfect on account of its very perfection, more even than others, suffers from its contact with the world, as the purest air suffers from the mere fact of its being breathed.

Whenever we can trace back a religion to its first beginnings, we find it free from many of the blemishes that offend us in its later phases. The founders of the ancient religions of the world, as far as we can judge, were minds of a high stamp, full of noble aspirations, yearning for truth, devoted to the welfare of their neighbors, examples of purity and unselfishness. What they desired to found upon earth was but seldom realized, and their sayings, if preserved in their original form, offer often a strange contrast to the practice of those who profess to be their disciples. As soon as a religion is established, and more particularly when it has become the religion of a powerful state, the foreign and worldly elements encroach more and more on the original foundation, and human interests mar the simplicity and purity of the plan which the founder had conceived in his own heart, and matured in his communings with his God. Even those who lived with Buddha misunderstood his words, and at the Great Council which had to settle the Buddhist canon, Asoka, the Indian Constantine, had to remind the assembled priests that "what had been said by Buddha, that alone was well said"; and that certain works ascribed to Buddha, as, for instance, the instruction given to his son, Râhula, were apocryphal, if not heretical.[6]

With every century, Buddhism, when it was accepted by nations differing as widely as Mongols and Hindus, when its sacred writings were translated into languages as wide apart as Sanskrit and Chinese, assumed widely different aspects, till at last the Buddhism of the Shamans in the steppes of Tartary is as different from the teaching of the original Samana as the Christianity of the leader of the Chinese rebels is from the teaching of Christ. If missionaries could show to the Brahmans, the Buddhists, the Zoroastrians, nay, even to the Mohammedans, how much their present faith differs from the faith of their forefathers and founders; if they could place in their hands and read with them in a kindly spirit the original documents on which these various religions profess to be founded, and enable them to distinguish between the doctrines of their own sacred books and the additions of later ages; an important advantage would be gained, and the choice between Christ and other Masters would be rendered far more easy to many a truth seeking soul. But for that purpose it is necessary that we too should see the beam in our own eyes, and learn to distinguish between the Christianity of the nineteenth century and the religion of Christ. If we find that the Christianity of the nineteenth century does not win as many hearts in India and China as it ought, let us remember that it was the Christianity of the first century in all its dogmatic simplicity, but with its overpowering love of God and man, that conquered the world and superseded religions and philosophies, more difficult to conquer than the religious and philosophical systems of Hindus and Buddhists. If we can teach something to the Brahmans in reading with them their sacred hymns, they too can teach us something when reading with us the Gospel of Christ.

Never shall I forget the deep despondency of a Hindu convert, a real martyr to his faith, who had pictured to himself from the pages of the New Testament what a Christian country must be, and who when he came to Europe found everything so different from what he had imagined in his lonely meditations at Benares! It was the Bible only that saved him from returning to his old religion, and helped him to discern beneath theological futilities, accumulated during nearly two thousand years, beneath pharisaical hypocrisy, infidelity, and want of charity, the buried, but still living seed, committed to the earth by Christ and His Apostles. How can a missionary in such circumstances meet the surprise and questions of his pupils, unless he may point to that seed, and tell them what Christianity was meant to be; unless he may show that, like all other religions, Christianity, too, has had its history; that the Christianity of the nineteenth century is not the Christianity of the Middle Ages, that the Christianity of the Middle Ages was not that of the early Councils, that the Christianity of the early Councils was not that of the Apostles, and "that what has been said by Christ, that alone was well said?"

The advantages, however, which missionaries and other defenders of the faith will gain from a comparative study of religions, though important hereafter, are not at present the chief object of these researches. In order to maintain their scientific character, they must be independent of all extraneous considerations: they must aim at truth, trusting that even unpalatable truths, like unpalatable medicine, will reinvigorate the system into which they enter. To those, no doubt, who value the tenets of their religion as the miser values his pearls and precious stones, thinking their value lessened if pearls and stones of the same kind are found in other parts of the world, the Science of Religion will bring many a rude shock; but to the true believer, truth, wherever it appears, is welcome, nor will any doctrine seem the less true or the less precious, because it was seen, not only by Moses or Christ, but likewise by Buddha or Lao-tse. Nor should it be forgotten that while a comparison of ancient religions will certainly show that some of the most vital articles of faith are the common property of the whole of mankind, at least of all who seek the Lord, if haply they might feel after Him, and find Him, the same comparison alone can possibly teach us what is peculiar to Christianity, and what has secured to it that preëminent position which now it holds in spite of all obloquy. The gain will be greater than the loss, if loss there be, which I, at least, shall never admit.

There is a strong feeling, I know, in the minds of all people against any attempt to treat their own religion as a member of a class, and, in one sense, that feeling is perfectly justified. To each individual, his own religion, if he really believes in it, is something quite inseparable from himself, something unique, that cannot be compared to anything else, or replaced by anything else. Our own religion is, in that respect, something like our own language. In its form it may be like other languages; in its essence and in its relation to ourselves, it stands alone and admits of no peer or rival.

But in the history of the world, our religion, like our own language, is but one out of many; and in order to understand fully the position of Christianity in the history of the world, and its true place among the religions of mankind, we must compare it, not with Judaism only, but with the religious aspirations of the whole world, with all, in fact, that Christianity came either to destroy or to fulfill. From this point of view Christianity forms part, no doubt, of what people call profane history, but by that very fact, profane history ceases to be profane, and regains throughout that sacred character of which it had been deprived by a false distinction. The ancient Fathers of the Church spoke on these subjects with far greater freedom than we venture to use in these days. Justin Martyr, in his *Apology* (A.D. 139), has this memorable passage:

> One article of our faith then is, that Christ is the first begotten of God, and we have already proved Him to be the very Logos (or universal Reason), of which mankind are all partakers; and therefore those who live according to the Logos are Christians, notwithstanding they may pass with you for Atheists; such among the Greeks were Sokrates and Herakleitos and the like; and such among the Barbarians were Abraham, and Ananias, and Azarias, and Misael, and Elias, and many others, whose actions, nay whose very names, I know, would be tedious to relate, and therefore shall pass them over. So, on the other side, those who have lived in former times in defiance of the Logos or Reason, were evil, and enemies to Christ and murderers of such as lived according to the Logos; but *they who have made or make the Logos or Reason the rule of their actions are Christians*, and men without fear and trembling.[7]

"God," says Clement (A.D. 200), "is the cause of all that is good: only of some good gifts He is the primary cause, as of the Old and New Testaments; of others the secondary, as of (Greek) philosophy. But even philosophy may have been given primarily by Him to the Greeks, before the Lord had called the Greeks also. For that philosophy, like a schoolmaster, has guided the Greeks also, as the Law did Israel, towards Christ. Philosophy, therefore, prepares and opens the way to those who are made perfect by Christ."[8]

And again: "It is clear that the same God to whom we owe the Old and New Testaments, gave also to the Greeks their Greek philosophy, by which the Almighty is glorified among the Greeks."[9]

And Clement was by no means the only one who spoke thus freely and fearlessly, though, no doubt, his knowledge of Greek philosophy qualified him better than many of his contemporaries to speak with authority on such subjects.

St. Augustine writes:

> If the Gentiles also had possibly something divine and true in their doctrines, our Saints did not find fault with it, although for their superstition, idolatry, and pride, and other evil habits, they had to be detested, and, unless they improved, to be punished by divine judgment. For the Apostle Paul, when he said something about God among the Athenians, quoted the testimony of some of the Greeks who had said something of the same kind: and this, if they came to Christ, would be acknowledged in them, and not blamed. St. Cyprian, too, uses such witnesses against the Gentiles. For when he speaks of the Magians, he says that the chief among them, Hostanes, maintains that the true God is invisible, and that true angels sit at His throne; and that Plato agrees with this, and believes in One God, considering the others to be angels or demons; and that Hermes Trismegistus also speaks of One God, and confesses that He is incomprehensible.[10]

Every religion, even the most imperfect and degraded, has something that ought to be sacred to us, for there is in all religions a secret yearning after the true, though unknown God. Whether we see the Papua squatting in dumb meditation before his fetich, or whether we listen to Firdusi exclaiming: "The height and the depth of the whole world have their centre in Thee, O my God! I do not know Thee what Thou art: but, I know that Thou art what Thou alone canst be,"—we ought to feel that the place whereon we stand is holy ground. There are philosophers, no doubt, to whom both Christianity and all other religions are exploded errors, things belonging to the past, and to be replaced by more positive knowledge. To them the study of the religions of the world could only have a pathological interest, and their hearts could never warm at the sparks of truth that light up, like stars, the dark yet glorious night of the ancient world. They tell us that the world has passed through the phases of religious and metaphysical errors, in order to arrive at the safe haven of positive knowledge of facts. But if they would but study positive facts, if they would but read, patiently and thoughtfully, the history of the world, as it is, not as it might have been: they would see that, as in geology, so in the history of human thought, theoretic uniformity does not exist, and that the past is never altogether lost. The oldest formations of thought crop out everywhere, and if we dig but deep enough, we shall find that even the sandy desert in which we are asked to live, rests everywhere on the firm foundation of that primeval, yet indestructible granite of the human soul—religious faith.

There are other philosophers, again, who would fain narrow the limits of the Divine government of the world to the history of the Jewish and of the Christian nations, who would grudge the very name of religion to the ancient creeds of the world, and to whom the name of natural religion has almost become a term of reproach. To them, too, I should like to say that if they would but study positive facts, if they would but read their own Bible, they would find that the greatness of

Divine Love cannot be measured by human standards, and that God has never forsaken a single human soul that has not first forsaken Him. "He hath made of one blood all nations of men, for to dwell on all the face of the earth; and hath determined the times before appointed, and the bounds of their habitation: that they should seek the Lord, if haply they might feel after Him, and find Him, though He be not far from every one of us." If they would but dig deep enough, they too would find that what they contemptuously call natural religion is in reality the greatest gift that God has bestowed on the children of man, and that without it, revealed religion itself would have no firm foundation, no living roots in the heart of man.

If by the essays here collected I should succeed in attracting more general attention towards an independent, yet reverent study of the ancient religions of the world, and in dispelling some of the prejudices with which so many have regarded the yearnings after truth embodied in the sacred writings of the Brahmans, the Zoroastrians, and the Buddhists, in the mythology of the Greeks and Romans, nay, even in the wild traditions and degraded customs of Polynesian savages, I shall consider myself amply rewarded for the labor which they have cost me. That they are not free from errors, in spite of a careful revision to which they have been submitted before I published them in this collection, I am fully aware, and I shall be grateful to any one who will point them out, little concerned whether it is done in a seemly or unseemly manner, as long as some new truth is elicited, or some old error effectually exploded. Though I have thought it right in preparing these essays for publication, to alter what I could no longer defend as true, and also, though rarely, to add some new facts that seemed essential for the purpose of establishing what I wished to prove, yet in the main they have been left as they were originally published. I regret that, in consequence, certain statements of facts and opinions are repeated in different articles in almost the same words; but it will easily be seen that this could not have been avoided without either breaking the continuity of an argument, or rewriting large portions of certain essays. If what is contained in these repetitions is true and right, I may appeal to a high authority "that in this country true things and right things require to be repeated a great many times." If otherwise, the very repetition will provoke criticism and insure refutation, I have added to all the articles the dates when they were written, these dates ranging over the last fifteen years; and I must beg my readers to bear these dates in mind when judging both of the form and the matter of these contributions towards a better knowledge of the creeds and prayers, the legends and customs of the ancient world.

Notes

1. August., *Retr.* 1, 13.
2. Abel Rémusat, *Mélanges*, p. 162.
3. The modern pandit's reply to the missionary who accuses him of polytheism is: "O, these are only various manifestations of the one God; the same as, though the sun be one in the heavens, yet he appears in multiform reflections upon the lake. The various sects are only different entrances to the one city." See W. W. Hunter, *Annals of Rural Bengal*, p. 116.
4. Basilius, *De legendis Græc.* libris, c. v.
5. Joguth Chundra Gangooly, a native convert, says: "I know from personal experience that the Hindu Scriptures have a great deal of truth. ... If you go to India, and examine the

common sayings of the people, you will be surprised to see what a splendid religion the Hindu religion must be. Even the most ignorant women have proverbs that are full of the purest religion. Now I am not going to India to injure their feelings by saying, 'Your Scripture is all nonsense, is good for nothing; anything outside the Old and New Testament is a humbug.' No; I tell you I will appeal to the Hindu philosophers, and moralists, and poets, at the same time bringing to them my light, and reasoning with them in the spirit of Christ. That will be my work." "A Brief Account of Joguth Chundra Gangooly, a Brahmin and a Convert to Christianity," *Christian Reformer*, August, 1860.

6. See Burnouf, *Lotus de la bonne Loi*, Appendice, No. x. § 4.

7. *Apol.*, i. 46.

8. Clem. Alex., *Strom.*, lib. I, cap. v. § 28.

9. *Strom.*, lib. VI, cap. v. § 42.

10. Augustinis, *De Baptismo contra Donatistas*, lib. VI, cap. xliv.

CHAPTER 4
BUDDHIST NIHILISM (1869)

I may be mistaken, but my belief is that the subject which I have chosen for my discourse cannot be regarded as alien to the general interests of this assembly.

Buddhism in its numerous varieties continues still the religion of the majority of mankind, and will therefore always occupy a very prominent place in a comparative study of the religions of the world. But the science of comparative theology, although the youngest branch on the tree of human knowledge, will, for an accurate and fruitful study of antiquity, soon become as indispensable as comparative philology. For how can we truly understand and properly appreciate a people, its literature, art, politics, morals, and philosophy, its entire conception of life, without having comprehended its religion, not only in its outer aspect, but in its innermost being, in its deepest far-reaching roots?

What our great poet once said almost prophetically of languages, may also be said of religions: "*He who knows only one knows none.*" As the true knowledge of a language requires a knowledge of languages, thus a true knowledge of religion requires a knowledge of religions. And however bold the assertion may sound, that all the languages of mankind have an Oriental origin, true it is that all religions, like the suns, have risen from the East.

Here, therefore, in treating religions scientifically (those of the Aryan as well as those of the Semitic races) the Oriental scholar lawfully enters into the "plenum" of philology, if philology still is, as our president told us yesterday, what it once intended and wished to be, namely, the true Humanitas, which, like an emperor of yore, could say of itself: "humani nihil a me alienum puto."[1]

Now it has been the peculiar fate of the religion of Buddha, that among all the so-called false or heathenish religions, it almost alone has been praised by all and everybody for its elevated, pure, and humanizing character. One hardly trusts one's eyes on seeing Catholic and Protestant missionaries vie with each other in their praises of the Buddha; and even the attention of those who are indifferent to all that concerns religion must be arrested for a moment, when they learn from statistical accounts that no religion, not even the Christian, has exercised so powerful an influence on the diminution of crime as the old simple doctrine of the Ascetic of Kapilavastu. Indeed no better authority can be brought forward in this respect than that of a still-living Bishop of the Roman Catholic Church. In his interesting work on the life of Buddha, the author, the Bishop of Ramatha, the Apostolic Vicar of Ava and Pegu, speaks with so much candor of the merits of the Buddhist religion, that we are often at a loss which most to admire, his courage or his learning. Thus he says in one place: "There are many moral precepts equally commanded and enforced in common by both creeds. It will not be deemed rash to assert that most of the moral truths, prescribed by the gospel, are to be met with in the Buddhistic scriptures."[2] In another place Bishop Bigandet says: "In reading the particulars of the life of the last Buddha Gaudama, it is impossible not to feel reminded of many circumstances relating to our Saviour's life, such as it has been sketched out by the Evangelists."[3]

I might produce many even stronger testimonies in honor of Buddha and Buddhism, but the above suffice for my purpose.

But then, on the other hand, it appears as if people had only permitted themselves to be so liberal in their praise of Buddha and Buddhism, because they could, in the end, condemn a religion which, in spite of all its merits, culminated in Atheism and Nihilism. Thus we are told by Bishop Bigandet: "It may be said in favor of Buddhism, that no philosophico-religious system has ever upheld, to an equal degree, the notions of a savior and deliverer, and the necessity of his mission, for procuring the salvation of man, in a Buddhist sense. The *role* of Buddha, from beginning to end, is that of a deliverer, who preaches a law designed to secure to man the deliverance from all the miseries he is laboring under. But by an inexplicable and deplorable eccentricity, the pretended savior, after having taught man the way to deliver himself from the tyranny of his passions, leads him, after all, into the bottomless gulf of a total annihilation."[4]

This language may have a slightly episcopal tinge, yet we find the same judgment, in almost identical words, pronounced by the most eminent scholars who have written on Buddhism. The warm discussions on this subject, which have recently taken place at the Académie des Inscriptions et Belles-Lettres of Paris, are probably known to many of those who are here present; but better still, the work of the man whose place has not yet been filled, either in the French Academy, or on the Council Board of German Science—the work of Eugène Burnouf, the true founder of a scientific study of Buddhism. Burnouf, too, in his researches arrives at the same result, namely, that Buddhism, as known to us from its canonical books, in spite of its great qualities, ends in Atheism and Nihilism.

Now, as to Atheism, it cannot be denied that, if we call the old gods of the Veda—Indra, and Agni, and Yama—gods, Buddha was an Atheist. He does not believe in the divinity of these deities. What is noteworthy is that he does not by any means deny their bare existence, just as little as St. Augustine and other Fathers of the Church endeavored to sublimize, or entirely explain away the existence of the Olympian deities. The founder of Buddhism treats the old gods as superhuman beings, and promises the believers that they shall after death be reborn into the world of the gods, and shall enjoy divine bliss with the gods. Similarly he threatens the wicked that after death they shall meet with their punishment in the subterranean abodes and hells, where the Asuras, Sarpas, Nâgas, and other evil spirits dwell, beings whose existence was more firmly rooted in the popular belief and language, than that even the founder of a new religion could have dared to reason them away. But although Buddha assigned to these mediatized gods and devils, palaces, gardens, and a court—not second to their former ones—he yet deprived them of all their sovereign rights. Although, according to Buddha, the worlds of the gods last for millions of years, they must perish, at the end of every kalpa, with the gods and with the spirits who in the circle of births have raised themselves to the world of the gods. Indeed, the reorganization of the spirit-world goes further still. Already, before Buddha, the Brahmans had surmounted the low stand-point of mythological poly-theism, and supplanting it by the idea of the Brahman, as the absolute divine or super-divine power.

What, then, does Buddha decree? To this Brahman also he assigns a place in his universe. Over and above the world of the gods with its six paradises, he heaps up sixteen Brahma-worlds, not to be attained through virtue and piety only, but through inner contemplation, through knowledge and enlightenment. The dwellers in these worlds are already purely spiritualized beings, without body, without weight, without desire, far above men and gods. Indeed, the Buddhist architect rises to a still more towering height, heaping upon the Brahma-world four still higher worlds, which he calls the world of the formless. All these worlds are open to man, and the beings ascend and descend in the circle of time, according to the works they have performed, according to the truths they have recognized. But in all these worlds the law of change obtains; in none is there exemption from birth, age, and death. The world of the gods will perish like that of men, even the world of the formless will not last forever; but the Buddha, the Enlightened and truly Free, stands higher, and will not be affected or disturbed by the collapse of the Universe: "Si fractus illabatur orbis, impavidum ferient ruinae."[5]

Now, however, we meet with a vein of irony, which one would hardly have expected in Buddha. Gods and devils he had located; to all mythological and philo-sophical acquisitions of the past he had done justice as far as possible. Even fabulous beings, such as Nâgas, Gandharvas, and Garudas, had escaped the process of disso-lution, which was to reach them later only at the hands of comparative mythology. There is only *one* idea, the idea of a personal creator, in regard to which Buddha is relentless.

It is not only denied, but even its origin, like that of an ancient myth, is carefully explained by him in its minutest details. This is done in the Brahmagâla-sûtra. Let us bear in mind that a destruction of the worlds occurs at the end of every kalpa, a destruction which not only annihilates earth and hell, but also all the worlds of the gods, and even the three lowest of the Brahma-worlds. A description of the duration of a kalpa can only be given in the language of Buddhism. Take a rock forming a cube of about fourteen miles, touch it once in a hundred years with a piece of fine cloth, and the rock will sooner be reduced to dust than a kalpa will have attained its end. It is said that at the end of the kalpa, after all the lower stories of the universe had been destroyed and a new world had again been slowly formed, the spirits dwelling in the higher Brahma-worlds had remained inviolate. Then one of these Spirits, a being without body, without weight, omnipresent and blessed within himself, descended, when his time had arrived, from the higher Brahma-world to the new-formed nether Brahma-world. There he first dwelt alone; but, by and by, the desire arose in him not to remain alone any longer. At the moment of the awaken-ing of this desire within him, a second being accidentally descended from the higher into the lower Brahma-world. Then and there the thought originated in the first being, "I am the Brahma, the great Brahma, the Highest, the Unconquerable, the Omniscient, the Lord and King of All. I am the Creator of all things, the Father of

All. *This* being has also been created by me; for as soon as I desired not to remain alone, my desire brought forth this second being." The other beings as they gradually descended from the higher worlds likewise believed that the first comer had been their Creator, for was he not older and mightier and handsomer than they?

But this is not all; for although it would explain how one spirit could consider himself the creator of other spirits, it would leave unexplained the circumstances of men on earth believing in such a creator. This is explained in the following manner: "In the course of time one of these higher beings sank lower and lower, and was finally born as a man on earth. There, by penances and deep meditation, he attained a state of inner enlightenment, which gives to man the faculty of remembering his former existences. He remembered the above narrated occurrences in the newly originated Brahma-world, and announced to mankind that there was a Creator, a Brahman, who had been prior to all other beings; that this Creator was eternal and immutable, while all beings created by him were mutable and mortal."

There is in this explanation, I believe, an unmistakable note of animosity, otherwise so alien to the character of Buddha, and the question naturally arises whether this can have been the doctrine of the founder of Buddhism himself. And herewith we at once approach our principal problem: "Is it possible to distinguish between Buddhism and the personal teaching of Buddha?" We possess the Buddhist canon and have a right to consider all that we find in this canon as orthodox Buddhist doctrine. But as there has been no lack of efforts in Christian theology to distinguish between the doctrine of the founder of our religion and that of the writers of the Gospels, to go beyond the canon of the New Testament, and to make the λόγια of the Master the only valid rule of our faith, so the same want was already felt at a very early period, among the followers of Buddha. King Asoka, the Indian Constantine, had to remind the assembled priests at the great council which had to settle the Buddhist canon, *that what had been said by Buddha that alone was well said.*[6] Works attributed to Buddha, but declared as apocryphal, or even as heterodox, already existed at that time.

Thus we are not by any means without an authority for distinguishing between Buddhism and the teaching of Buddha; the question is only whether such a separation is still practicable for us?

My belief is that all honest inquirers must oppose a No to this question. Burnouf never ventured to cast a glance beyond the boundaries of the Buddhist canon. What he finds in the canonical books, in the so-called Three Baskets, is to him the doctrine of Buddha, similarly as we must accept, as the doctrine of Christ, what is contained in the four Gospels.

Still the question ought to be asked again, and again, whether, at least with regard to certain doctrines or facts, it may not be possible to make a step further in advance, even with the conviction that it cannot lead us to results of apodictic certainty. For if, as happens frequently, we find in the different parts of the canon, views, not only differing from, but even contradictory to each other, it follows, I think, that one only of them can belong to Buddha personally, and I believe that in such a case we have the right to choose, and the liberty to accept *that* view as the original one, the one peculiar to Buddha, which *least* harmonizes with the later system of orthodox Buddhism.

As regards the denial of a Creator, or Atheism in the ordinary acceptation of the term, I do not think that any one passage from the books of the canon known to us, can be quoted which contradicts it, or which in any way presupposes the belief in a personal God or a Creator. All that may be urged are the words said to have been spoken by Buddha at the moment when he became the Enlightened, the Buddha. They are as follows: "Without ceasing shall I run through a course of many births, looking for the maker of this tabernacle—and painful is birth again and again. But now, maker of the tabernacle, thou hast been seen; thou shalt not make up this tabernacle again. All thy rafters are broken, thy ridge-pole is sundered; the mind, being sundered, has attained to the extinction of all desires."

Here in the maker of the tabernacle, i.e., the body, one might be tempted to see a Creator. But he who is acquainted with the general run of thought in Buddhism soon finds that this architect of the house is only a poetical expression, and that whatever meaning may underlie it, it evidently signifies a force subordinated to the Buddha, the Enlightened.

But whilst we have no ground for exonerating the Buddha personally from the accusation of Atheism, the matter stands very differently as regards the charge of Nihilism. Buddhist Nihilism has always been much more incomprehensible than mere Atheism. A kind of religion is still conceivable, when there is something firm somewhere, when a something, eternal and self-dependent, is recognized, if not *without* and *above* man, at least *within* him. But if, as Buddhism teaches, the soul, after having passed through all the phases of existence, all the worlds of the gods and of the higher spirits, attains finally Nirvâna as its highest aim and last reward, i.e., becomes quite extinct, then religion is not any more what it ought to be—a bridge from the finite to the infinite, but a trap-bridge hurling man into the abyss, at the very moment when he thought he had arrived at the stronghold of the Eternal. According to the metaphysical doctrine of Buddhism, the soul cannot dissolve itself in a higher being, or be absorbed in the absolute substance, as was taught by the Brahmans and other mystics of ancient and modern times. For Buddhism knew not the Divine, the Eternal, the Absolute, and the soul, even as the I, or as the mere Self, the Atman, as called by the Brahmans, was represented in the orthodox Metaphysics of Buddhism as transient, as futile, as a mere phantom.

No person who reads with attention the metaphysical speculations on the Nirvâna contained in the Buddhist canon, can arrive at any other conviction than that expressed by Burnouf, namely, that Nirvâna, the highest aim, the *summum bonum* of Buddhism, is the absolute nothing.

Burnouf adds, however, that this doctrine, in its crude form, appears only in the third part of the canon, the so-called Abhidharma, but not in the first and second parts, in the Sûtras, the sermons, and the Vinaya, the ethics, which together bear the name of Dharma or Law. He next points out that, according to some ancient authorities, this entire part of the canon was designated as "not pronounced by Buddha."[7] These are, at once, two important limitations. I add a third, and maintain that sayings of the Buddha occur in the first and second parts of the canon, which are in open contradiction to this metaphysical Nihilism.

Now as regards the soul, or the self, the existence of which, according to the orthodox metaphysics, is purely phenomenal, a sentence attributed to the Buddha says, "Self is the Lord of Self, who else could be the Lord?" And again, "A man who controls himself enters the untrodden land through his own self-controlled self." And this untrodden land is the Nirvâna.

Nirvâna certainly means extinction, whatever its later arbitrary interpretations may have been, and seems therefore to imply, even etymologically, a real blowing out or passing away. But Nirvâna occurs also in the Brahmanic writings, as synonymous with Moksha, Nirvritti, and other words, all designating the highest stage of spiritual liberty and bliss, but not annihilation. Nirvâna may mean the extinction of many things—of selfishness, desire, and sin, without going so far as the extinction of subjective consciousness. Further, if we consider that Buddha himself, after he had already seen Nirvâna, still remains on earth until his body falls a prey to death; that Buddha appears, in the legends, to his disciples even after his death, it seems to me that all these circumstances are hardly reconcilable with the orthodox metaphysical doctrine of Nirvâna.

What does it mean when Buddha calls reflection the path of immortality, and thoughtlessness the path of death? Buddhaghosha, a learned man of the fifth century, here explains immortality by Nirvâna, and that this was also Buddha's thought is clearly established by a passage following immediately after: "These wise people, meditative, steady, always possessed of strong powers, attain to Nirvâna, the highest happiness." Can this be annihilation?; and would such expressions have been used by the founder of this new religion, if what he called immortality had, in his own idea, been annihilation?

I could quote many more such passages did I not fear to tire you. Nirvâna occurs even in the purely moral sense of quietness and absence of passion. "When a man can bear everything without uttering a sound," says Buddha, "he has attained Nirvâna." Quiet long-suffering he calls the highest Nirvâna; he who has conquered passion and hatred is said to enter into Nirvâna.

In other passages, Nirvâna is described as the result of just knowledge. There we read: "Hunger or desire is the worst ailment, the body the greatest of all evils; where this is properly known, there is Nirvâna, the greatest happiness."

When it is said in one passage that Rest (Sânti) is the highest bliss, it is said in another that Nirvâna is the highest bliss.

Buddha says: "The sages who injure nobody, and who always control their body, they will go to the unchangeable place (Nirvâna), where, if they have gone, they will suffer no more."

Nirvâna is called the quiet place, the immortal place, even simply that which is immortal; and the expression occurs, that the wise dived into this immortal. As, according to Buddha, everything that was made, everything that was put together, passes away again, and resolves itself into its component parts, he calls in contradistinction, that which is not made, i.e., the uncreated and eternal, Nirvâna. He says: "When you have understood the destruction of all that was made, you will understand that which was not made." Whence it appears that even for him a certain something exists, which is not made, which is eternal and imperishable.

On considering such sayings, to which many more might be added, one recognizes in them a conception of Nirvâna, altogether irreconcilable with the Nihilism of the third part of the Buddhist canon. The question in such matters is not a more or less, but an *aut-aut* [either-or]. If these sayings have maintained themselves, in spite of their contradiction to orthodox metaphysics, the only explanation, in my opinion, is that they were too firmly fixed in the tradition which went back to Buddha and his disciples. What Bishop Bigandet and others represent as the popular view of the Nirvâna, in contradistinction to that of the Buddhist divines, was, if I am not mistaken, the conception of Buddha and his disciples. It represented the entrance of the soul into rest, a subduing of all wishes and desires, indifference to joy and pain, to good and evil, an absorption of the soul in itself, and a freedom from the circle of existences from birth to death, and from death to a new birth. This is still the meaning which educated people attach to it, whilst, to the minds of the larger masses,[8] Nirvâna suggests rather the idea of a Mohammedan paradise or of blissful Elysian fields.

Only in the hands of the philosophers, to whom Buddhism owes its metaphysics, the Nirvâna, through constant negations, carried to an indefinite degree, through the excluding and abstracting of all that is not Nirvâna, at last became an empty Nothing, a philosophical myth. There is no lack of such philosophical myths either in the East or in the West. What has been fabled by philosophers of a Nothing, and of the terrors of a Nothing, is as much a myth as the myth of Eos and Tithonus. There is no more a Nothing than there is an Eos or a Chaos. All these are sickly, dying, or dead words, which, like shadows and ghosts, continue to haunt language, and succeed in deceiving for a while even the healthiest understanding.

Even modern philosophy is not afraid to say that there is a Nothing. We find passages in the German mystics, such as Eckhart and Tauler, where the abyss of the Nothing is spoken of quite in a Buddhist style. If Buddha had said, like St. Paul, "that what no eye hath seen, nor ear heard, neither has it entered into the heart of man," was prepared in the Nirvâna for those who had advanced to the highest degree of spiritual perfection, such expressions would have been quite sufficient to serve as a proof to the philosophers by profession that this Nirvâna, which could not become an object of perception by the senses, nor of conception by the categories of the understanding, could be nothing more nor less than the Nothing. Could we dare with Hegel to distinguish between a Nothing (Nichts) and a Not (Nicht), we might say that the Nirvâna had through a false dialectical process become from a relative Nothing an absolute Not. This was the work of the theologians and of the orthodox philosophers. But a religion has never been founded by such teaching, and a man like Buddha, who knew mankind, must have known that he could not with such weapons overturn the tyranny of the Brahmans. Either we must bring ourselves to believe that Buddha taught his disciples two diametrically opposed doctrines on Nirvâna, say an exoteric and esoteric one, or we must allow *that* view of Nirvâna to have been the original view of the founder of this marvelous religion, which corresponds best with the simple, clear, and practical character of Buddha.

I have now said all that can be said in vindication of Buddha within the brief time allowed to these discourses. But I should be sorry if you carried away the impression that Buddhism contained nothing but empty, useless speculations; permit me, therefore, to read to you, in conclusion, a short Buddhist parable, which will show you Buddhism in a more human form. It is borrowed from a work which will soon appear, and which contains the translation of the parables used by the Buddhists to obtain acceptance for their doctrines amongst the people. I shall only omit some technical expressions and minor details which are of no importance:[9]

Some time after this, Kisâgotamî gave birth to a son. When the boy was able to walk by himself, he died. The young girl, in her love for it, carried the dead child clasped to her bosom, and went about from house to house, asking if any one would give her some medicine for it. When the neighbors saw this, they said, "Is the young girl mad that she carries about on her breast the dead body of her son!" But a wise man thinking to himself, "Alas!, this Kisâgotamî does not understand the law of death, I must comfort her," said to her, "My good girl, I cannot myself give medicine for it, but I know of a doctor who can attend to it." The young girl said, "If so, tell me who it is." The wise man continued "Buddha can give medicine, you must go to him."

Kisâgotamî went to Buddha and, doing homage to him, said, "Lord and master, do you know any medicine that will be good for my boy?" Buddha replied, "I know of some." She asked, "What medicine do you require?" He said, "I want a handful of mustard seed." The girl promised to procure it for him, but Buddha continued, "I require some mustard seed taken from a house where no son, husband, parent, or slave has died." The girl said, "Very good," and went to ask for some at the different houses, carrying the dead body of her son astride on her hip. The people said, "Here is some mustard seed, take it." Then she asked, "In my friend's house has there died a son, a husband, a parent, or a slave?" They replied, "Lady, what is this that you say! *The living are few, but the dead are many.*" Then she went to other houses, but one said, "I have lost a son"; another, "I have lost my parents"; another, "I have lost my slave." At last, not being able to find a single house where no one had died, from which to procure the mustard seed, she began to think, "This is a heavy task that I am engaged in. I am not the only one whose son is dead. In the whole of the Sâvatthi country, everywhere children are dying, parents are dying." Thinking thus, she was seized by fear, and putting away her affection for her child, she summoned up resolution, and left the dead body in a forest; then she went to Buddha and paid him homage. He said to her, "Have you procured the handful of mustard seed?" "I have not," she replied; "the people of the village told me, '*The living are few, but the dead are many.*'" Buddha said to her, "You thought that you alone had lost a son; the law of death is that among all living creatures there is no permanence." When Buddha had finished preaching the law, Kisâgotamî was established in the reward of the noviciate; and all the assembly who heard the law were established in the same reward.

Some time afterwards, when Kisâgotamî was one day engaged in the performance of her religious duties, she observed the lights (in the houses) now shining, now extinguished, and began to reflect, "My state is like these lamps." Buddha, who was then in the Gandhakutî building, sent his sacred appearance to her, which said to her, just as if he himself were preaching, "All living beings resemble the flame of these lamps, one moment lighted, the next extinguished; those only who have arrived at Nirvâna are at rest." Kisâgotamî, on hearing this, reached the stage of a saint possessed of intuitive knowledge.

Gentlemen, this is a specimen of the true Buddhism; this is the language, intelligible to the poor and the suffering, which has endeared Buddhism to the hearts of millions,—not the silly, metaphysical phantasmagorias of worlds of gods and worlds of Brahma, or final dissolution of the soul in Nirvâna,—no, the beautiful, the tender, the humanly true, which, like pure gold, lies buried in all religions, even in the sand of the Buddhist canon.

Notes

1. "Nothing that relates to man is alien to me."—The Editor.
2. Page 494.
3. Page 495.
4. Page viii.
5. "Were the world to come crashing down, he would remain undaunted amid the ruins."—The Editor.
6. See Max Müller's *Chips from a German Workshop*, second edition, vol. 1, p. xxiv.
7. Max Müller's *Chips*, second edition, vol. 1, p. 285, note.
8. Bigandet, *The Life or Legend of Gaudama, the Buddha of the Burmese*, with Annotations. The ways to Neibban, and notice on the Phongyies, or Burmese Monks. 8vo, sewed. London, Trübner & Co., pp. xi, 538, and v.
9. *Buddhaghosha's Parables*. Translated from the Burmese by Captain H. T. Rogers, R. E. With an Introduction containing "Buddha's Dhammapada, or the Path of Virtue." Translated from Pâli by Professor F. Max Müller. London, Trübner & Co.

CHAPTER 5

ON FALSE ANALOGIES IN COMPARATIVE THEOLOGY (1870)

Very different from the real similarities that can be discovered in nearly all the religions of the world and which, owing to their deeply human character, in no way necessitate the admission that one religion borrowed from the other, are those minute coincidences between the Jewish and the Pagan religions which have so often been discussed by learned theologians, and which were intended by them as proof positive, either that the Pagans borrowed their religious ideas direct from the Old Testament, or that some fragments of a primeval revelation, granted to the ancestors of the whole race of mankind, had been preserved in the temples of Greece and Italy.

Bochart, in his *Geographia Sacra,* considered the identity of Noah and Saturn so firmly established as hardly to admit of the possibility of a doubt. The three sons of Saturn—Jupiter, Neptune, and Pluto—he represented as having been originally the three sons of Noah: Jupiter being Ham; Neptune, Japhet; and Shem, Pluto. Even in the third generation the two families were proved to have been one, for Phut, the son of Ham, or of Jupiter Hammon, could be no other than Apollo Pythius; Canaan no other than Mercury; and Nimrod no other than Bacchus, whose original name was supposed to have been Bar-chus, the son of Cush. G. J. Vossius, in his learned work, *De Origine et Progressu Idolatriæ* (1688), identified Saturn with Adam, Janus with Noah, Pluto with Ham, Neptune with Japhet, Minerva with Naamah, Vulcan with Tubal Cain, Typhon with Og. Huet, the friend of Bochart, and the colleague of Bossuet, went still farther; and in his classical work, the *Demonstratio Evangelica,* he attempted to prove that the whole theology of the heathen nations was borrowed from Moses, whom he identified not only with ancient law-givers, like Zoroaster and Orpheus, but with gods and demi-gods, such as Apollo, Vulcan, Faunus, and Priapus.

All this happened not more than two hundred years ago; and even a hundred years ago, nay, even after the discovery of Sanskrit and the rise of Comparative Philology, the troublesome ghost of Huet was by no means laid at once [to rest]. On the contrary, as soon as the ancient language and religion of India became known in Europe, they were received by many people in the same spirit. Sanskrit, like all other languages, was to be derived from Hebrew, the ancient religion of the Brahmans from the Old Testament.

There was at that time an enthusiasm among Oriental scholars, particularly at Calcutta, and an interest for Oriental antiquities in the public at large, of which we in these days of apathy for Eastern literature can hardly form an adequate idea. Everybody wished to be first in the field, and to bring to light some of the treasures which were supposed to be hidden in the sacred literature of the Brahmans. Sir William Jones, the founder of the Asiatic Society at Calcutta, published in the first volume of the *Asiatic Researches* his famous essay, "On the Gods of Greece, Italy, and India"; and he took particular care to state that his essay, though published only in 1788, had been written in 1784. In that essay he endeavored to show that there existed an intimate connection, not only between the mythology of India and that of Greece and Italy, but likewise between the legendary stories of the Brahmans and the accounts of certain historical events as recorded in the Old Testament. No doubt, the temptation was great. No one could look down for a moment into the rich mine of religious and mythological lore that was suddenly opened before the eyes of scholars and theologians without being struck by a host of similarities, not only in the languages, but also in the ancient traditions of the Hindus, the Greeks, and the Romans; and if at that time the Greeks and Romans were still supposed to have borrowed their language and their religion from Jewish quarters, the same conclusion could hardly be avoided with regard to the language and the religion of the Brahmans of India.

The first impulse to look in the ancient religion of India for reminiscences of revealed truth seems to have come from missionaries rather than from scholars. It arose from a motive, in itself most excellent, of finding some common ground for those who wished to convert and those who were to be converted. Only, instead of looking for that common ground where it really was to be found—namely, in the broad foundations on which all religions are built up: the belief in a divine power, the acknowledgment of sin, the habit of prayer, the desire to offer sacrifice, and the hope of a future life—the students of Pagan religion as well as Christian missionaries were bent on discovering more striking and more startling coincidences, in order to use them in confirmation of their favorite theory that some rays of a primeval revelation, or some reflection of the Jewish religion, had reached the uttermost ends of the world. This was a dangerous proceeding—dangerous because superficial, dangerous because undertaken with a foregone conclusion; and very soon the same arguments that had been used on one side in order to prove that all religious truth had been derived from the Old Testament were turned against Christian scholars and Christian missionaries, in order to show that it was not Brahmanism and Buddhism which had borrowed from the Old and New Testament, but that the Old and the New Testament had borrowed from the more ancient religions of the Brahmans and Buddhists.

This argument was carried out, for instance, in Holwell's *Original Principles of the Ancient Brahmans,* published in London as early as 1779, in which the author maintains that "the Brahmanic religion is the first and purest product of supernatural revelation," and "that the Hindu scriptures contain to a moral certainty the original doctrines and terms of restoration delivered from God himself, by the mouth of his first created Birmah, to mankind, at his first creation in the form of man."

Sir William Jones tells us that one or two missionaries in India had been absurd enough, in their zeal for the conversion of the Gentiles, to urge "that the Hindus were even now almost Christians, because their Brahma, Vishnu, and Mahesa were no other than the Christian Trinity"; a sentence in which, he adds, we can only doubt whether folly, ignorance, or impiety predominates.[1]

Sir William Jones himself was not likely to fall into that error. He speaks against it most emphatically. "Either," he says, "the first eleven chapters of Genesis—all due allowance being made for a figurative Eastern style—are true, or the whole fabric of our national religion is false; a conclusion which none of us, I trust, would wish to be drawn. But it is not the truth of our national religion as such that I have at heart; it is truth itself; and if any cool, unbiassed [sic] reasoner will clearly convince me that Moses drew his narrative through Egyptian conduits from the primeval fountains of Indian literature, I shall esteem him as a friend for having weeded my mind from a capital error, and promise to stand amongst the foremost in assisting to circulate the truth which he has ascertained."

But though he speaks so strongly against the uncritical proceedings of those who would derive anything that is found in the Old Testament from Indian sources, Sir William Jones himself was really guilty of the same want of critical caution in his own attempts to identify the gods and heroes of Greece and Rome with the gods and heroes of India. He begins his essay, "On the Gods of Greece, Italy, and India," with the following remarks:

> We cannot justly conclude, by arguments preceding the proof of facts, that one idolatrous people must have borrowed their deities, rites, and tenets from another, since gods of all shapes and dimensions may be framed by the boundless powers of imagination, or by the frauds and follies of men, in countries never connected; but when features of resemblance, too strong to have been accidental, are observable in different systems of polytheism, without fancy or prejudice to colour them and improve the likeness, we can scarce help believing that some connection has immemorially subsisted between the several nations who have adopted them. It is my design in this essay to point out such a resemblance between the popular worship of the old Greeks and Italians and that of the Hindus; nor can there be any room to doubt of a great similarity between their strange religions and that of Egypt, China, Persia, Phrygia, Phœnice, and Syria; to which, perhaps, we may safely add some of the southern kingdoms, and even islands of America; while the Gothic system which prevailed in the northern regions of Europe was not merely similar to those of Greece and Italy, but almost the same in another dress, with an embroidery of images apparently Asiatic. From all this, if it be satisfactorily proved, we may infer a general union or affinity between the most distinguished inhabitants of the primitive world at the time when they deviated, as they did too early deviate, from the rational adoration of the only true God.[2]

Here, then, in an essay written nearly a hundred years ago by Sir W. Jones, one of the most celebrated Oriental scholars in England, it might seem as if we should find the first outlines of that science which is looked upon as but of to-day or yesterday—the outlines of Comparative Mythology. But in such an expectation we are disappointed. What we find is merely a superficial comparison of the mythology of India and that of other nations, both Aryan and Semitic, without any scientific value, because carried out without any of those critical tests which alone keep Comparative Mythology from running riot. This is not intended as casting a slur on Sir W. Jones. At his time the principles which have now been established by the students of the science of language were not yet known, and as with words, so with the names of deities, similarity of sound, the most treacherous of all sirens, was the only guide in such researches.

It is not pleasant to have to find fault with a man possessed of such genius, taste, and learning as Sir W. Jones, but no one who is acquainted with the history of these researches will be surprised at my words. It is the fate of all pioneers, not only to be left behind in the assault which they had planned, but to find that many of their approaches were made in a false direction, and had to be abandoned. But as the authority of their names continues to sway the public at large, and is apt to mislead even painstaking students and to entail upon them repeated disappointments, it is necessary that those who know should speak out, even at the risk of being considered harsh or presumptuous.

A few instances will suffice to show how utterly baseless the comparisons are which Sir W. Jones instituted between the gods of India, Greece, and Italy. He compares the Latin Janus with the Sanskrit deity Ganesa. It is well known that Janus is connected with the same root that has yielded the names of Jupiter, Zeus, and Dyaus, while Ganesa is a compound, meaning lord of hosts, lord of the companies of gods.

Saturnus is supposed to have been the same as Noah, and is then identified by Sir W. Jones with the Indian Manu Satyavrata, who escaped from the flood. Ceres is compared with the goddess Sri, Jupiter or Diespiter with Indra or Divaspati; and though etymology is called a weak basis for historical inquiries, the three syllables Jov in Jovis, Zeu in Zeus, and Siv in Siva are placed side by side, as possibly containing the same root, only differently pronounced. Now the s of Siva is a palatal s, and no scholar who has once looked into a book on Comparative Philology need be told that such an s could never correspond to a Greek Zeta or a Latin J.

In Krishna, the lovely shepherd-god, Sir W. Jones recognizes the features of Apollo Nomius, who fed the herds of Admetus, and slew the dragon Python, and he leaves it to etymologists to determine whether Gopâla—i.e., the cow-herd—may not be the same word as Apollo. We are also assured, on the authority of Colonel Vallancey, that Krishna in Irish means the sun, and that the goddess Kâlî, to whom human sacrifices were offered, as enjoined in the Vedas (?) was the same as Hekate. In conclusion, Sir W. Jones remarks, "I strongly incline to believe that Egyptian priests have actually come from the Nile to the Gangâ and Yamunâ, and that they visited the Sarmans of India, as the sages of Greece visited them, rather to acquire than to impart knowledge."

The interest that had been excited by Sir William Jones's researches did not subside, though he himself did not return to the subject, but devoted his great powers to more useful labors. Scholars, both in India and in Europe, wanted to know more of the ancient religion of India. If Jupiter, Apollo, and Janus had once been found in the ancient pantheon of the Brahmans; if the account of Noah and the deluge could be traced back to the story of Manu Satyavrata, who escaped from the flood, more discoveries might be expected in this newly-opened mine, and people rushed to it with all the eagerness of gold-diggers. The idea that everything in India was of extreme antiquity had at that time taken a firm hold on the minds of all students of Sanskrit; and, as there was no one to check their enthusiasm, everything that came to light in Sanskrit literature was readily accepted as more ancient than Homer, or even than the Old Testament.

It was under these influences that Lieutenant Wilford, a contemporary of Sir William Jones at Calcutta, took up the thread which Sir William Jones had dropped, and determined at all hazards to solve the question which at that time had excited a world-wide interest. Convinced that the Brahmans possessed in their ancient literature the originals, not only of Greek and Roman mythology, but likewise of the Old Testament history, he tried every possible means to overcome their reserve and reticence. He related to them, as well as he could, the principal stories of classical mythology, and the leading events in the history of the Old Testament; he assured them that they would find the same things in their ancient books, if they would but look for them; he held out the hopes of ample rewards for any extracts from their sacred literature containing the histories of Adam and Eve, of Deukalion and Prometheus; and at last he succeeded. The coyness of the Pandits yielded; the incessant demand created a supply; and for several years essay after essay appeared in the *Asiatic Researches,* with extracts from Sanskrit manuscripts, containing not only the names of Deukalion, Prometheus, and other heroes and deities of Greece, but likewise the names of Adam and Eve, of Abraham and Sarah, and all the rest.

Great was the surprise, still greater the joy, not only in Calcutta, but in London, at Paris, and all the universities of Germany. The Sanskrit manuscripts, from which Lieutenant Wilford quoted, and on which his theories were based, had been submitted to Sir W. Jones and other scholars; and though many persons were surprised, and for a time even incredulous, yet the fact could not be denied that all was found in these Sanskrit manuscripts as stated by Lieutenant Wilford. Sir W. Jones, then president of the Asiatic Society, printed the following declaration at the end of the third volume of the *Asiatic Researches*:

> Since I am persuaded that the learned essay on Egypt and the Nile has afforded you equal delight with that which I have myself received from it, I cannot refrain from endeavoring to increase your satisfaction by confessing openly that I have at length abandoned the greatest part of the natural distrust and incredulity which had taken possession of my mind before I had examined the sources from which our excellent associate, Lieutenant Wilford, has drawn so great a variety of new and interesting opinions. Having lately read again and again, both alone and with a Pandit, the numerous original passages in the Purânas, and other Sanskrit books, which the writer of the dissertation adduces in support of his assertions, I am happy in bearing testimony to his perfect good faith and general accuracy, both in his extracts and in the translation of them.

Sir W. Jones then proceeds himself to give a translation of some of these passages. "The following translation," he writes, "of an extract from the Padma-purâna is minutely exact":

1. To Satyavarman, the sovereign of the whole earth, were born three sons: the eldest, Sherma; then Charma; and thirdly, Jyapeti.
2. They were all men of good morals, excellent in virtue and virtuous deeds, skilled in the use of weapons to strike with, or to be thrown, brave men, eager for victory in battle.
3. But Satyavarman, being continually delighted with devout meditation, and seeing his sons fit for dominion, laid upon them the burden of government,
4. Whilst he remained honoring and satisfying the gods, and priests, and kine. One day, by the act of destiny, the king, having drunk mead,

5. Became senseless, and lay asleep naked; then was he seen by Charma, and by him were his two brothers called.
6. To whom he said: What now has befallen? In what state is this our sire? By those two was he bidden with clothes, and called to his senses again and again.
7. Having recovered his intellect, and perfectly knowing what had passed, he cursed Charma, saying, Thou shalt be the servant of servants:
8. And since thou wast a laugher in their presence, from laughter shalt thou acquire a name. Then he gave to Sherma the wide domain on the south of the snowy mountains.
9. And to Jyapeti he gave all on the north of the snowy mountains; but he, by the power of religious contemplation, obtained supreme bliss.

After this testimony from Sir W. Jones—wrung from him, as it would seem, against his own wish and will—Lieutenant Wilford's essays became more numerous and more startling every year.

At last, however, the coincidences became too great. The manuscripts were again carefully examined; and then it was found that a clever forgery had been committed, that leaves had been inserted in ancient manuscripts, and that on these leaves the Pandits, urged by Lieutenant Wilford to disclose their ancient mysteries and traditions, had rendered in correct Sanskrit verse all that they had heard about Adam and Abraham from their inquisitive master. Lieutenant (then Colonel) Wilford did not hesitate for one moment to confess publicly that he had been imposed upon; but in the meantime the mischief had been done, his essays had been read all over Europe, they retained their place in the volumes of the *Asiatic Researches,* and to the present day some of his statements and theories continue to be quoted authoritatively by writers on ancient religion.

Such accidents, and, one might almost say, such misfortunes, will happen, and it would be extremely unfair were we to use unnecessarily harsh language with regard to those to whom they have happened. It is perfectly true that at present, after the progress that has been made in an accurate and critical study of Sanskrit, it would be unpardonable if any Sanskrit scholar accepted such passages as those translated by Sir W. Jones as genuine. Yet it is by no means certain that a further study of Sanskrit will not lead to similar disenchantments, and deprive many a book in Sanskrit literature which now is considered as very ancient of its claims to any high antiquity. Certain portions of the Veda even, which, as far as our knowledge goes at present, we are perfectly justified in referring to the tenth or twelfth century before our era, may some day or other dwindle down from their high estate, and those who have believed in their extreme antiquity will then be held up to blame or ridicule, like Sir W. Jones or Colonel Wilford. This cannot be avoided, for science is progressive, and does not acknowledge, even in the most distinguished scholars, any claims to infallibility. One lesson only may we learn from the disappointment that befell Colonel Wilford, and that is to be on our guard against anything which in ordinary language would be called "too good to be true."

Comparative Philology has taught us again and again that when we find a word exactly the same in Greek and Sanskrit, we may be certain that it cannot be the same word; and the same applies to Comparative Mythology. The same god or the same hero cannot have exactly the same name in Sanskrit and Greek, for the simple reason that Sanskrit and Greek have deviated from each other, have both followed their own way, have both suffered their own phonetic corruptions; and hence, if they do possess the same word, they can only possess it either in its Greek or its Sanskrit disguise. And if that caution applies to Sanskrit and Greek, members of the same family of language, how much more strongly must it apply to Sanskrit and Hebrew! If the first man were called in Sanskrit Âdima, and in Hebrew Adam, and if the two were really the same word, then Hebrew and Sanskrit could not be members of two different families of speech, or we should be driven to admit that Adam was borrowed by the Jews from the Hindus for it is in Sanskrit only that âdima means the first, whereas in Hebrew it has no such meaning.

The same remark applies to a curious coincidence pointed out many years ago by Mr. Ellis in his *Polynesian Researches* (London, 1829, vol. ii, p. 38). We there read:

A very generally received Tahitian tradition is that the first human pair were made by Taaroa, the principal deity formerly acknowledged by the nation. On more than one occasion I have listened to the details of the people respecting his work of creation. They say that, after Taaroa had formed the world, he created man out of araea, red earth, which was also the food of man until bread first was made. In connection with this some relate that Taaroa one day called for the man by name. When he came, he caused him to fall asleep, and, while he slept, he took out one of his *ivi*, or bones, and with it made a woman, whom he gave to the man as his wife, and they became the progenitors of mankind.

"This," Mr. Ellis continues,

always appeared to me a mere recital of the Mosaic account of creation, which they had heard from some European, and I never placed any reliance on it, although they have repeatedly told me it was a tradition among them before any foreigners arrived. Some have also stated that the woman's name was *Ivi*, which would be by them pronounced as if written *Eve*. *Ivi* is an aboriginal word, and not only signifies a bone, but also a widow, and a victim slain in war. Notwithstanding the assertion of the natives, I am disposed to think that *Ivi*, or *Eve*, is the only aboriginal part of the story, as far as it respects the mother of the human race. Should more careful and minute inquiry confirm the truth of this declaration, and prove that their account was in existence among them prior to their intercourse with Europeans, it will be the most remarkable and valuable oral tradition of the origin of the human race yet known.

In this case, I believe the probability is that the story of the creation of the first woman from the bone of a man[3] existed among the Tahitians before their intercourse with Christians, but I need hardly add that the similarity between the Polynesian name for bone, *ivi*, even when it was used as the name of the first woman, and the English corruption of the Hebrew, Chavah, Eve, could be the result of accident only. Whatever Chavah meant in Hebrew, whether life or living or anything else, it never meant bone, while the Tahitian *ivi*, the Maori *wheva*,[4] meant bone, and bone only.

These principles and these cautions were hardly thought of in the days of Sir William Jones and Colonel Wilford, but they ought to be thought of at present. Thus, before Bopp had laid down his code of phonetic laws, and before Burnouf had written his works on Buddhism, one cannot be very much surprised that Buddha should have been identified with Minos and Lamech; nay, that even the Babylonian deity Belus, and the Teutonic deity Wodan or Odin, should have been supposed to be connected with the founder of Buddhism in India. As Burnouf said in his *Introduction à l'Histoire du Buddhisme*, p. 70: "On avait même fait du Buddha une planète; et je ne sais pas si quelques savants ne se plaisent pas encore aujourd'hui à retrouver ce sage paisible sous les traits du belliqueux Odin."[5] But we did not expect that we should have to read again, in a book published in 1869, such statements as these:

There is certainly a much greater similarity between the Buddhism of the Topes and the Scandinavian mythology than between it and the Buddhism of the books; but still the gulf between the two is immense; and if any traces of the doctrines of the gentle ascetic (Buddha) ever existed in the bosom of Odin or his followers, while dwelling near the roots of the Caucasus, all that can be said is, that they suffered fearful ship-wreck among the rocks of the savage superstitions of the North, and sank, never again to appear on the surface of Scandinavian mythology. If the two religions come anywhere in contact, it is at their base, for underlying both there existed a strange substratum of Tree and Serpent Worship; on this the two structures seem to have been raised, though they afterwards diverged into forms so strangely dissimilar (p. 34).[6]

Or again:

We shall probably not err far if we regard these traces of serpent worship as indicating the presence in the Northeast of Scotland of the head of that column of migration, or of propagandism, which, under the myth of Wodenism, we endeavored in a previous chapter to trace from the Caucasus to Scandinavia. (p. 32);

The arbors under which two of the couples are seated are curious instances of that sort of summer-house which may be found adorning tea-gardens in the neighborhood of London to the present day. It is scenes like these that make us hesitate before asserting that there could not possibly be any connection between Buddhism and Wodenism. (p. 140);

One of the most tempting nominal similarities connected with this subject is suggested by the name of Mâyâ. The mother of Buddha was called Mâyâ. The mother of Mercury was also Maia, the daughter of Atlas. The Romans always called Wodin, Mercury, and *dies Mercurii* and *Wodensday* alike designated the fourth day of the week. ... These and other similarities have been frequently pointed out and insisted upon, and they are too numerous and too distinct not to have some foundation in reality. (p. 186, note)

Statements like these cannot be allowed to pass unnoticed or uncontradicted, particularly if supported by the authority of a great name; and after having spoken so freely of the unscientific character of the mythological comparisons instituted by scholars like Sir William Jones and Lieutenant Wilford, who can no longer defend themselves, it would be mere cowardice to shrink from performing the same unpleas-ant duty in the case of a living writer, who has shown that he knows how to wield the weapons both of defence and attack.

It is perfectly true that the mother of Buddha was called Mâyâ but it is equally true that the Sanskrit Mâyâ cannot be the Greek Maia. It is quite true, also, that the fourth day of the week is called *dies Mercurii* in Latin, and Wednesday in English;

nay, that in Sanskrit the same day is called Budha-dina or Budha-vâra. But the origin of all these names falls within perfectly historical times, and can throw no light whatever on the early growth of mythology and religion.

First of all, we have to distinguish between Budha and Buddha. The two names, though so like each other, and therefore constantly mistaken one for the other, have nothing in common but their root. Buddha with two *d*s, is the participle of budh, and means awakened, enlightened.[7] It is the name given to those who have reached the highest stage of human wisdom, and it is known most generally as the title of Gotama, Sâkya-muni, the founder of Buddhism, whose traditional era dates from 543 B.C. Budha, on the contrary, with one *d*, means simply knowing, and it became in later times, when the Hindus received from the Greeks a knowledge of the planets, the name of the planet Mercury.

It is well known that the names of the seven days of the week are derived from the names of the planets,[8] and it is equally well known that in Europe the system of weeks and week-days is comparatively of very modern origin. It was not a Greek, nor a Roman, nor a Hindu, but a Jewish or Babylonian invention. The Sabbath (Sabbata) was known and kept at Rome in the first century B.C. with many superstitious practices. It is mentioned by Horace, Ovid, Tibullus (*dies Saturni*), Persius, Juvenal. Ovid calls it a day *"rebus minus apta gerendis."* Augustus (Suet., *Aug.*, c. 76) evidently imagined that the Jews fasted on their Sabbath, for he said, "Not even a Jew keeps the fast of the Sabbath so strictly as I have kept this day." In fact, Josephus (*Contra Apion.*, ii. 39) was able to say that there was no town, Greek or not Greek, where the custom observing the seventh day had not spread.[9] It is curious that we find the seventh day, the Sabbath, even under its new Pagan name, as *dies Saturni* or *Kronike,* mentioned by Roman and Greek writers, before the names of the other days of the week made their appearance. Tibullus speaks of the day of Saturn, *dies Saturni*; Julius Frontinus (under Nerva, 96–98) says that Vespasian attacked the Jews on the day of Saturn, *dies Saturni*; and Justin Martyr (died 165) states that Christ was crucified the day before the day of Kronos, and appeared to his disciples the day after the day of Kronos. He does not use the names of Friday and Sunday. Sunday, as *dies Solis,* is mentioned by Justin Martyr (*Apolog.*, i. 67), and by Tertullian (died 220), the usual name of that day amongst Christians being the Lord's-day, Κυριακή, *dominica* or *dominicus.* Clemens of Alexandria (died 220) seems to have been the first who used the names of Wednesday and Friday, Ἑρμοῦ καὶ Ἀφροδίτης ἡμέρα.

It is generally stated, on the authority of Cassius Dio, that the system of counting by weeks and week-days was first introduced in Egypt, and that at his time, early in the third century, the Romans had adopted it, though but recently. Be this as it may, it would seem that, if Tibullus could use the name of *dies Saturni* for Saturday, the whole system of week-days must have been settled and known at Rome in his time. Cassius Dio tells us that the names were assigned to each day διὰ τεσσάρων, by fours; or by giving the first hour of the week to Saturn, then giving one hour to each planet in succession, till the twenty-fifth hour became again the first of the next day. Both systems lead to the same result, as will be seen from the following table [5.1].

After the names of the week-days had once been settled, we have no difficulty in tracing their migration towards the East and towards the West. The Hindus had their own peculiar system of reckoning days and months, but they adopted at a later time

Table 5.1

Planets	Latin	French	Sanskrit
1 Saturn 1	Dies Saturni	Samedi (dies sabbati)	Sani-vâra
2 Jupiter 6	Dies Solis	Dimanche (dominicus)	Ravi-vâra
3 Mars 4	Dies Lunæ	Lundi	Soma-vâra
4 Sun 2	Dies Martis	Mardi	Bhauma-vâra
5 Venus 7	Dies Mercurii	Mercredi	Budha-vâra
6 Mercury 5	Dies Jovis	Jeudi	Brihaspati-vâra
7 Moon 3	Dies Veneris	Vendredi	Sukra-vâra

	Old Norse	Anglo-Saxon	English
1 Saturn 1	laugardagr (washing day)	sätres däg	Saturday
2 Jupiter 6	sunnudagr	sunnan däg	Sunday
3 Mars 4	mânadagr	monan däg	Monday
4 Sun 2	tysdagr	tives däg	Tuesday
5 Venus 7	odhinsdagr	vôdenes däg	Wednesday
6 Mercury 5	thôrsdagr	thunores däg	Thursday
7 Moon 3	friadagr	frige däg	Friday

	Old-High German	Middle-High German	German
1 Saturn 1	sambaztag (sunnûn âband)	samztac (sunnen âbent)	Samstag (Sonnabend)
2 Jupiter 6	sunnûn dag	sunnen tac	Sonntag
3 Mars 4	mânin tac (?)	mân tac	Montag
4 Sun 2	ziuwes tac (cies dac)	zies tac (eritic)	Dienstag
5 Venus 7	wuotanes tac (?) (mittawecha)	mittwoch	Mittwoch
6 Mercury 5	donares tac	donres tac	Donnerstag
7 Moon 3	fria dag	frîtac	Freitag

the foreign system of counting by weeks of seven days, and assigning a presiding planetary deity to each of the seven days, according to the system described above. As the Indian name of the planet Mercury was Budha, the *dies Mercurii* was naturally called Budha-vâra but never Buddha-vâra; and the fact that the mother of Mercury was called Maia, and the mother of Buddha Mâyâ, could, therefore, have had no bearing whatever on the name assigned to the Indian Wednesday.[10] The very Buddhists, in Ceylon, distinguish between buddha, the enlightened, and budha, wise, and call Wednesday the day of Budha, not of Buddha.[11] Whether the names of the planets were formed in India independently, or after Greek models, is difficult to settle. The name of Budha, the knowing or the clever, given to the planet Mercury, seems, however, inexplicable except on the latter hypothesis.

Having traced the origin of the Sanskrit name of the *dies Mercurii,* Budha-vâra, let us now see why the Teutonic nations, though perfectly ignorant of Buddhism, called the same day the day of Wodan.

That the Teutonic nations received the names of the week-days from their Greek and Roman neighbors admits of no doubt. For commercial and military arrangements between Romans and Germans, some kind of *lingua franca* must soon have sprung up, and in it the names of the week-days must have found their place. There would have been little difficulty in explaining the meaning of Sun-day and Mon-day to the Germans, but in order to make them understand the meaning of the other names, some explanations must have been given on the nature of the different deities, in order to enable the Germans to find corresponding names in their own language. A Roman would tell his German friend that *dies Veneris* meant the day of a goddess who represented beauty and love, and on hearing this the German would at once have thought of his own goddess of love, *Freyja,* and have called the *dies Veneris* the day of *Freyja* or Friday.[12]

If *Jupiter* was described as the god who wields the thunderbolt, his natural representative in German would be *Donar,*[13] the Anglo-Saxon *Thunar,* the Old Norse *Thor,* and hence the *dies Jovis* would be called the day of *Thor,* or Thursday. If the fact that Jupiter was the king of the gods had been mentioned, his proper representative in German would, no doubt, have been *Wuotan* or Odin.[14] As it was, *Wuotan* or *Odin* was chosen as the nearest approach to *Mercury,* the character which they share in common, and which led to their identification, being most likely their love of travelling through the air,[15] also their granting wealth and fulfilling the wishes of their worshippers, in which capacity Wuotan is known by the name of *Wunsch*[16] or Wish. We can thus understand how it happened that father and son changed places, for while *Mercurius* is the son of *Jupiter, Wuotan* is the father of *Donar. Mars,* the god of war, was identified with the German *Tiu* or *Ziu,* a name which, though originally the same as *Zeus* in Greek or *Dyaus* in Sanskrit, took a peculiarly national character among the Germans, and became their god of war.[17]

There remained thus only the *dies Saturni,* the day of Saturn, and whether this was called so in imitation of the Latin name, or after an old German deity of a similar name and character, is a point which for the present we must leave unsettled.

What, however, is not unsettled is this, that if the Germans, in interpreting these names of Roman deities as well as they could, called the *dies Mercurii,* the same day which the Hindus had called the day of Budha (with one *d*), their day of *Wuotan,* this was not because "the doctrines of the gentle ascetic existed in the bosom of Odin or his followers, while dwelling near the roots of the Caucasus," but for very different and much more tangible reasons.

But, apart from all this, by what possible process could Buddha and Odin have ever been brought together in the flesh? In the history of ancient religions, Odin belongs to the same stratum of mythological thought as *Dyaus* in India, *Zeus* in Greece, *Jupiter* in Italy. He was worshipped as the supreme deity during a period long anterior to the age of the Veda and of Homer. His travels in Greece, and even in Tyrkland,[18] and his half-historical character as a mere hero and a leader of his people, are the result of the latest Euhemerism. Buddha, on the contrary, is not a mythological, but a personal and historical character, and to think of a meeting of Buddha and

Odin, or even of their respective descendants, at the roots of Mount Caucasus would be like imagining an interview between Cyrus and Odin, between Mohammed and Aphrodite.

A comparative study of ancient religions and mythologies, as will be seen from these instances, is not a subject to be taken up lightly. It requires not only an accurate acquaintance with the minutest details of comparative philology, but a knowledge of the history of religions which can hardly be gained without a study of original documents. As long, however, as researches of this kind are carried on for their own sake, and from a mere desire of discovering truth, without any ulterior objects, they deserve no blame, though, for a time, they may lead to erroneous results. But when coincidences between different religions and mythologies are searched out simply in support of preconceived theories, whether by the friends or enemies of religion, the sense of truth, the very life of all science, is sacrificed, and serious mischief will follow without fail. Here we have a right, not only to protest, but to blame.

There is on this account a great difference between the books we have hitherto examined, and a work lately published in Paris by M. Jacolliot, under the sensational title of *La Bible dans l'Inde, Vie de Jeseus Christna*. If this book had been written with the pure enthusiasm of Lieutenant Wilford, it might have been passed by as a mere anachronism. But when one sees how its author shuts his eyes against all evidence that would tell against him, and brings together, without any critical scruples, whatever seems to support his theory that Christianity is a mere copy of the ancient religion of India, mere silence would not be a sufficient answer. Besides, the book has lately been translated into English, and will be read, no doubt, by many people who cannot test the evidence on which it professes to be founded. We learn that M. Jacolliot was some years ago appointed President of the Court of Justice at Chandernagore, and that he devoted the leisure left him from the duties of his position to studying Sanskrit and the holy books of the Hindus. He is said to have put himself in communication with the Brahmans, who had obtained access to a great number of manuscripts carefully stored up in the depths of the pagodas. "The purport of his book is" (I quote from a friendly critic), "that our civilization, our religion, our legends, our gods, have come to us from India, after passing in succession through Egypt, Persia, Judea, Greece, and Italy." This statement, we are told, is not confined to M. Jacolliot, but has been admitted by almost all Oriental scholars. The Old and New Testaments are found again in the Vedas, and the texts quoted by M. Jacolliot in support of his theory are said to leave it without doubt. Brahma created Âdima (in Sanskrit, the first man) and gave him for companion Heva (in Sanskrit, that which completes life). He appointed the island of Ceylon for their residence. What follows afterwards is so beautifully described that I may be pardoned for quoting it. Only I must warn my readers, lest the extract should leave too deep an impression on their memory, that what M. Jacolliot calls a simple translation from Sanskrit is, as far as I can judge, a simple invention of some slightly mischievous Brahman, who, like the Pandits of Lieutenant Wilford, took advantage of the zeal and credulity of a French judge:

> Having created the Man and the Woman (*simultaneously,* not one after the other), and animated them with the divine afflatus—the Lord said unto them: 'Behold, your mission is to people this beautiful Island [Ceylon], where I have gathered together everything

pleasant and needful for your subsistence—the rest of the Earth is as yet uninhabitable, but should your progeny so increase as to render the bounds of paradise too narrow a habitation, let them inquire of me by sacrifice and I will make known my will.'

And thus saying, the Lord disappeared....

Then Adam and Eve dwelt together for a time in perfect happiness; but ere long a vague disquietude began to creep upon them.... The Spirit of Evil, jealous of their felicity and of the work of Brahma, inspired them with disturbing thoughts;—'Let us wander through the Island,' said Adam to his companion, 'and see if we may not find some part even more beautiful than this.'...

And Eve followed her husband... wandering for days and for months;... but as they advanced the woman was seized with strange and inexplicable terrors: 'Adam,' said she, 'let us go no farther: it seems to me that we are disobeying the Lord; have we not already quitted the place which he assigned us for a dwelling and forbade us to leave?'

'Fear not,' replied Adam; 'this is not that fearful wilderness of which he spake to us.'...

And they wandered on....

Arriving at last at the extremity of the Island, they beheld a smooth and narrow arm of the sea, and beyond it a vast and apparently boundless country, connected with their Island only by a narrow and rocky pathway arising from the bosom of the waters.

The two wanderers stood amazed: the country before them was covered with stately trees, birds of a thousand colors flitting amidst their foliage....

'Behold, what beautiful things!' cried Adam, 'and what good fruit such trees must produce;... let us go and taste them, and if that country is better than this, we will dwell there.'

Eve, trembling, besought Adam to do nothing that might irritate the Lord against them. 'Are we not well here? Have we not pure water and delicious fruits? Wherefore seek other things?'

'True,' replied Adam, 'but we will return; what harm can it be to visit this unknown country that presents itself to our view?'... And as he approached the rocks, Eve, trembling, followed.

Placing his wife upon his shoulders, he proceeded to cross the space that separated him from the object of his desires, but no sooner did he touch the shore than trees, flowers, fruits, birds, all that they had perceived from the opposite side, in an instant vanished amidst terrific clamor;... the rocks by which they had crossed sunk beneath the waters, a few sharp peaks alone remaining above the surface, to indicate the place of the bridge which had been destroyed by Divine displeasure.

The vegetation which they had seen from the opposite shore was but a delusive mirage raised by the Spirit of Evil to tempt them to disobedience.

Adam fell, weeping, upon the naked sands,... but Eve throwing herself into his arms, besought him not to despair;... 'let us rather pray to the Author of all things to pardon us.'...

And as she spake there came a voice from the clouds, saying, 'Woman!, *thou* hast only sinned from love to thy husband, whom I commanded thee to love, and thou hast hoped in me.

'I therefore pardon thee—and I pardon him also for *thy* sake:... but ye may no more return to paradise, which I had created for your happiness;... through your disobedience to my commands the Spirit of Evil has obtained possession of the Earth.... Your children reduced to labor and to suffer by your fault will become corrupt and forget me....

'But I will send Vishnu, who will be born of a woman, and who will bring to all the hope of a reward in another life, and the means by prayer of softening their sufferings.'

The translator from whom I have quoted exclaims at the end, as well he might: "What grandeur and what simplicity is this Hindu legend!, and at the same time how simply logical!... Behold here the veritable Eve—the true woman."

But much more extraordinary things are quoted by M. Jacolliot, from the Vedas and the commentaries.

On p. 63, we read that Manu, Minos, and Manes, had the same name as Moses; on p. 73, the Brahmans who invaded India are represented as the successors of a great reformer called Christna. The name of Zoroaster is derived from the Sanskrit Sûryastara (p. 110), meaning "he who spreads the worship of the Sun." After it has been laid down (p. 116) that Hebrew was derived from Sanskrit, we are assured that there is little diffi- culty in deriving Jehovah from Zeus.[19] Zeus, Jezeus, Jesus, and Isis are all declared to be the same name, and later on (p. 130) we learn that "at present the Brahmans who offi- ciate in the pagodas and temples give this title of Jeseus—i.e., the pure essence, the divine emanation—to Christna only, who alone is recognized as the Word, the truly incarnated, by the worshippers of Vishnu and the freethinkers among the Brahmans."

We are assured that the Apostles, the poor fishermen of Galilee, were able to read the Veda (p. 356); and it was their greatest merit that they did not reject the miraculous accounts of the Vedic period, because the world was not yet ripe for freedom of thought. Kristna, or Christna, we read on p. 360, signified in Sanskrit, sent by God, promised by God, holy; and as the name of Christ or *Christos* is not Hebrew, whence could it have been taken except from Krishna, the son of Devakî, or, as M. Jacolliot writes, Devanaguy?

It is difficult, nay, almost impossible, to criticise or refute such statements, and yet it is necessary to do so; for such is the interest, or I should rather say the feverish curiosity, excited by anything that bears on ancient religion, that M. Jacolliot's book has produced a very wide and very deep impression. It has been remarked with some surprise that Vedic scholars in Europe had failed to discover these important passages in the Veda which he has pointed out, or, still worse, that they had never brought them to the knowledge of the public. In fact, if anything was wanting to show that a general knowledge of the history of ancient religion ought to form part of our education, it was the panic created by M. Jacolliot's book. It is simply the story of Lieutenant Wilford over again, only far less excusable now than a hundred years ago. Many of the words which M. Jacolliot quotes as Sanskrit are not Sanskrit at all; others never have the meaning which he assigns to them; and as to the passages from the Vedas (including our old friend the Bhagaveda-gîta), they are not from the Veda, they are not from any old Sanskrit writer—they simply belong to the second half of the nineteenth century. What happened to Lieutenant Wilford has happened again to M. Jacolliot. He tells us the secret himself: "One day," he says,

when we were reading the translation of Manu, by Sir W. Jones, a note led us to consult the Indian commentator, Kullûka Bhatta, when we found an allusion to the sacrifice of a son by his father prevented by God himself after he had commanded it. We then had only one *idée fixe*—namely, to find again in the dark mass of the religious books of the Hindu, the original account of that event. We should never have succeeded but for 'the complaisance' of a Brahman with whom we were reading Sanskrit, and who, yielding to our request, brought us from the library of his pagoda the works of the theologian Ramatsariar, which have yielded us such precious assistance in this volume (p. 280).

As to the story of the son offered as a sacrifice by his father, and released at the command of the gods, M. Jacolliot might have found the original account of it from the Veda, both text and translation, in my *History of Ancient Sanskrit Literature*. He would soon have seen that the story of Sunahsepa being sold by his father in order to be sacrificed in the place of an Indian prince, has very little in common with the intended sacrifice of Isaac by Abraham. M. Jacolliot has, no doubt, found out by this time that he has been imposed upon; and if so, he ought to follow the example of Colonel Wilford, and publicly state what has happened. Even then, I doubt not that his statements will continue to be quoted for a long time, and that Âdima and Heva, thus brought to life again, will make their appearance in many a book and many a lecture-room.

Lest it be supposed that such accidents happen to Sanskrit scholars only, or that this fever is bred only in the jungles of Indian mythology, I shall mention at least one other case which will show that this disease is of a more general character, and that want of caution will produce it in every climate.

Before the discovery of Sanskrit, China had stood for a long time in the place which was afterwards occupied by India. When the ancient literature and civilization of China became first known to the scholars of Europe, the Celestial Empire had its admirers and prophets as full of enthusiasm as Sir W. Jones and Lieutenant Wilford, and there was nothing, whether Greek philosophy or Christian morality, that was not supposed to have had its first origin among the sages of China. The proceedings of the Jesuit missionaries in China were most extraordinary. They had themselves admitted the antiquity of the writings of Confucius and Lao-tse, both of whom lived in the sixth century B.C.[20] But in their zeal to show that the sacred books of the Chinese contained numerous passages borrowed from the Bible, nay, even some of the dogmas of the later Church, they hardly perceived that, taking into account the respective dates of these books, they were really proving that a kind of anticipated Christianity had been accorded to the ancient sages of the Celestial Empire. The most learned advocate of this school was Father Prémare. Another supporter of the same view, Montucci,[21] speaking of Lao-tse's Tao-te-king, says:

> We find in it so many sayings clearly referring to the triune God, that no one who has read this book can doubt that the mystery of the most holy Trinity was revealed to the Chinese more than five centuries before the advent of Christ. Everybody, therefore, who knows the strong feeling of the Chinese for their own teachers, will admit that nothing more efficient could be found in order to fix the dogmas of the Christian religion in the mind of the Chinese than the demonstration that these dogmas agree with their own books. The study, therefore, and the translation of this singular book (the Tao-te-king) would prove most useful to the missionaries, in order to bring to a happy issue the desired gathering in of the Apostolic harvest.

What followed is so extraordinary that, though it has often been related, it deserves to be related again, more particularly as the whole problem which was supposed to have been solved once for all by M. Stanislas Julien, has of late been opened again by Dr. von Strauss, in the *Journal of the German Oriental Society,* 1869.

There is a passage at the beginning of the fourteenth chapter of the Tao-te-king in which Father Amyot felt certain that the three Persons of the Trinity could be recognized. He translated it:

> He who is as it were visible, but cannot be seen, is called *Khi.*
> He whom we cannot hear, and who does not speak to our ear, is called *Hi.*
> He who is as it were tangible, but cannot be touched, is called *Wei.*

Few readers, I believe, would have been much startled by this passage, or would have seen in it what Father Amyot saw. But more startling revelations were in store. The most celebrated Chinese scholar of his time, Abel Rémusat, took up the subject; and after showing that the first of the three names had to be pronounced, not *Khi*, but *I*, he maintained that the three syllables *I-Hi-Wei*, were meant for Je-ho-vah. According to him, the three characters employed in this name have no meaning in Chinese; they are only signs of sounds foreign to the Chinese language; and they were intended to render the Greek 'Ιαώ, the name which, according to Diodorus Siculus, the Jews gave to their God. Rémusat goes on to remark that Lao-tse had really rendered this Hebrew name more accurately than the Greeks, because he had preserved the aspiration of the second syllable, which was lost in Greek. In fact, he entertained no doubt that this word, occurring in the work of Lao-tse, proves an intellectual communication between the West and China, in the sixth century B.C.

Fortunately, the panic created by this discovery did not last long. M. Stanislas Julien published in 1842 a complete translation of this difficult book; and here all traces of the name of Jehovah have disappeared.

"The three syllables," he writes,

which Abel Rémusat considered as purely phonetic and foreign to the Chinese language, have a very clear and intelligible meaning, and have been fully explained by Chinese commentators. The first syllable, *I*, means without color; the second, *Hi*, without sound or voice; the third, *Wei*, without body. The proper translation therefore is:

You look (for the Tao, the law) and you see it not: it is colorless.
You listen and you hear it not: it is voiceless.
You wish to touch it and you reach it not: it is without body.

Until, therefore, some other traces can be discovered in Chinese literature proving an intercourse between China and Judæa in the sixth century B.C., we can hardly be called upon to believe that the Jews should have communicated this one name, which they hardly trusted themselves to pronounce at home, to a Chinese philosopher; and we must treat the apparent similarity between *I-Hi-Wei* and Jehovah as an accident, which ought to serve as a useful warning, though it need in no way discourage a careful and honest study of Comparative Theology.

Notes

1. *Asiatic Researches,* vol. 1, p. 272; *Life of Sir W. Jones,* vol. 2, p. 240 seq.
2. *Asiatic Researches,* vol. 1, p. 221.
3. See *Introduction to the Science of Religion,* p. 48.
4. The Rev. W. W. Gill tells me that the Maori word for bone is *iwi,* but he suspects a foreign origin for the fable founded on it.
5. "We have even created a planet out of Buddha, and I don't know if some scholars wouldn't enjoy discovering again today this peaceful, wise man under the guise of the aggressive Odin."—Sondra Bacharach
6. *Tree and Serpent Worship,* by James Fergusson. London, 1868. Very similar opinions had been advocated by Rajendralal Mitra, in a paper published in 1858 in the *Journal of the Royal Asiatic Society,* "Buddhism and Odinism, illustrated by extracts from Professor

Holmboe's Memoir on the *Traces du Buddhisme en Norvège.*" How much mischief is done by opinions of this kind when they once find their way into the general public, and are supported by names which carry weight, may be seen by the following extracts from the *Pioneer* (July 30, 1878), a native paper published in India. Here we read that the views of Holmboe, Rajendralal Mitra, and Fergusson, as to a possible connection between Buddha and Wodan, between Buddhism and Wodenism, have been adopted and preached by an English bishop, in order to convince his hearers, who were chiefly Buddhists, that the religion of the gentle ascetic came originally, if not from the Northeast of Scotland, at all events from the Saxons. "Gotama Buddha," he maintained, "was a Saxon," coming from "a Saxon family which had penetrated into India." And again: "The most convincing proof to us Anglo-Indians lies in the fact that the Purânas named Varada and Matsy distinctly assert that the White Island in the West—meaning England—was known in India as Sacana, having been conquered at a very early period by the Sacas or Saks." After this the bishop takes courage, and says: "Let me call your attention to the Pâli word Nibban, called in Sanskrit Nirvâna. In the Anglo-Saxon you have the identical word—Nabban, meaning 'not to have,' or 'to be without a thing.'"

7. See *Buddhaghosha's Parables,* translated by Captain Rogers, with an Introduction containing Buddha's Dhammapada, translated from Pâli, by M. M., 1870, p. 110, note.
8. Hare, "On the Names of the Days of the Week" (*Philol. Museum,* Nov. 1831); Ideler, *Handbuch der Chronologie,* p. 177; Grimm, *Deutsche Mythologie,* p. 111.
9. A writer in the *Index* objects to my representation of what Josephus said with regard to the observance of the seventh day in Greek and barbarian towns. He writes:

WASHINGTON, Nov. 9, 1872
The article by Max Müller in the *Index* of this week contains, I think, one error, caused doubtless by his taking a false translation of a passage from Josephus instead of the original. 'In fact,' says Professor Müller, 'Josephus (*Contra Apion.* ii. 39) was able to say that there was no town, Greek or not Greek, where the custom of observing the seventh day had not spread.' Mr. Wm. B. Taylor, in a discussion of the Sabbath question with the Rev. Dr. Brown, of Philadelphia, in 1853 (*Obligation of the Sabbath,* p. 120), gives this rendering of the passage: 'Nor is there anywhere any city of the Greeks, nor a single barbarian notion, whither the institution of the Hebdomade (*which we mark by resting*) has not travelled'; then in a note Mr. Taylor gives the original Greek of part of the passage and adds: 'Josephus does not say that the Greek and barbarian rested, but that *we* [the Jews] observe it by rest.'

The corrected translation only adds strength to Max Müller's position in regard to the very limited extent of Sabbath observance in ancient times; and Mr. Taylor brings very strong historical proof to maintain the assertion (p. 24) that 'throughout all history we discover no trace of a Sabbath among the nations of antiquity.'

It seems to me that if we read the whole of Josephus's work, *On the Antiquity of the Jews,* we cannot fail to perceive that what Josephus wished to show towards the end of the second book was that other nations had copied or were trying to copy the Jewish customs. ... He ... says that the early Greek philosophers, though apparently original in their theoretic speculations, followed the Jewish laws with regard to practical and moral precepts. ... Standing where it [ἑβδομάς (hebdomas)] stands, the sentence ... can only mean that "there is no town of Greeks nor of barbarians, nor one single people, where the custom of the seventh day, on which we rest, has not spread, and where fastings, and lighting of lamps, and much of what is forbidden to us with regard to food are not observed. They try to imitate our mutual concord also, etc." Hebdomas, which originally meant the week, is here clearly used in the sense of the seventh day, and though Josephus may exaggerate, what he says is certainly "that there was no town, Greek or not Greek, where the custom of observing the seventh day had not spread."

10. Grimm, *Deutsche Mythologie*, p. 118, note.
11. In Singalese Wednesday is Badâ, in Tamil Budau. See Kennet, in *Indian Antiquary*, 1874, p. 90; D'Alwis, *Journal of Ceylon Branch of the Royal Asiatic Society*, 1870, p. 17.
12. Grimm, *Deutsche Mythologie*, p. 276.
13. Ibid., p. 151.
14. Ibid., p. 120.
15. Grimm, *Deutsche Mythologie*, pp. 137–148.
16. Ibid., p. 126. Oski in Icelandic, the god Wish, one of the names of the highest god.
17. Tacit. *Hist.*, iv. 64: "Communibus Diis et præcipuo Deorum Marti grates agimus."
18. Grimm, loc. cit., p. 148.
19. P. 125. "Pour quiconque s'est occupé d'études philologiques, Jéhova dérivé de Zeus est facile à admettre." ["For whoever has been concerned with philological studies, it is easy to accept that Jehovah is derived from Zeus."—Sondra Bacharach]
20. Stanislas Julien, *Le Livre de la Voie et de la Vertu*. Paris, 1842, p. iv.
21. Montucci, *De Studiis Sinicis*. Berolini, 1808.

CHAPTER 6

THE SCIENCE OF RELIGION: LECTURE ONE

from *Lectures on the Science of Religion* (1870)

When I undertook for the first time to deliver a course of lectures in this institution, I chose for my subject the *Science of Language*. What I then had at heart was to show to you, and to the world at large, that the comparative study of the principal languages of mankind was based on principles sound and scientific, and that it had brought to light results which deserved a larger share of public interest than they had as yet received. I tried to convince, not only scholars by profession, but historians, theologians, and philosophers, nay everybody who had once felt the charm of gazing inwardly upon the secret workings of his own mind, veiled and revealed as they are in the flowing forms of language, that the discoveries made by comparative philologists could no longer be ignored with impunity; and I submitted that after the progress achieved in a scientific study of the principal branches of the vast realm of human speech, our new science, the Science of Language, might claim by right its seat at the round-table of the intellectual chivalry of our age.

Such was the goodness of the cause I had then to defend, that, however imperfect my own pleading, the verdict of the public has been immediate and almost unanimous.

During the years that have elapsed since the delivery of my first course of lectures, the Science of Language has had its full share of public recognition. Whether we look at the number of books that have been published for the advancement and elucidation of our science, or at the excellent articles in the daily, weekly, fortnightly, monthly, or quarterly reviews, or at the frequent notices of its results scattered about in works on philosophy, theology, and ancient history, we may well rest satisfied. The example set by France and Germany, in founding chairs of Sanskrit and Comparative Philology, has been followed of late in nearly all the universities of England, Ireland, and Scotland. We need not fear for the future of the Science of Language. A career so auspiciously begun, in spite of strong prejudices that had to be encountered, will lead on from year to year to greater triumphs. Our best public schools, if they have not done so already, will soon have to follow the example set by the universities. It is but fair that school-boys who are made to devote so many hours every day to the laborious acquisition of languages, should now and then be taken by a safe guide to enjoy from a higher point of view that living panorama of human speech which has been surveyed and carefully mapped out by patient explorers and bold discoverers; nor is there any longer an excuse why, even in the most elementary lessons, nay I should say, why more particularly in these elementary lessons, the dark and dreary passages of Greek and Latin, of French and German grammar, should not be lighted up by the electric light of Comparative Philology.

When last year I travelled in Germany I found that lectures on Comparative Philology are now attended in the universities by all who study Greek and Latin. At Leipzig alone the lectures of the professor of Sanskrit were attended by more than fifty under-graduates, who first acquire that amount of knowledge of Sanskrit which is absolutely necessary before entering upon a study of Comparative Grammar. The introduction of Greek into the universities of Europe in the fifteenth century could hardly have caused a greater revolution than the discovery of Sanskrit and the study of Comparative Philology in the nineteenth century. Very few indeed now take their degree of Master of Arts in Germany, or would be allowed to teach at a public school, without having been examined in the principles of Comparative Philology, nay in the elements of Sanskrit Grammar. Why should it be different in England? The intellectual fibre, I know, is not different in the youth of England and in the youth of Germany, and if there is but a fair field and no favor, Comparative Philology, I feel convinced, will soon hold in England, too, that place which it ought to hold at every public school, in every university, and in every classical examination.

In beginning to-day a course of lectures on the *Science of Religion,*—or I should rather say on some preliminary points that have to be settled before we can enter upon a truly scientific study of the religions of the world—I feel as I felt when first pleading in this very place for the Science of Language.

I know that I shall have to meet determined antagonists who will deny the possibility of a scientific treatment of religions as they denied the possibility of a scientific treatment of languages. I foresee even a far more serious conflict with familiar prejudices and deep-rooted convictions; but I feel at the same time that I am prepared to meet my antagonists; and I have such faith in their honesty of purpose, that I doubt not of a patient and impartial hearing on their part, and of a verdict influenced by nothing but by the evidence that I shall have to place before them.

In these our days it is almost impossible to speak of religion without giving offense either on the right or on the left. With some, religion seems too sacred a subject for scientific treatment; with others it stands on a level with alchemy and astrology, a mere tissue of errors or hallucinations, far beneath the notice of the man of science. In a certain sense, I accept both these views. Religion is a sacred subject, and whether in its most perfect or in its most imperfect form, it has a right to our highest reverence. No one—this I can promise—who attends these lectures, be he Christian or Jew, Hindu or Mohammedan, shall hear his own way of serving God spoken of irreverently. But true reverence does not consist in declaring a subject, because it is dear to us, to be unfit for free and honest inquiry; far from it! True reverence is shown in treating every subject, however sacred, however dear to us, with perfect confidence; without fear and without favor; with tenderness and love, by all means, but, before all, with an unflinching and uncompromising loyalty to truth. I also admit that religion has stood in former ages, and stands even in our own age, if we look abroad, aye, even if we look into some dark places at home, on a level with alchemy and astrology; but for the discovery of truth there is nothing so useful as the study of errors, and we know that in alchemy there lay the seed of chemistry, and that astrology was more or less a yearning and groping after the true science of astronomy.

But although I shall be most careful to avoid giving offense, I know perfectly well that many a statement I shall have to make, and many an opinion I shall have to express, will sound strange and startling to some of my hearers. The very title of the Science of Religion jars on the ears of many persons, and a comparison of all the religions of the world, in which none can claim a privileged position, must seem to many reprehensible in itself, because ignoring that peculiar reverence which everybody, down to the mere fetich [sic] worshipper, feels for his *own* religion and for his *own* God. Let me say then at once that I myself have shared these misgivings, but that I have tried to overcome them, because I would not and could not allow myself to surrender either what I hold to be the truth, or what I hold still dearer than the truth, the right tests of truth. Nor do I regret it. I do not say that the Science of Religion is all gain. No; it entails losses, and losses of many things which we hold dear. But this I will say, that, as far as my humble judgment goes, it does not entail the loss of anything that is essential to true religion, and that if we strike the balance honestly, the gain is immeasurably greater than the loss.

One of the first questions that was asked by classical scholars when invited to consider the value of the Science of Language, was, "What shall we gain by a comparative study of languages?" Languages, it was said, are wanted for practical purposes, for speaking and reading; and by studying too many languages at once, we run the risk of losing the firm grasp which we ought to have on the few that are really important. Our knowledge, by becoming wider, must needs, it was thought, become shallower, and the gain, if there is any, in knowing the structure of dialects which have never produced any literature at all, would certainly be outweighed by the loss in accurate and practical scholarship.

If this could be said of a comparative study of languages, with how much greater force will it be urged against a comparative study of religions! Though I do not expect that those who study the religious books of Brahmans and Buddhists, of Confucius and Lao-tse, of Mohammed and Nanak, will be accused of cherishing in their secret heart the doctrines of those ancient masters, or of having lost the firm hold on their own religious convictions, yet I doubt whether the practical utility of wider studies in the vast field of the religions of the world will be admitted with greater readiness by professed theologians than the value of a knowledge of Sanskrit, Zend, Gothic, or Celtic for a thorough mastery of Greek and Latin, and for a real appreciation of the nature, the purpose, the laws, the growth and decay of language was admitted, or is even now admitted, by some of our most eminent professors and teachers.

People ask, What is gained by comparison? Why, all higher knowledge is gained by comparison, and rests on comparison. If it is said that the character of scientific research in our age is preëminently comparative; this really means that our researches are now based on the widest evidence that can be obtained, on the broadest inductions that can be grasped by the human mind. What can be gained by comparison? Why, look at the study of languages. If you go back but a hundred years and examine the folios of the most learned writers upon questions connected with language, and then open a book written by the merest tyro in Comparative Philology, you will see what can be gained, what has been gained, by the comparative method.

A few hundred years ago, the idea that Hebrew was the original language of mankind was accepted as a matter of course, even as a matter of faith, the only problem being to find out by what process Greek, or Latin, or any other language could have been developed out of Hebrew. The idea, too, that language was revealed, in the scholastic sense of that word, was generally accepted, although, as early as the fourth century, St. Gregory, the learned Bishop of Nyssa, had strongly protested against it. The grammatical frame-work of a language was either considered as the result of a conventional agreement, or the terminations of nouns and verbs were supposed to have sprouted forth like buds from the roots and stems of language; and the vaguest similarity in the sound and meaning of words was taken to be a sufficient criterion for testing their origin and their relationship.

Of all this philological somnambulism we hardly find a trace in works published since the days of Humboldt, Bopp, and Grimm. Has there been any loss here? Has it not been pure gain? Does language excite admiration less because we know that though the faculty of speaking is the work of Him who has so framed our nature, the invention of words for naming each object was left to man, and was achieved through the working of the human mind? Is Hebrew less carefully studied because it is no longer believed to be a revealed language sent down from heaven, but a language closely allied to Arabic, Syriac, and ancient Babylonian, and receiving light from these cognate, and in some respects more primitive languages, for the explanation of many of its grammatical forms, and for the exact interpretation of many of its obscure and difficult words? Is the grammatical articulation of Greek and Latin less instructive because, instead of seeing in the termination of nouns and verbs merely arbitrary signs to distinguish the singular from the plural, or the present from the future, we can now perceive an intelligible principle in the gradual production of formal out of the material elements of language? And are our etymologies less important because, instead of being suggested by superficial similarities, they are now based on honest historical and physiological research? Lastly, has our own language ceased to hold its own peculiar place? Is our love for our own native tongue at all impaired? Do men speak less boldly or pray less fervently in their own mother-tongue, because they know its true origin and its unadorned history; or because they have discovered that in all languages, even in the jargons of the lowest savages, there is order and wisdom; there is in them something that makes the world akin?

Why, then, should we hesitate to apply the comparative method, which has produced such great results in other spheres of knowledge, to a study of religion? That it will change many of the views commonly held about the origin, the character, the growth, and decay of the religions of the world, I do not deny; but unless we hold that fearless progression in new inquiries, which is our bounden duty and our honest pride in all other branches of knowledge, is dangerous in the study of religions, unless we allow ourselves to be frightened by the once famous dictum, that whatever is new in theology is false, this ought to be the very reason why a comparative study of religions should no longer be neglected or delayed.

When the students of Comparative Philology boldly adopted Goethe's paradox, "*He who knows one language, knows none,*" people were startled at first, but they soon began to feel the truth which was hidden beneath the paradox. Could Goethe have meant that Homer did not know Greek, or that Shakespeare did not know English, because neither of them knew more than his own mother-tongue? No!, what was meant was that neither Homer nor Shakespeare knew what that language really was which he handled with so much power and cunning. Unfortunately the old verb "to can," from which "canny" and "cunning," is lost in English, otherwise we should be able in two words to express our meaning, and to keep apart the two kinds of knowledge of which we are here speaking. As we say in German *können* is not *kennen,* we might say in English *to can,* that is to be cunning, is not *to ken,* that is to know; and it would then become clear at once, that the most eloquent speaker and the most gifted poet, with all their command of words and skillful mastery of expression, would have but little to say if asked what language really is! The same applies to religion. *He who knows one, knows none.*

There are thousands of people whose faith is such that it could move mountains, and who yet, if they were asked what religion really is, would remain silent, or would speak of outward tokens rather than of the inward nature, or of the faculty of faith.

It will easily be perceived that religion means at least two very different things. When we speak of the Jewish, or the Christian, or the Hindu religion, we mean a body of doctrines handed down by tradition, or in canonical books, and containing all that constitutes the faith of Jew, Christian, or Hindu. Using religion in that sense, we may say that a man has changed his religion, that is, that he has adopted the Christian instead of the Brahmanical body of religious doctrines, just as a man may learn to speak English instead of Hindustani. But religion is also used in a different sense. As there is a faculty of speech, independent of all the historical forms of language, so we may speak of a faculty of faith in man, independent of all historical religions. If we say that it is religion which distinguishes man from the animal, we do not mean the Christian or Jewish religions only; we do not mean any special religion, but we mean a mental faculty, that faculty which, independent of, nay in spite of sense and reason, enables man to apprehend the Infinite under varying disguises. Without that faculty, no religion, not even the lowest worship of idols and fetiches, would be possible; and if we will but listen attentively, we can hear in all religions a groaning of the spirit, a struggle to conceive the inconceivable, to utter the unutterable, a longing after the Infinite, a love of God. Whether the etymology which the ancients gave of the Greek word ἄνθρωπος [anthropos], man, be true or not (they derived it from ὁ ἄνω ἀθρῶν, he who looks upward): certain it is that what makes man to be man, is that he alone can turn his face to heaven; certain it is that he alone yearns for something that neither sense nor reason can supply.

If then there is a philosophical discipline which examines into the conditions of sensuous perception, and if there is another philosophical discipline which examines into the conditions of rational conception, there is clearly a place for a third philosophical discipline that has to examine into the conditions of that third faculty of man, coördinate with sense and reason, the faculty of perceiving the Infinite, which is at the root of all religions. In German we can distinguish that third faculty by the name of *Vernuft,* as opposed to *Verstand,* reason, and *Sinne,* sense. In English I know no better name for it than the faculty of faith, though it will have to be guarded by careful definition, and to be restricted to those objects only, which cannot be supplied either by the evidence of the senses, or by the evidence of reason. No simply historical fact can ever fall under the cognizance of faith.

If we look at the history of modern thought, we find that the dominant school of philosophy, previous to Kant, had reduced all intellectual activity to *one* faculty, that of the senses. "Nihil in intellectu quod non ante fuerit in sensu," "Nothing exists in the intellect but what has before existed in the senses," was their watch-word; and Leibnitz [sic] answered it epigrammatically, but most profoundly, "Nihil—nisi intellectus." "Yes, nothing but the intellect." Then followed Kant, who, in his great work written ninety years ago, but not yet antiquated, proved that our knowledge requires the admission of two independent faculties, the intuitions of the senses, and the categories, or, as we might call them, the necessities of reason. But satisfied with having established the independent faculty of reason, as coördinate with the faculty of sense, or, to use his own technical language, satisfied with having proved the possibility of apodictic judgments à priori, Kant declined to go further, and denied to the intellect the power of transcending the finite, the faculty of approaching the Divine. He closed the ancient gates through which man had gazed into Infinity, but, in spite of himself, he was driven, in his *Critique of Practical Reason,* to open a side-door through which to admit the sense of the Divine. This is the vulnerable point in Kant's philosophy, and if philosophy has to explain what is, not what ought to be, there will be and can be no rest till we admit, what cannot be denied, that there is in man a third faculty, which I call simply the faculty of apprehending the Infinite, not only in religion, but in all things; a power independent of sense and reason, a power in a certain sense contradicted by sense and reason, but yet, I suppose, a very real power, if we see how it has held its own from the beginning of the world, how neither sense nor reason have been able to overcome it, while it alone is able to overcome both reason and sense.

According to the two meanings of the word religion, then, the Science of Religion is divided into two parts; the former, which has to deal with the historical forms of religion, is called *Comparative Theology;* the latter, which has to explain the conditions under which religion, in its highest or lowest form, is possible, is called *Theoretic Theology.*

We shall at present have to deal with the former only; nay, it will be my object to show that the problems which chiefly occupy theoretic theology, ought not to be taken up till all the evidence that can possibly be gained from a comparative study of the religions of the world has been fully collected, classified, and analyzed.

It may seem strange that while theoretical theology, or the analysis of the inward and outward conditions under which faith is possible, has occupied so many thinkers, the study of comparative theology has never as yet been seriously taken in hand. But the explanation is very simple. The materials on which alone a comparative study of the religions of mankind could have been founded were not accessible in former days, while in our own days they have come to light in such profusion as almost to challenge these more comprehensive inquiries in a voice that cannot be disobeyed.

It is well known that the Emperor Akbar had a passion for the study of religions, so that he invited to his court Jews, Christians, Mohammedans, Brahmans, and Fire-worshippers, and had as many of their sacred books as he could get access to, translated for his own study. Yet, how small was the collection of sacred books that even an emperor of India could command not more than two hundred and fifty years ago, compared to what may now be found in the library of every poor scholar! We have the original text of the Veda, which neither the bribes nor the threats of Akbar could extort from the Brahmans. The translation of the Veda which he is said to have obtained, was a translation of the so-called Atharva-veda, and comprised most likely the Upanishads only, mystic and philosophical treatises, very interesting, very important in themselves, but as far removed from the ancient poetry of the Veda as the Talmud is from the Old Testament, as Sufism is from the Koran. We have the Zend-Avesta, the sacred writings of the so-called fire-worshippers, and we possess the translation of it, far more complete and far more correct than any that the Emperor Akbar could have obtained. The religion of Buddha, certainly in many respects more important than either Brahmanism, or Zoroastrianism, or Mohammedanism, is never mentioned in the religious discussions that took place one evening in every week at the imperial court of Delhi. Abufazl, it is said, the minister of Akbar, could find no one to assist him in his inquiries respecting Buddhism. We possess the whole sacred canon of the Buddhists in various languages, in Pâli, in Sanskrit, in Burmese, Siamese, Tibetan, Mongolian, and Chinese, and it is our fault entirely, if as yet there is no complete translation in any European tongue of this important collection of sacred books. The ancient religions of China again, that of Confucius and that of Lao-tse, may now be studied in excellent translations of their sacred books by anybody interested in the ancient faith of mankind.

But this is not all. We owe to missionaries particularly, careful accounts of the religious belief and worship among tribes far lower in the scale of civilization than the poets of the Vedic hymns, or the followers of Confucius. Though the belief of African and Melanesian savages is more recent in point of time, it represents an earlier and far more primitive phase in point of growth, and is therefore as instructive to the student of religion as the study of uncultivated dialects has proved to the student of language.

Lastly, and this, I believe, is the most important advantage which we enjoy as students of the history of religion, we have been taught the rules of critical scholarship. No one would venture nowadays, to quote from any book, whether sacred or profane, without having asked these simple and yet momentous questions: When was it written?, Where?, and by whom? Was the author an eye-witness, or does he only relate what he has heard from others? And if the latter, were his authorities at least contemporaneous with the events which they relate, and were they under the sway of party feeling or any other disturbing influence? Was the whole book written at once, or does it contain portions of an earlier date; and if so, is it possible for us to separate these earlier documents from the body of the book?

A study of the original documents on which the principal religions of the world profess to be founded, carried out in this spirit, has enabled some of our best living scholars to distinguish in each religion between what is really ancient and what is comparatively modern; what was the doctrine of the founders and their immediate disciples, and what were the afterthoughts and, generally, the corruptions of later ages. A study of these later developments, of these later corruptions, or, it may be, improvements, is not without its own peculiar charms, and full of practical lessons; yet, as it is essential that we should know the most ancient forms of every language, before we proceed to any comparisons, it is indispensable that we should have a clear conception of the most primitive form of every religion before we proceed to determine its own value, and to compare it with other forms of religious faith. Many an orthodox Mohammedan, for instance, will relate miracles wrought by Mohammed; but in the Koran, Mohammed says distinctly that he is a man like other men. He disdains to work miracles, and appeals to the great works of Allah, the rising and setting of the sun, the rain that fructifies the earth, the plants that grow, and the living souls that are born into the world—who can tell whence?—as the real signs and wonders in the eyes of a true believer.

The Buddhist legends teem with miserable miracles attributed to Buddha and his disciples—miracles which in wonderfulness certainly surpass the miracles of any other religion; yet in their own sacred canon a saying of Buddha's is recorded, prohibiting his disciples from working miracles, though challenged by the multitudes who required a sign that they might believe. And what is the miracle that Buddha commands his disciples to perform? "Hide your good deeds," he says, "and confess before the world the sins you have committed."

Modern Hinduism rests on the system of caste as on a rock which no arguments can shake; but in the Veda, the highest authority of the religious belief of the Hindus, no mention occurs of the complicated system of castes, such as we find it in Manu; nay, in one place, where the ordinary classes of the Indian, or any other society, are alluded to, namely, the priests, the warriors, the citizens, and the slaves, all are represented as sprung alike from Brahman, the source of all being.

It would be too much to say that the critical sifting of the authorities for a study of each religion has been already fully carried out. There is work enough still to be done. But a beginning, and a very successful beginning, has been made, and the results thus brought to light will serve as a wholesome caution to everybody who is engaged in religious researches. Thus, if we study the primitive religion of the Veda, we have to distinguish most carefully, not only between the hymns of the Rig-veda on one side, and the hymns collected in the Sama-veda, Yagur-veda, and Atharva-veda on the other, but critical scholars would distinguish with equal care between the more ancient and the more modern hymns of the Rig-veda, as far as even the faintest indications of language, of grammar, or metre enable them to do so.

In order to gain a clear insight into the motives and impulses of the founder of the worship of Ahura-mazda, we must chiefly, if not entirely, depend on those portions of the Zend-Avesta which are written in the Gatha dialect, a more primitive dialect than that of the rest of the sacred code of the Zoroastrians.

In order to do justice to Buddha, we must not mix the practical portions of the Tripitaka, the Dharma, with the metaphysical portions, the Abhidharma. Both, it is true, belong to the sacred canon of the Buddhists; but their original sources lie in very different latitudes of religious thought.

We have in the history of Buddhism an excellent opportunity for watching the process by which a canon of sacred books is called into existence. We see here, as elsewhere, that during the life-time of the teacher, no record of events, no sacred code containing the sayings of the master was wanted. His presence was enough, and thoughts of the future, and more particularly of future greatness, seldom entered the minds of those who followed him. It was only after Buddha had left the world to enter into Nirvâna, that his disciples attempted to recall the sayings and doings of their departed friend and master. At that time everything that seemed to redound to the glory of Buddha, however extraordinary and incredible, was eagerly welcomed,

while witnesses who would have ventured to criticize or reject unsupported statements, or to detract in any way from the holy character of Buddha, had no chance
of even being listened to. And when, in spite of all this, differences of opinion arose,
they were not brought to the test by a careful weighing of evidence, but the names
of "unbeliever" and "heretic" (*nastika, pashanda*) were quickly invented in India as
elsewhere, and bandied backwards and forwards between contending parties, till at
last, when the doctors disagreed, the help of the secular power had to be invoked,
and kings and emperors convoked councils for the suppression of schism, for the
settlement of an orthodox creed, and for the completion of a sacred canon. We know
of King Asoka, the contemporary of Seleucus, sending his royal missive to the assembled elders, and telling them what to do, and what to avoid, warning them also in
his own name of the apocryphal or heretical character of certain books which, as he
thinks, ought not to be admitted into the sacred canon.

We here learn a lesson, which is confirmed by the study of other religions, that
canonical books, though they furnish in most cases the most ancient and most
authentic information within the reach of the student of religion, are not to be
trusted implicitly, nay, that they must be submitted to a more searching criticism and
to more stringent tests than any other historical books. For that purpose the Science
of Language has proved in many cases a most valuable auxiliary. It is not easy to
imitate ancient language so as to deceive the practiced eye of the grammarian, even
if it were possible to imitate ancient thought that should not betray to the historian
its modern origin. A forged book, like the Ezour Veda, which deceived even Voltaire,
and was published by him as "the most precious gift for which the West was indebted
to the East," could hardly impose again on any Sanskrit scholar of the present day.
This most precious gift from the East to the West, is about the silliest book that can
be read by the student of religion, and all one can say in its defense is that the original writer never meant it as a forgery, never intended it for the purpose for which it
was used by Voltaire. I may add that a book which has lately attracted considerable
attention, *La Bible dans l'Inde,* by M. Jacolliot, belongs to the same class of books.
Though the passages from the sacred books of the Brahmans are not given in the
original, but only in a very poetical French translation, no Sanskrit scholar would
hesitate for one moment to say that they are forgeries, and that M. Jacolliot, the
President of the Court of Justice at Chandernagore, has been deceived by his native
teacher. We find many childish and foolish things in the Veda, but when we read the
following line, as an extract from the Veda: *La femme c'est l'ame de l'humanité;* it is
not difficult to see that this is the folly of the nineteenth century, and not of the
childhood of the human race. M. Jacolliot's conclusions and theories are such as
might be expected from his materials.

With all the genuine documents for studying the history of the religions of
mankind that have lately been brought to light, and with the great facilities which a
more extensive study of Oriental languages has afforded to scholars at large for inves-
tigating the deepest springs of religious thought all over the world, a comparative
study of religions has become a necessity. A science of religion, based on a comparison of all, or, at all events, of the most important religions of mankind, is now only
a question of time. It is demanded by those whose voice cannot be disregarded. Its
title, though implying as yet a promise rather than a fulfillment, has become more

or less familiar in Germany, France, and America; its great problems have attracted the eyes of many inquirers, and its results have been anticipated either with fear or delight. It becomes the duty of those who have devoted their life to the study of the principal religions of the world in their original documents, and who value religion and reverence it in whatever form it may present itself, to take possession of this new territory in the name of true science, and thus to protect its sacred precincts from the inroads of mere babblers. Those who would use a comparative study of religions as a means for debasing Christianity by exalting the other religions of mankind, are to my mind as dangerous allies as those who think it necessary to debase all other religions in order to exalt Christianity. Science wants no partisans. I make no secret that true Christianity seems to me to become more and more exalted the more we appreciate the treasures of truth hidden in the despised religions of the world. But no one can honestly arrive at that conviction, unless he uses honestly the same measure for all religions. It would be fatal for any religion to claim an exceptional treatment, most of all for Christianity. Christianity enjoyed no privileges and claimed no immunities when it boldly confronted and confounded the most ancient and the most powerful religions of the world. Even at present it craves no mercy, and it receives no mercy from those whom our missionaries have to meet face to face in every part of the world; and unless our religion has ceased to be what it was, its defenders should not shrink from this new trial of strength, but should encourage rather than depreciate the study of Comparative Theology.

And let me remark this, in the very beginning, that no other religion, with the exception, perhaps, of early Buddhism, would have favored the idea of an impartial comparison of the principal religions of the world—would have tolerated our science. Nearly every religion seems to adopt the language of the Pharisee rather than of the publican. It is Christianity alone which, as the religion of humanity, as the religion of no caste, of no chosen people, has taught us to respect the history of humanity, as a whole, to discover the traces of a divine wisdom and love in the government of all the races of mankind, and to recognize, if possible, even in the lowest and crudest forms of religious belief, not the work of demoniacal agencies, but something that indicates a divine guidance, something that makes us perceive, with St. Peter, "that God is no respecter of persons, but that in every nation he that feareth Him and worketh righteousness is accepted with Him."

In no religion was there a soil so well prepared for the cultivation of Comparative Theology as in our own. The position which Christianity from the very beginning took up with regard to Judaism, served as the first lesson in Comparative Theology, and directed the attention, even of the unlearned, to a comparison of two religions, differing in their conception of the Deity, in their estimate of humanity, in their motives of morality, and in their hope of immortality, yet sharing so much in common that there are but few of the psalms and prayers in the Old Testament in which a Christian cannot heartily join even now, and but few rules of morality which he ought not even now to obey. If we have once learned to see in the exclusive religion of the Jews a preparation of what was to be the all-embracing religion of humanity, we shall feel much less difficulty in recognizing in the mazes of other religions a hidden purpose; a wandering in the desert, it may be, but a preparation also for the land of promise.

A study of these two religions, the Jewish and the Christian, such as it has long been carried on by some of our most learned divines, simultaneously with the study of Greek and Roman mythology, has, in fact, served as a most useful preparation for wider inquiries. Even the mistakes that have been committed by earlier scholars have proved useful to those who followed after; and, once corrected, they are not likely to be committed again. The opinion, for instance, that the pagan religions were mere corruptions of the religion of the Old Testament, once supported by men of high authority and great learning, is now as completely surrendered as the attempts of explaining Greek and Latin as corruptions of Hebrew. The theory again that there was a primeval preternatural revelation granted to the fathers of the human race, and that the grains of truth which catch our eye when exploring the temples of heathen idols, are the scattered fragments of that sacred heirloom—the seeds that fell by the wayside or upon stony places—would find but few supporters at present; no more, in fact, than the theory that there was in the beginning one complete and perfect primeval language, broken up in later times into the numberless languages of the world.

Some other principles, too, have been established within this limited sphere by a comparison of Judaism and Christianity with the religions of Greece and Rome, which will prove extremely useful in guiding us in our own researches. It has been proved, for instance, that the language of antiquity is not like the language of our own times; that the language of the East is not like the language of the West; and that, unless we make allowance for this, we cannot but misinterpret the utterances of the most ancient teachers and poets of the human race. The same words do not mean the same thing in Anglo-Saxon and English, in Latin and French; much less can we expect that the words of any modern language should be the exact equivalents of an ancient Semitic language, such as the Hebrew of the Old Testament.

Ancient words and ancient thoughts, for both go together, have not yet arrived at that stage of abstraction in which, for instance, active powers, whether natural or supernatural, can be represented in any but a personal and more or less human form. When we speak of a temptation from within or from without, it was more natural for the ancients to speak of a tempter, whether in a human or in an animal form; when we speak of the ever-present help of God, they call the Lord their rock, and their fortress, their buckler, and their high tower; what with us is a heavenly message, or a godsend, was to them a winged messenger; what we call divine guidance, they speak of as a pillar of a cloud to lead them the way, and a pillar of light to give them light, a refuge from the storm, and a shadow from the heat. What is really meant is no doubt the same, and the fault is ours, not theirs, if we willfully misinterpret the language of ancient prophets, if we persist in understanding their words in their outward and material aspect only, and forget that before language had sanctioned a distinction between the concrete and the abstract, between the purely spiritual as opposed to the coarsely material, the intention of the speakers comprehends both the concrete and the abstract, both the material and the spiritual, in a manner which has become quite strange to us, though it lives on in the language of every true poet. Unless we make allowance for this mental parallax, all our readings in the ancient skies will be, and must be erroneous. Nay, I believe it can be proved that more than half of the difficulties in the history of religious thought owe their origin to this constant misinterpretation of ancient language by modern language, of ancient thought by modern thought.

That much of what seems to us, and seemed to the best among the ancients, irrational and irreverent in the mythologies of India, Greece, and Italy, can thus be removed, and that many of their childish fables can thus be read again in their original child-like sense, has been proved by the researches of Comparative Mythologists. The phase of language which gives rise, inevitably, we may say, to these misunderstandings, is earlier than the earliest literary documents. Its work in the Aryan languages was done before the time of the Veda, before the time of Homer, though its influence continues to be felt to a much later period.

Is it likely that the Semitic languages, and, more particularly, Hebrew, should, as by a miracle, have escaped the influence of a process which is inherent in the very nature and growth of language, which, in fact, may rightly be called an infantine disease, against which no precautions can be of any avail?

And if it is not, are we likely to lose anything if we try to get at the most ancient, the most original intention of sacred traditions, instead of being satisfied with their later aspect, their modern misinterpretations? Have we lost anything if, while reading the story of Hephaestos splitting open with his axe the head of Zeus, and Athene springing from it full[y] armed, we perceive behind this savage imagery, Zeus as the bright Sky, his forehead as the East, Hephaestos as the young, not yet risen Sun, and Athene as the Dawn, the daughter of the Sky, stepping forth from the fountain-head of light:

GLAUKOPIS (Γλαυκῶπις), with eyes like an owl (and beautiful they are);
PARTHENOS (Παρθένος), pure as a virgin;
CHRYSEA (Χρύσεα), the golden;
AKRIA (Ἀκρία), lighting up the tops of the mountains, and her own
   glorious Parthenon in her own favorite town of Athens;
PALLAS (Παλλάς), whirling the shafts of light;
ALEA (Ἀλέα), the genial warmth of the morning;
PROMACHOS (Πρόμαχος), the foremost champion in the battle
   between night and day;
PANOPLOS (Πάνοπλος), in full armor, in her panoply of light,
   driving away the darkness of night, and awakening men to a bright life,
   to bright thoughts, to bright endeavors.

Would the Greeks have had less reverence for their gods if, instead of believing that Apollo and Artemis murdered the twelve children of Niobe, they had perceived that Niobe was, in a former period of language, a name of snow and winter; and that no more was intended by the ancient poet than that Apollo and Artemis, the vernal deities, must slay every year with their darts the brilliant and beautiful but doomed children of the Snow? Is it not something worth knowing, worth knowing even to us after the lapse of four or five thousand years, that before the separation of the Aryan race, before the existence of Sanskrit, Greek, or Latin, before the gods of the Veda had been worshipped, and before there was a sanctuary of Zeus among the sacred oaks of Dodona, one supreme deity had been found, had been named, had been invoked by the ancestors of our race, and had been invoked by a name which has never been excelled by any other name?

No; if a critical examination of the ancient language of the Jews leads to no worse results than those which have followed from a careful interpretation of the petrified language of ancient India and Greece, we need not fear; we shall be gainers, not losers. Like an old precious medal, the ancient religion, after the rust of ages has been removed, will come out in all its purity and brightness; and the image which it discloses will be the image of the Father, the Father of all the nations upon earth; and the superscription, when we can read it again, will be, not only in Judæa, but in the languages of all the races of the world, the Word of God, revealed, where alone it can be revealed—revealed in the heart of man.

CHAPTER 7

ON THE MIGRATION OF FABLES

A Lecture Delivered at the Royal Institution, on Friday, June 3, 1870

"Count not your chickens before they be hatched" is a well-known proverb in English, and most people, if asked what was its origin, would probably appeal to La Fontaine's delightful fable, *La Laitière et le Pot au Lait*.[1] We all know Perrette, lightly stepping along from her village to the town, carrying the milk-pail on her head, and in her day-dreams selling her milk for a good sum, then buying a hundred eggs, then selling the chickens, then buying a pig, fattening it, selling it again, and buying a cow with a calf. The calf frolics about, and kicks up his legs—so does Perrette, and, alas!, the pail falls down, the milk is spilt, her riches gone, and she only hopes when she comes home that she may escape a flogging from her husband.

Did La Fontaine invent this fable?; or did he merely follow the example of Sokrates, who, as we know from the Phædon,[2] occupied himself in prison, during the last days of his life, with turning into verse some of the fables, or, as he calls them, the myths of Æsop?

La Fontaine published the first six books of his fables in 1668,[3] and it is well known that the subjects of most of these early fables were taken from Æsop, Phædrus, Horace, and other classical fabulists, if we may adopt this word "fabuliste," which La Fontaine was the first to introduce into French.

In 1678 a second edition of these six books was published, enriched by five books of new fables, and in 1694 a new edition appeared, containing one additional book, thus completing the collection of his charming poems.

The fable of Perrette stands in the seventh book, and was published, therefore, for the first time in the edition of 1678. In the preface to that edition La Fontaine says: "It is not necessary that I should say whence I have taken the subjects of these new fables. I shall only say, from a sense of gratitude, that I owe the largest portion of them to Pilpay the Indian sage."

If, then, La Fontaine tells us himself that he borrowed the subjects of most of his new fables from Pilpay, the Indian sage, we have clearly a right to look to India in order to see whether, in the ancient literature of that country, any traces can be discovered of Perrette with the milk-pail.

Sanskrit literature is very rich in fables and stories; no other literature can vie with it in that respect; nay, it is extremely likely that fables, in particular animal fables, had their principal source in India. In the sacred literature of the Buddhists, fables held a most prominent place. The Buddhist preachers, addressing themselves chiefly to the people, to the untaught, the uncared for, the outcast, spoke to them, as we still speak to children, in fables, in proverbs and parables. Many of these fables and parables must have existed before the rise of the Buddhist religion; others, no doubt, were added on the spur of the moment, just as Sokrates would invent a myth or fable whenever that form of argument seemed to him most likely to impress and convince his hearers. But Buddhism gave a new and permanent sanction to this whole branch of moral mythology, and in the sacred canon, as it was settled in the third century before Christ, many a fable received, and holds to the present day, its recognized place. After the fall of Buddhism in India, and even during its decline, the Brahmans claimed the inheritance of their enemies, and used their popular fables for educational purposes. The best known of these collections of fables in Sanskrit is the Pañkatantra, literally the Pentateuch, or Pentamerone. From it and from other sources another collection was made, well known to all Sanskrit scholars by the name of Hitopadesa, i.e., Salutary Advice. Both these books have been published in England and Germany, and there are translations of them in English, German, French, and other languages.[4]

The first question which we have to answer refers to the date of these collections, and dates in the history of Sanskrit literature are always difficult points. Fortunately, as we shall see, we can in this case fix the date of the Pañkatantra at least, by means of a translation into ancient Persian, which was made about 550 years after Christ, though even then we can only prove that a collection somewhat like the Pañkatantra must have existed at that time; but we cannot refer the book, in exactly that form in which we now possess it, to that distant period.

If we look for La Fontaine's fable in the Sanskrit stories of the Pañkatantra, we do not find, indeed, the milkmaid counting her chickens before they are hatched, but we meet with the following story:

> There lived in a certain place a Brâhman, whose name was Svabhâvakripana, which means 'a born miser.' He had collected a quantity of rice by begging (this reminds us somewhat of the Buddhist mendicants), and after having dined off it, he filled a pot with what was left over. He hung the pot on a peg on the wall, placed his couch beneath, and looking intently at it all the night, he thought, 'Ah, that pot is indeed brimful of rice. Now, if there should be a famine, I should certainly make a hundred rupees by it. With this I shall buy a couple of goats. They will have young ones every six months, and thus I shall have a whole herd of goats. Then, with the goats, I shall buy cows. As soon as they have calved, I shall sell the calves. Then, with the cows, I shall buy buffaloes; with the buffaloes, mares. When the mares have foaled, I shall have plenty of horses; and when I sell them, plenty of gold. With that gold I shall get a house with four wings. And then a Brâhman will come to my house, and will give me his beautiful daughter, with a large dowry. She will have a son, and I shall call him Somasarman. When he is old enough to be danced on his father's knee, I shall sit with a book at the back of the stable, and while I am reading, the boy will see me, jump from his mother's lap, and run towards me to be danced on my knee. He will come too near the horse's hoof, and, full of anger, I shall call to my wife, "Take the baby; take him!" But she, distracted by some domestic work does not hear me. Then I get up, and give her such a kick with my foot.' While he thought this, he gave a kick with his foot, and broke the pot. All the rice fell over him, and made him quite white. Therefore, I say, 'He who makes foolish plans for the future will be white all over, like the father of Somasarman.'[5]

I shall at once proceed to read you the same story, though slightly modified, from the Hitopadesa.[6] The Hitopadesa professes to be taken from the Pañkatantra and some other books; and in this case it would seem as if some other authority had been followed. You will see, at all events, how much freedom there was in telling the old story of the man who built castles in the air:

> In the town of Devíkotta there lived a Brâhman of the name of Devasarman. At the feast of the great equinox he received a plate full of rice. He took it, went into a potter's shop, which was full of crockery, and, overcome by the heat, he lay down in a corner and began to doze. In order to protect his plate of rice, he kept a stick in his hand, and began to think, 'Now, if I sell this plate of rice, I shall receive ten cowries (*kapardaka*). I shall then, on the spot, buy pots and plates, and after having increased my capital again and again, I shall buy and sell betel nuts and dresses till I become enormously rich. Then I shall marry four wives, and the youngest and prettiest of the four I shall make a great pet of. Then the other wives will be so angry, and begin to quarrel. But I shall be in a great rage, and take a stick, and give them a good flogging.' ... While he said this, he flung his stick away; the plate of rice was smashed to pieces, and many of the pots in the shop were broken. The potter, hearing the noise, ran into the shop, and, when he saw his pots broken, he gave the Brâhman a good scolding, and drove him out of his shop. Therefore I say, 'He who rejoices over plans for the future will come to grief, like the Brâhman who broke the pots.'

In spite of the change of a Brahman into a milkmaid, no one, I suppose, will doubt that we have here in the stories of the Pañkatantra and Hitopadesa the first germs of La Fontaine's fable. But how did that fable travel all the way from India to France? How did it doff its Sanskrit garment and don the light dress of modern French? How was the stupid Brâhman born again as the brisk milkmaid, "*cotillon simple et souliers plats?*"

It seems a startling case of longevity that while languages have changed, while works of art have perished, while empires have risen and vanished again, this simple children's story should have lived on, and maintained its place of honor and its undisputed sway in every school-room of the East and every nursery of the West. And yet it is a case of longevity so well attested that even the most skeptical would hardly venture to question it. We have the passport of these stories *viséed* at every place through which they have passed, and, as far as I can judge, *parfaitement en règle.* The story of the migration of these Indian fables from East to West is indeed wonderful; more wonderful and more instructive than many of these fables themselves. Will it be believed that we, in this Christian country and in the nineteenth century, teach our children the first, the most important lessons of worldly wisdom, nay, of a more than worldly wisdom, from books borrowed from Buddhists and Brahmans, from heretics and idolaters, and that wise words, spoken a thousand, nay, two thousand years ago, in a lonely village of India, like precious seed scattered broadcast all over the world, still bear fruit a hundred and a thousand-fold in that soil which is the most precious before God and man the soul of a child? No lawgiver, no philosopher, has made his influence felt so widely, so deeply, and so permanently as the author of these children's fables. But who was he? We do not know. His name, like the name of many a benefactor of the human race, is forgotten. We only know he was an Indian—a "nigger," as some people would call him—and that he lived at least two thousand years ago.

No doubt, when we first hear of the Indian origin of these fables, and of their migration from India to Europe, we wonder whether it can be so; but the fact is, that the story of this Indo-European migration is not, like the migration of the Indo-European languages, myths, and legends, a matter of theory, but of history, and that it was never quite forgotten either in the East or in the West. Each translator, as he handed on his treasure, seems to have been anxious to show how he came by it.

Several writers who have treated of the origin and spreading of Indo-European stories and fables have mixed up two or three questions which ought to be treated each on its own merits.

The first question is whether the Aryans, when they broke up their pro-ethnic community, carried away with them not only their common grammar and diction-ary, but likewise some myths and legends which we find that Indians, Persians, Greeks, Romans, Celts, Germans, Slaves [sic], when they emerge into the light of history, share in common? That certain deities occur in India, Greece, and Germany, having the same names and the same character, is a fact that can no longer be denied. That certain heroes, too, known to Indians, Greeks, and Romans, point to one and the same origin, both by their name and by their history, is a fact by this time admit-ted by all whose admission is of real value. As heroes are in most cases gods in disguise, there is nothing very startling in the fact that nations, who had worshipped the same gods, should also have preserved some common legends of demi-gods or heroes, nay, even in a later phase of thought, of fairies and ghosts. The case, however, becomes much more problematical when we ask whether stories also, fables told with a decided moral purpose, formed part of that earliest Aryan inheritance? This is still doubted by many who have no doubts whatever as to common Aryan myths and legends, and even those who, like myself, have tried to establish by tentative argu-ments the existence of common Aryan fables, dating from before the Aryan separa-tion, have done so only by showing a possible connection between ancient popular saws and mythological ideas, capable of a moral application. To any one, for instance, who knows how in the poetical mythology of the Aryan tribes, the golden splendor of the rising sun leads to conceptions of the wealth of the Dawn in gold and jewels and her readiness to shower them upon her worshippers, the modern German proverb, *Morgenstunde hat Gold im Munde,* seems to have a kind of mythological ring, and the stories of benign fairies, changing everything into gold, sound likewise like an echo from the long-forgotten forest of our common Aryan home.

If we know how the trick of dragging stolen cattle backwards into their place of hiding, so that their footprints might not lead to the discovery of the thief, appears again and again in the mythology of different Aryan nations, then the pointing of the same trick as a kind of proverb, intended to convey a moral lesson and, illustrated by fables of the same or a very similar character in India and Greece, makes one feel inclined to suspect that here too the roots of these fables may reach to a pro-ethnic period. *Vestigia nulla retrorsum* is clearly an ancient proverb, dating from a nomadic period, and when we see how Plato (*Alcibiades,* i. 123) was perfectly, familiar with the Æsopian myth or fable—κατὰ τὸν Αἰσώπου μῦθον, he says—of the fox declining to enter the lions cave, because all footsteps went into it and none came out, and how the Sanskrit Pañkatantra (III.14), tells of a jackal hesitating to enter his own cave, because he sees the footsteps of a lion going in, but not coming out, we feel strongly inclined to admit a common origin for both fables. Here, however, the idea that the Greeks, like La Fontaine, had borrowed their fable from the Pañkatantra would be simply absurd, and it would be much more rational, if the process must be one of borrowing, to admit, as Benfey (*Pantschatantra,* i. 381) does, that the Hindus, after Alexander's discovery of India, borrowed this story from the Greeks. But if we consider that each of the two fables has its own peculiar tendency, the one deriving its lesson from the absence of backward footprints of the victims, the other from the absence of backward footprints of the lion himself, the admission of a common Aryan proverb such as "*vestigia nulla retrorsum,*" would far better explain the facts such as we find them. I am not ignorant of the difficulties of this explanation, and I would myself point to the fact that among the Hottentots, too, Dr. Bleek has found a fable of the jackal declining to visit the sick lion, "because the traces of the animals who went to see him did not turn back."[7]

Without, however, pronouncing any decided opinion on this vexed question, what I wish to place clearly before you is this, that the spreading of Aryan myths, legends, and fables, dating from a pro-ethnic period, has nothing whatever to do with the spreading of fables taking place in strictly historical times from India to Arabia, to Greece and the rest of Europe, not by means of oral tradition, but through more or less faithful translations of literary works. Those who like may doubt whether *Zeus* was Dyaus, whether *Daphne* was Ahanâ, whether *La Belle au Bois* was the mother of two children, called *L'Aurore* and *Le Jour,*[8] but the fact that a collection of fables was, in the sixth century of our era, brought from India to Persia, and by means of various translations naturalized among Persians, Arabs, Greeks, Jews, and all the rest, admits of no doubt or cavil. Several thousand years have passed between those two migrations, and to mix them up together, to suppose that Comparative Mythology has anything to do with the migration of such fables as that of Perrette, would be an anachronism of a portentous character.

There is a third question, viz., whether besides the two channels just mentioned, there were others through which Eastern fables could have reached Europe, or Æsopian and other European fables have been transferred to the East. There are such channels, no doubt. Persian and Arab stories, of Indian origin, were through the crusaders brought back to Constantinople, Italy, and France; Buddhist fables were through Mongolian[9] conquerors (thirteenth century) carried to Russia and the eastern parts of Europe. Greek stories may have reached Persia and India at the time of Alexander's conquests and during the reigns of the Diadochi, and even Christian legends may have found their way to the East through missionaries, travellers, or slaves.

Lastly, there comes the question, how far our common human nature is sufficient to account for coincidences in beliefs, customs, proverbs, and fables, which, at first sight, seem to require an historical explanation. I shall mention but one instance. Professor Wilson (*Essays on Sanskrit Literature,* i, p. 201) pointed out that the story of the Trojan horse occurs in a Hindu tale, only that instead of the horse we have an elephant. But he rightly remarked that the coincidence was accidental. In the one case, after a siege of nine years, the principal heroes of the Greek army are concealed in a wooden horse, dragged into Troy by a stratagem, and the story ends by their falling upon the Trojans and conquering the city of Priam. In the other story a king bent on securing a son-in-law, had an elephant constructed by able artists, and filled with armed men. The elephant was placed in a forest, and when the young prince came to hunt, the armed men sprang out, overpowered the prince and brought him to the king, whose daughter he was to marry. However striking the similarity may seem to one unaccustomed to deal[ing] with ancient legends, I doubt whether any comparative mythologist has postulated a common Aryan origin for these two stories. They feel that, as far as the mere construction of a wooden animal is concerned, all that was necessary to explain the origin of the idea in one place was present also in the other, and that while the Trojan horse forms an essential part of a mythological cycle, there is nothing truly mythological or legendary in the Indian story. The idea of a hunter disguising himself in the skin of an animal, or even of one animal assuming the disguise of another,[10] are familiar in every part of the world, and if that is so, then the step from hiding under the skin of a large animal to that of hiding in a wooden animal is not very great.

Every one of these questions, as I said before, must be treated on its own merits, and while the traces of the first migration of Aryan fables can be rediscovered only by the most minute and complex inductive processes, the documents of the latter are to be found in the library of every intelligent collector of books. Thus, to return to Perrette and the fables of Pilpay, Huet, the learned bishop of Avranches, the friend of La Fontaine, had only to examine the prefaces of the principal translations of the Indian fables in order to track their wanderings, as he did in his famous *Traite de l'Origine des Romans,* published at Paris in 1670, two years after the appearance of the first collection of La Fontaine's fables. Since his time the evidence has become more plentiful, and the whole subject has been more fully and more profoundly treated by Sylvestre de Sacy,[11] Loiseleur Deslongchamps,[12] and Professor Benfey.[13] But though we have a more accurate knowledge of the stations by which the Eastern fables reached their last home in the West, Bishop Huet knew as well as we do that they came originally from India through Persia by way of Bagdad and Constantinople.

In order to gain a commanding view of the countries traversed by these fables, let us take our position at Bagdad in the middle of the eighth century, and watch from that central point the movements of our literary caravan in its progress from the far East to the far West. In the middle of the eighth century, during the reign of the great Khalif Almansur, Abdallah ibn Almokaffa wrote his famous collection of fables, the *Kalila and Dimnah,* which we still possess. The Arabic text of these fables has been published by Sylvestre de Sacy, and there is an English translation of it by Mr. Knatchbull, formerly Professor of Arabic at Oxford. Abdallah ibn Almokaffa was a Persian by birth, who after the fall of the Omeyyades became a convert to Mohammedanism, and rose to high office at the court of the Khalifs. Being in possession of important secrets of state, he became dangerous in the eyes of the Khalif Almansur, and was foully murdered.[14] In the preface, Abdallah ibn Almokaffa tells us that he translated these fables from Pehlevi, the ancient language of Persia; and that they had been translated into Pehlevi (about two hundred years before his time) by Barzûyeh, the physician of Khosru Nushirvan, the king of Persia, the contemporary of the Emperor Justinian. The king of Persia had heard that there existed in India a book full of wisdom, and he had commanded his vizier, Buzurjmihr, to find a man acquainted with the languages both of Persia and India. The man chosen was Barzûyeh. He travelled to India, got possession of the book, translated it into Persian, and brought it back to the court of Khosru. Declining all rewards beyond a dress of honor, he only stipulated that an account of his own life and opinions should be added to the book. This account, probably written by himself, is extremely curious. It is a kind of *Religio Medici* of the sixth century, and shows us a soul dissatisfied with traditions and formularies, striving after truth, and finding rest only where many other seekers after truth have found rest before and after him, in a life devoted to alleviating the sufferings of mankind.

There is another account of the journey of this Persian physician to India. It has the sanction of Firdúsi, in the great Persian epic, the Shah Nâmeh, and it is considered by some[15] as more original than the one just quoted. According to it, the Persian physician read in a book that there existed in India trees or herbs supplying a medicine with which the dead could be restored to life. At the command of the king he went to India in search of those trees and herbs; but, after spending a year in vain researches, he consulted some wise people on the subject. They told him that the medicine of which he had read as having the power of restoring men to life had to be understood in a higher and more spiritual sense, and that what was really meant by it were ancient books of wisdom preserved in India, which imparted life to those who were dead in their folly and sins.[16] Thereupon the physician translated these books, and one of them was the collection of fables, the *Kalila and Dimnah*.

It is possible that both these stories were later inventions; the preface also by Ali, the son of Alshah Farési, in which the names of Bidpai and King Dabshelim are mentioned for the first time, is of later date. But the fact remains that Abdallah ibn Almokaffa, the author of the oldest Arabic collection of our fables, translated them from Pehlevi, the language of Persia at the time of Khosru Nushirvan, and that the Pehlevi text which he translated was believed to be a translation of a book brought from India in the middle of the sixth century. That Indian book could not have been the Pañkatantra, as we now possess it, but must have been a much larger collection of fables, for the Arabic translation, the *Kalilah and Dimnah,* contains eighteen chapters instead of the five of the Pañkatantra, and it is only in the fifth, the seventh, the eighth, the ninth, and the tenth chapters that we find the same stories which form the five books of the Pañkatantra in the *textus ornatior*. Even in these chapters the Arabic translator omits stories which we find in the Sanskrit text, and adds others which are not to be found there.

In this Arabic translation the story of the Brahman and the pot of rice runs as follows:

A religious man was in the habit of receiving every day from the house of a merchant a certain quantity of butter (oil) and honey, of which, having eaten as much as he wanted, he put the rest into a jar, which he hung on a nail in a corner of the room, hoping that the jar would in time be filled. Now, as he was leaning back one day on his couch, with a stick in his hand, and the jar suspended over his head, he thought of the high price of butter and honey, and said to himself, 'I will sell what is in the jar, and buy with the money which I obtain for it ten goats, which, producing each of them a young one every five months, in addition to the produce of the kids as soon as they begin to bear, it will not be long before there is a large flock.' He continued to make his calculations, and found that he should at this rate, in the course of two years, have more than four hundred goats. 'At the expiration of this term I will buy,' said he, 'a hundred black cattle, in the proportion of a bull or a cow for every four goats. I will then purchase land, and hire workmen to plough it with the beasts, and put it into tillage, so that before five years are over I shall, no doubt, have realized a great fortune by the sale of the milk which the cows will give and of the produce of my land. My next business will be to build a magnificent house, and engage a number of servants, both male and female; and, when my establishment is completed, I will marry the handsomest woman I can find, who, in due time becoming a mother, will present me with an heir to my possessions, who, as he advances in age, shall receive the best masters that can be procured; and, if the progress which he makes in learning is equal to my reasonable expectations, I shall be amply repaid for the pains and expense which I have bestowed upon him; but if, on the other hand, he disappoints my hopes, the rod which I have here shall be the instrument with which I will make him feel the displeasure of a justly-offended parent.' At these words he suddenly raised the hand which held the stick towards the jar, and broke it, and the contents ran down upon his head and face.[17]

You will have observed the coincidences between the Arabic and the Sanskrit versions, but also a considerable divergence, particularly in the winding up of the story. The Brahman and the holy man both build their castles in the air; but, while the former kicks his wife, the latter only chastises his son. How this change came to pass we cannot tell. One might suppose that, at the time when the book was translated from Sanskrit into Pehlevi, or from Pehlevi into Arabic, the Sanskrit story was exactly like the Arabic story, and that it was changed afterwards. But another explanation is equally admissible, viz., that the Pehlevi or the Arabic translator wished to avoid the offensive behavior of the husband kicking his wife, and therefore substituted the son as a more deserving object of castigation.

We have thus traced our story from Sanskrit to Pehlevi, and from Pehlevi to Arabic; we have followed it in its migrations from the hermitages of Indian sages to the court of the kings of Persia, and from thence to the residence of the powerful Khalifs at Bagdad. Let us recollect that the Khalif Almansur, for whom the Arabic translation was made, was the contemporary of Abderrhaman, who ruled in Spain, and that both were but little anterior to Harun al Rashid and Charlemagne. At that time, therefore, the way was perfectly open for these Eastern fables, after they had once reached Bagdad, to penetrate into the seats of Western learning, and to spread to every part of the new empire of Charlemagne. They may have done so, for all we know; but nearly three hundred years pass before these fables meet us again in the literature of Europe. The Carlovingian empire had fallen to pieces, Spain had been rescued from the Mohammedans, William the Conqueror had landed in England, and the Crusades had begun to turn the thoughts of Europe towards the East, when, about the year 1080, we hear of a Jew of the name of Symeon, the son of Seth, who translated these fables from Arabic into Greek. He states in his preface that the book came originally from India, that it was brought to the King Chosroes of Persia, and then translated into Arabic. His own translation into Greek must have been made from an Arabic manuscript of the *Kalila and Dimna,* in some places more perfect, in others less perfect, than the one published by De Sacy. The Greek text has been published, though very imperfectly, under the title of *Stephanites and Ichnelates.*[18] Here, our fable is told as follows:

It is said that a beggar kept some honey and butter in a jar close to where he slept. One night he thus thought within himself: 'I shall sell this honey and butter for however small a sum; with it I shall buy ten goats, and these in five months will produce as many again. In five years they will become four hundred. With them I shall buy one hundred cows, and with them I shall cultivate some land. And what with their calves and the harvests, I shall become rich in five years, and build a house with four wings,[19] ornamented with gold, and buy all kinds of servants, and marry a wife. She will give me a child, and I shall call him Beauty. It will be a boy, and I shall educate him properly; and if I see him lazy, I shall give him such a flogging with this stick. ... ' With these words he took a stick that was near him, struck the jar, and broke it, so that the honey and milk ran down on his beard. (p. 337)

This Greek translation might, no doubt, have reached La Fontaine; but as the French poet was not a great scholar, least of all a reader of Greek manuscripts, and as the fables of Symeon Seth were not published till 1697, we must look for other channels through which the old fable was carried along from East to West.

There is, first of all, an Italian translation of the *Stephanites and Ichnelates*, which was published at Ferrara in 1583. The title is *Del Governo de' Regni. Sotto morali essempi di animali ragionanti tra loro. Tratti prima di lingua Indiana in Agarena da Lelo Demno Saraceno. Et poi dall' Agarena nella Greca da Simeone Setto, philosopho Antiocheno. Et hora tradotti di Greco in Italiano.* This translation was probably the work of Giulio Nuti.

There is, besides, a Latin translation, or rather a free rendering of the Greek translation by the learned Jesuit, Petrus Possinus, which was published at Rome in 1666. This may have been, and, according to some authorities, has really been one of the sources from which La Fontaine drew his inspirations. But though La Fontaine may have consulted this work for other fables, I do not think that he took from it the fable of Perrette and the milk-pail.

The fact is, these fables had found several other channels through which, as early as the thirteenth century, they reached the literary market of Europe, and became familiar as household words, at least among the higher and educated classes. We shall follow the course of some of these channels. First, then, a learned Jew, whose name seems to have been Joel, translated our fables from Arabic into Hebrew (1250?). His work has been preserved in one manuscript at Paris, but has not yet been published, except the tenth book, which was communicated by Dr. Neubauer to Benfey's journal, *Orient und Occident* (vol. i, p. 658). This Hebrew translation was translated by another converted Jew, Johannes of Capua, into Latin. His translation was finished between 1263 and 1278, and, under the title of *Directorium Humanæ Vitæ,* it became very soon a popular work with the select reading public of the thirteenth century. In the *Directorium,* and in Joel's translation, the name of Sendebar is substituted for that of Bidpay. The *Directorium* was translated into German at the command of Eberhard, the great Duke of Würtemberg, and both the Latin text and the German translation occur, in repeated editions, among the rare books printed between 1480 and the end of the fifteenth century.[20] A Spanish translation, founded both on the German and the Latin texts, appeared at Burgos in 1493[21]; and from these different sources flowed in the sixteenth century the Italian renderings of Firenzuola (1548)[22] and Doni (1552).[23] As these Italian translations were repeated in French[24] and English, before the end of the sixteenth century, they might no doubt have supplied La Fontaine with subjects for his fables.

But, as far as we know, it was a third channel that really brought the Indian fables to the immediate notice of the French poet. A Persian poet, of the name of Nasr

Allah, translated the work of Abdallah ibn Almokaffa into Persian about 1150. This Persian translation was enlarged in the fifteenth century by another Persian poet, Husain ben Ali called el Vaez, under the title of *Anvári Suhaili*.[25] This name will be familiar to many members of the Indian Civil Service, as being one of the old Haileybury class-books which had to be construed by all who wished to gain high honors in Persia[n]. This work, or at least the first books of it, were translated into French by David Sahid of Ispahan, and published at Paris in 1644, under the title of *Livre des Lumières, ou, la Conduite des Rois, composé par le Sage Pilpay, Indien*. This translation, we know, fell into the hands of La Fontaine, and a number of his most charming fables were certainly borrowed from it.

But Perrette with the milk-pail has not yet arrived at the end of her journey, for if we look at the *Livre des Lumières*, as published at Paris, we find neither the milkmaid nor her prototype, the Brahman who kicks his wife, or the religious man who flogs his boy. That story occurs in the later chapters, which were left out in the French translation; and La Fontaine, therefore, must have met with his model elsewhere.

Remember that in all our wanderings we have not yet found the milkmaid, but only the Brahman or the religious man. What we want to know is who first brought about this metamorphosis.

No doubt La Fontaine was quite the man to seize on any jewel which was contained in the Oriental fables, to remove the cumbersome and foreign-looking setting, and then to place the principal figure in that pretty frame in which most of us have first become acquainted with it. But in this case the charmer's wand did not belong to La Fontaine, but to some forgotten worthy, whose very name it will be difficult to fix upon with certainty.

We have, as yet, traced three streams only, all starting from the Arabic translation of Abdallah ibn Almokaffa, one in the eleventh, another in the twelfth, a third in the thirteenth century, all reaching Europe, some touching the very steps of the throne of Louis XIV, yet none of them carrying the leaf which contained the story of Perrette, or of the Brahman, to the threshold of La Fontaine's home. We must, therefore, try again.

After the conquest of Spain by the Mohammedans, Arabic literature had found a new home in Western Europe, and among the numerous works translated from Arabic into Latin or Spanish, we find towards the end of the thirteenth century (1289) a Spanish translation of our fables, called *Calila é Dymna*. In this the name of the philosopher is changed from Bidpai to Bundobel. This, or another translation from Arabic, was turned into Latin verse by Raimond de Béziers in 1313 (not published).

Lastly, we find in the same century another translation from Arabic straight into Latin verse, by Baldo, which became known under the name of *Æsopus alter*.

From these frequent translations, and translations of translations, in the eleventh, twelfth, and thirteenth centuries, we see quite clearly that these Indian fables were extremely popular, and were, in fact, more widely read in Europe than the Bible, or any other book. They were not only read in translations, but having been introduced into sermons, homilies, and works on morality, they were improved upon, acclimatized, localized, moralized, till at last it is almost impossible to recognize their Oriental features under their homely disguises.

I shall give you one instance only.

Rabelais, in his *Gargantua,* gives a long description how a man might conquer the whole world. At the end of this dialogue, which was meant as a satire on Charles V, we read:

> There was there present at that time an old gentleman well experienced in the wars, a stern soldier, and who had been in many great hazards, named Echephron, who, hearing this discourse, said: 'J'ay grand peur que toute ceste entreprise sera semblable à la farce *du pot au laict* duquel un cordavanier se faisoit riche par resverie, puis le pot cassé, n'eut de quoy disner.'[26]

This is clearly our story, only the Brahman has, as yet, been changed into a shoemaker only, and the pot of rice or the jar of butter and honey into a pitcher of milk. Now it is perfectly true that if a writer of the fifteenth century changed the Brahman into a shoemaker, La Fontaine might, with the same right, have replaced the Brahman by his milkmaid. Knowing that the story was current, was, in fact, common property in the fifteenth century, nay, even at a much earlier date, we might really be satisfied after having brought the germs of Perrette within easy reach of La Fontaine. But, fortunately, we can make at least one step further, a step of about two centuries. This step backwards brings us to the thirteenth century, and there we find our old Indian friend again, and this time really changed into a milkmaid.

The book I refer to is written in Latin, and is called *Dialogus Creaturarum optime moralizatus;* in English, the *Dialogue of Creatures moralysed.* It was a book intended to teach the principles of Christian morality by examples taken from ancient fables. It was evidently a most successful book, and was translated into several modern languages. There is an old translation of it in English, first printed by Rastell,[27] and afterwards repeated in 1816. I shall read you from it the fable in which, as far as I can find, the milkmaid appears for the first time on the stage, surrounded already by much of that scenery which, four hundred years later, received its last touches at the hand of La Fontaine:

> DIALOGO C. (p. ccxxiii): For as it is but madnesse to trust to moche in surete, so it is but foly to hope to moche of vanyteys, for vayne be all erthly thinges longynge to men, as sayth Davyd, Psal. xciiii: Wher of it is tolde in fablys that a lady upon a tyme delyvered to her mayden a *galon of mylke* to sell at a cite, and by the way, as she sate and restid her by a dyche side, she began to thinke that with the money of the mylke she wold bye an henne, the which shulde bringe forth chekyns, and when they were growyn to hennys she wolde sell them and by piggis, and eschaunge them in to shepe, and the shepe in to oxen, and so whan she was come to richesse she sholde be maried right worshipfully unto some worthy man, and thus she reioycid. And whan she was thus mervelously comfortid and ravisshed inwardly in her secrete solace, thinkynge with howe greate ioye she shuld be ledde towarde the chirche with her husbond on horsebacke, she sayde to her self: 'Goo we, goo we.' Sodaynlye she smote the ground with her fote, myndynge to spurre the horse, but her fote slypped, and she fell in the dyche, and there lay all her mylke, and so she was farre from her purpose, and never had that she hopid to have.[28]

Here we have arrived at the end of our journey. It has been a long journey across fifteen or twenty centuries, and I am afraid our following Perrette from country to country, and from language to language, may have tired some of my hearers. I shall, therefore, not attempt to fill the gap that divides the fable of the thirteenth century from La Fontaine. Suffice it to say, that the milkmaid, having once taken the place of the Brahman, maintained it against all comers. We find her as Dona Truhana, in the famous *Conde Lucanor*, the work of the Infante Don Juan Manuel who died in 1347, the grandson of St. Ferdinand, the nephew of Alfonso the Wise, though himself not a king, yet more powerful than a king; renowned both by his sword and by his pen, and possibly not ignorant of Arabic, the language of his enemies. We find her again in the *Contes et Nouvelles* of Bonaventure des Periers, published in the sixteenth century, a book which we know that La Fontaine was well acquainted with. We find her after La Fontaine in all the languages of Europe.[29]

You see now before your eyes the bridge on which our fables came to us from East to West. The same bridge which brought us Perrette brought us hundreds of fables, all originally sprung up in India, many of them carefully collected by Buddhist priests, and preserved in their sacred canon, afterwards handed on to the Brahminic writers of a later age, carried by Barzûyeh from India to the court of Persia, then to the courts of the Khalifs at Bagdad and Cordova, and of the emperors at Constantinople. Some of them, no doubt, perished on their journey, others were mixed up together, others were changed till we should hardly know them again. Still, if you once know the eventful journey of Perrette, you know the journey of all the other fables that belong to this Indian cycle. Few of them have gone through so many changes, few of them have found so many friends, whether in the courts of kings or in the huts of beggars. Few of them have been to places where Perrette has not also been. This is why I selected her and her passage through the world as the best illustration of a subject which otherwise would require a whole course of lectures to do it justice.

But though our fable represents one large class or cluster of fables, it does not represent all. There were several collections, besides the Pañkatantra, which found their way from India to Europe. The most important among them is the *Book of the Seven Wise Masters, or the Book of Sindbad,* the history of which has lately been written, with great learning and ingenuity, by Signor Comparetti.[30]

These large collections of fables and stories mark what may be called the high roads on which the literary products of the East were carried to the West. But there are, beside these high roads, some smaller, less trodden paths on which single fables, sometimes mere proverbs, similes, or metaphors, have come to us from India, from Persepolis, from Damascus and Bagdad. I have already alluded to the powerful influence which Arabic literature exercised on Western Europe through Spain. Again, a most active interchange of Eastern and Western ideas took place at a later time during the progress of the Crusades. Even the inroads of Mongolian tribes into Russia and the east of Europe kept up a literary bartering between Oriental and Occidental nations. [see Chart, p. 135]

But few would have suspected a Father of the Church as an importer of Eastern fables. Yet so it is.

At the court of the same Khalif Almansur, where Abdallah ibn Almokaffa translated the fables of Calila and Dimna from Persian into Arabic, there lived a Christian of the name of Sergius, who for many years held the high office of treasurer to the

OLD COLLECTION OF INDIAN FABLES

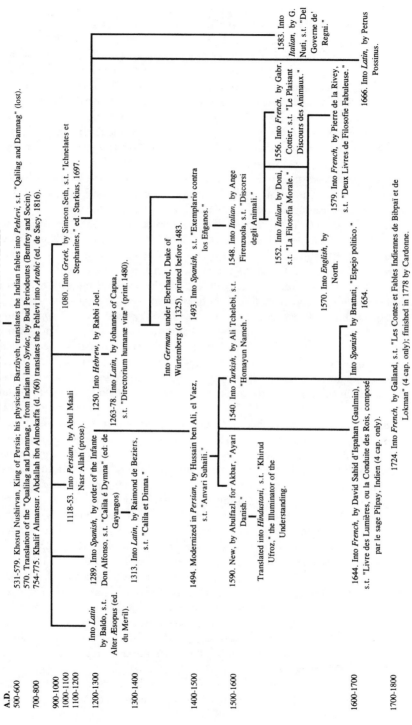

A.D.

500-600

531-579. Khosru Nushirvan, King of Persia; his physician, Barzûyeh, translates the Indian fables into *Pehlevi*, s.t. "Qalilag and Damnag" (lost).

700-800

570. Translation of the "Qualilag and Damnag," from Indian into *Syriac*, by Bud Periodeutes (Benfrey and Socin).
754-775. Khalif Almansur. Abdallah ibn Almokaffa (d. 760) translates the Pehlevi into *Arabic* (ed. de Sacy, 1816).

900-1000
1000-1100
1100-1200

1080. Into *Greek*, by Simeon Seth, s.t. "Ichnelates et Stephanites," ed. Starkius, 1697.

Into *Latin* by Baldo, s.t. Alter Æsopus (ed. du Meril).

1118-53. Into *Persian*, by Abul Maali Nasr Allah (prose).

1250. Into *Hebrew*, by Rabbi Joel.

1200-1300

1289. Into *Spanish*, by order of the Infante Don Alfonso, s.t. "Calila é Dymna" (ed. de Gayangos)

1263-78. Into *Latin*, by Johannes of Capua, s.t. "Directorium humanæ vitæ" (print. 1480).

1300-1400

1313. Into *Latin*, by Raimond de Beziers, s.t. "Calila et Dimna."

Into *German*, under Eberhard, Duke of Wurtemberg (d. 1325). printed before 1483.

1493. Into *Spanish*, s.t. "Exemplario contra los Engaños."

1400-1500

1494. Modernized in *Persian*, by Hussain ben Ali, el Vaez, s.t. "Anvari Suhaili."

1540. Into *Turkish*, by Ali Tchelebi, s.t. "Homayun Nameh."

1548. Into *Italian*, by Ange Firenzuola, s.t. "Discorsi degli Animali."

1500-1600

1590. New, by Abulfazl, for Akbar, "Ayari Danish."
Translated into *Hindustani*, s.t. "Khirud Ufroz," the Illuminator of the Understanding.

1552. Into *Italian*, by Doni, s.t. "La Filosofia Morale."

1570. Into *English*, by North.

1556. Into *French*, by Gabr. Cottier, s.t. "Le Plaisant Discours des Animaux."

1579. Into *French*, by Pierre de la Rivey, s.t. "Deux Livres de Filosofie Fabuleuse."

1583. Into *Italian*, by G. Nuti, s.t. "Del Governe de' Regni."

1666. Into *Latin*, by Petrus Possinus.

Into *Spanish*, by Brattuti, "Espejo politico." 1654.

1600-1700

1644. Into *French*, by David Sahid d'Ispahan (Gaulmin), s.t. "Livre des Lumières, ou la Conduite des Rois, composé par le sage Pilpay, Indien (4 cap. only).

1700-1800

1724. Into *French*, by Galland, s.t. "Les Contes et Fables Indiennes de Bibpai et de Lokman" (4 cap. only): finished in 1778 by Cardonne.

Khalif. He had a son to whom he gave the best education that could then be given, his chief tutor being one Cosmas, an Italian monk, who had been taken prisoner by the Saracens, and sold as a slave at Bagdad. After the death of Sergius, his son succeeded him for some time as chief councillor (πρωτοσύμβουλος) to the Khalif Almansur. Such, however, had been the influence of the Italian monk on his pupil's mind, that he suddenly resolved to retire from the world, and to devote himself to study, meditation, and pious works. From the monastery of St. Saba, near Jerusalem, this former minister of the Khalif issued the most learned works on theology, particularly his *Exposition of the Orthodox Faith*. He soon became the highest authority on matters of dogma in the Eastern Church, and he still holds his place among the saints both of the Eastern and Western Churches. His name was Joannes, and from being born at Damascus, the former capital of the Khalifs, he is best known in history as Joannes Damascenus, or St. John of Damascus. He must have known Arabic, and probably Persian; but his mastery of Greek earned him, later in life, the name of Chrysorrhoas, or Gold-flowing. He became famous as the defender of the sacred images, and as the determined opponent of the Emperor Leo the Isaurian, about 726. It is difficult in his life to distinguish between legend and history, but that he had held high office at the court of the Khalif Almansur, that he boldly opposed the iconoclastic policy of the Emperor Leo, and that he wrote the most learned theological works of his time, cannot be easily questioned.

Among the works ascribed to him is a story called *Barlaam and Joasaph*.[31] There has been a fierce controversy as to whether he was the author of it or not. Though for our own immediate purposes it would be of little consequence whether the book was written by Joannes Damascenus or by some less distinguished ecclesiastic, I must confess that the arguments hitherto adduced against his authorship seem to me very weak.

The Jesuits did not like the book, because it was a religious novel. They pointed to a passage in which the Holy Ghost is represented as proceeding from the Father "and the Son," as incompatible with the creed of an Eastern ecclesiastic. That very passage, however, has now been proved to be spurious; and it should be borne in mind, besides, that the controversy on the procession of the Holy Ghost from the Father and the Son, or from the Father through the Son, dates a century later than Joannes. The fact, again, that the author does not mention Mohammedanism,[32] proves nothing against the authorship of Joannes, because, as he places Barlaam and Joasaph in the early centuries of Christianity, he would have ruined his story by any allusion to Mohammed's religion, then only a hundred years old. Besides, he had written a separate work, in which the relative merits of Christianity and Mohammedanism are discussed. The prominence given to the question of the worship of images shows that the story could not have been written much before the time of Joannes Damascenus, and there is nothing in the style of our author that could be pointed out as incompatible with the style of the great theologian. On the contrary, the author of *Barlaam and Joasaph* quotes the same authors whom Joannes Damascenus quotes most frequently—e.g., Basilius and Gregorius Nazianzenus. And no one but Joannes could have taken long passages from his own works without saying where he borrowed them.[33]

The story of Barlaam and Joasaph—or, as he is more commonly called, Josaphat—may be told in a few words: "A king in India, an enemy and persecutor of the Christians, has an only son. The astrologers have predicted that he would embrace the new doctrine. His father, therefore, tries by all means in his power to keep him ignorant of the miseries of the world, and to create in him a taste for pleasure and enjoyment. A Christian hermit, however, gains access to the prince, and instructs him in the doctrines of the Christian religion. The young prince is not only baptized, but resolves to give up all his earthly riches; and after having converted his own father and many of his subjects, he follows his teacher into the desert."

The real object of the book is to give a simple exposition of the principal doctrines of the Christian religion. It also contains a first attempt at Comparative Theology, for in the course of the story there is a disputation on the merits of the principal religions of the world—the Chaldaean, the Egyptian, the Greek, the Jewish, and the Christian. But one of the chief attractions of this manual of Christian theology consisted in a number of fables and parables with which it is enlivened. Most of them have been traced to an Indian source. I shall mention one only which has found its way into almost every literature of the world:[34]

A man was pursued by a unicorn, and while he tried to flee from it, he fell into a pit. In falling he stretched out both his arms, and laid hold of a small tree that was growing on one side of the pit. Having gained a firm footing, and holding to the tree, he fancied he was safe, when he saw two mice, a black and a white one, busy gnawing the root of the tree to which he was clinging. Looking down into the pit, he perceived a horrid dragon with his mouth wide open, ready to devour him, and when examining the place on which his feet rested, the heads of four serpents glared at him. Then he looked up, and observed drops of honey falling down from the tree to which he clung. Suddenly the unicorn, the dragon, the mice, and the serpents were all forgotten, and his mind was intent only on catching the drops of sweet honey trickling down from the tree.

An explanation is hardly required. The unicorn is Death, always chasing man; the pit is the world; the small tree is man's life, constantly gnawed by the black and the white mouse—i.e., by night and day; the four serpents are the four elements which compose the human body; the dragon below is meant for the jaws of hell. Surrounded by all these horrors, man is yet able to forget them all, and to think only of the pleasures of life, which, like a few drops of honey, fall into his mouth from the tree of life.[35]

But what is still more curious is, that the author of Barlaam and Josaphat has evidently taken his very hero, the Indian Prince Josaphat, from an Indian source. In the Lalita Vistara—the life, though no doubt the legendary life, of Buddha— the father of Buddha is a king. When his son is born, the Brahman Asita predicts that he will rise to great glory, and become either a powerful king, or, renouncing the throne and embracing the life of a hermit, become a Buddha.[36] The great object of his father is to prevent this. He therefore keeps the young prince, when he grows up, in his garden and palaces, surrounded by all pleasures which might turn his mind from contemplation to enjoyment. More especially he is to know nothing of illness, old age, and death, which might open his eyes to the misery and unreality of life. After a time, however, the prince receives permission to drive out; and then follow the four drives, so famous in Buddhist history.[37] The places where these drives took place were commemorated by towers still standing in the time of Fa Hian's visit to

India, early in the fifth century after Christ, and even in the time of Hiouen Thsang, in the seventh century. I shall read you a short account of the three drives:[38]

One day when the prince with a large retinue was driving through the eastern gate of the city, on the way to one of his parks, he met on the road an old man, broken and decrepit. One could see the veins and muscles over the whole of his body, his teeth chattered, he was covered with wrinkles, bald, and hardly able to utter hollow and unmelodious sounds. He was bent on his stick, and all his limbs and joints trembled. 'Who is that man?' said the prince to his coachman. 'He is small and weak, his flesh and his blood are dried up, his muscles stick to his skin, his head is white, his teeth chatter, his body is wasted away; leaning on his stick, he is hardly able to walk, stumbling at every step. Is there something peculiar in his family, or is this the common lot of all created beings?'

'Sir,' replied the coachman, 'that man is sinking under old age, his senses have become obtuse, suffering has destroyed his strength, and he is despised by his relations. He is without support and useless, and people have abandoned him, like a dead tree in a forest. But this is not peculiar to his family. In every creature youth is defeated by old age. Your father, your mother, all your relations, all your friends, will come to the same state; this is the appointed end of all creatures.'

'Alas!,' replied the prince, 'are creatures so ignorant, so weak and foolish as to be proud of the youth by which they are intoxicated, not seeing the old age which awaits them? As for me, I go away. Coachman, turn my chariot quickly. What have I, the future prey of old age—what have I to do with pleasure?' And the young prince returned to the city without going to the park.

Another time the prince was driving through the southern gate to his pleasure-garden, when he perceived on the road a man suffering from illness, parched with fever, his body wasted, covered with mud, without a friend, without a home, hardly able to breathe, and frightened at the sight of himself, and the approach of death. Having questioned his coachman, and received from him the answer which he expected, the young prince said, 'Alas!, health is but the sport of a dream, and the fear of suffering must take this frightful form. Where is the wise man who, after having seen what he is, could any longer think of joy and pleasure?' The prince turned his chariot, and returned to the city.

A third time he was driving to his pleasure-garden through the western gate, when he saw a dead body on the road, lying on a bier and covered with a cloth. The friends stood about crying, sobbing, tearing their hair, covering their heads with dust, striking their breasts, and uttering wild cries. The prince, again, calling his coachman to witness this painful scene, exclaimed, 'Oh, woe to youth, which must be destroyed by old age! Woe to health, which must be destroyed by so many diseases! Woe to this life, where a man remains so short a time! If there were no old age, no disease, no death; if these could be made captive forever!' Then, betraying for the first time his intentions, the young prince said, 'Let us turn back, I must think how to accomplish deliverance.'

A last meeting put an end to hesitation. He was driving through the northern gate on the way to his pleasure-gardens, when he saw a mendicant, who appeared outwardly calm, subdued, looking downwards, wearing with an air of dignity his religious vestment, and carrying an alms-bowl.

'Who is that man?' asked the prince.

'Sir,' replied the coachman, 'this man is one of those who are called Bhikshus, or mendicants. He has renounced all pleasures, all desires, and leads a life of austerity. He tries to conquer himself. He has become a devotee. Without passion, without envy, he walks about asking for alms.'

'This is good and well said,' replied the prince. 'The life of a devotee has always been praised by the wise. It will be my refuge, and the refuge of other creatures; it will lead us to a real life, to happiness and immortality.'

With these words the young prince turned his chariot, and returned to the city.

If we now compare the story of Joannes of Damascus, we find that the early life of Josaphat is exactly the same as that of Buddha. His father is a king, and after the birth of his son, an astrologer predicts that he will rise to glory; not, however, in his own kingdom, but in a higher and better one; in fact, that he will embrace the new and persecuted religion of the Christians. Everything is done to prevent this. He is kept in a beautiful palace, surrounded by all that is enjoyable; and great care is taken to keep him in ignorance of sickness, old age, and death. After a time, however, his father gives him leave to drive out. On one of his drives he sees two men, one maimed, the other blind. He asks what they are, and is told that they are suffering from disease. He then inquires whether all men are liable to disease, and whether it is known beforehand who will suffer from disease and who will be free; and when he hears the truth, he becomes sad, and returns home. Another time, when he drives out, he meets an old man with wrinkled face and shaking legs, bent down, with white hair, his teeth gone, and his voice faltering. He asks again what all this means, and is told that this is what happens to all men; and that no one can escape old age, and that in the end all men must die. Thereupon he returns home to meditate on death, till at last a hermit appears,[39] and opens before his eyes a higher view of life, as contained in the Gospel of Christ.

No one, I believe, can read these two stories without feeling convinced that one was borrowed from the other; and as Fa Hian, three hundred years before John of Damascus, saw the towers which commemorated the three drives of Buddha still standing among the ruins of the royal city of Kapilavastu, it follows that the Greek father borrowed his subject from the Buddhist scriptures. Were it necessary, it would be easy to point out still more minute coincidences between the life of Josaphat and of Buddha, the founder of the Buddhist religion. Both in the end convert their royal fathers, both fight manfully against the assaults of the flesh and the devil, both are regarded as saints before they die. Possibly even a proper name may have been transferred from the sacred canon of the Buddhists to the pages of the Greek writer. The driver who conducts Buddha when he flees by night from his palace where he leaves his wife, his only son, and all his treasures, in order to devote himself to a contemplative life, is called Chandaka, in Burmese, Sanna.[40] The friend and companion of Barlaam is called Zardan.[41] Reinaud in his *Mémoire sur l'Inde* (1849, p. 91), was the first, it seems, to point out that Youdasf, mentioned by Massoudi as the founder of the Sabæan religion, and Youasaf, mentioned as the founder of Buddhism by the author of the *Kitáb-al-Fihrist*, are both meant for Bodhisattva, a corruption quite intelligible with the system of transcribing that name with Persian letters. Professor Benfey has identified Theudas, the sorcerer in *Barlaam and Joasaph*, with the Devadatta of the Buddhist scriptures.[42]

How palpable these coincidences are between the two stories is best shown by the fact that they were pointed out, independently of each other, by scholars in France, Germany, and England. I place France first, because in point of time M. Laboulaye was the first who called attention to it in one of his charming articles in the *Débats*.[43] A more detailed comparison was given by Dr. Liebrecht.[44] And, lastly, Mr. Beal, in his translation of the *Travels of Fa Hian*,[45] called attention to the same fact—viz., that the story of Josaphat was borrowed from the "Life of Buddha." I could mention the names of two or three scholars besides who happened to read the two books, and who could not help seeing, what was as clear as daylight, that Joannes Damascenus took the principal character of his religious novel from the *Lalita Vistara,* one of the sacred books of the Buddhists; but the merit of having been the first belongs to M. Laboulaye.

This fact is, no doubt, extremely curious in the history of literature; but there is another fact connected with it which is more than curious, and I wonder that it has never been pointed out before. It is well known that the story of *Barlaam and Josaphat* became a most popular book during the Middle Ages. In the East it was translated into Syriac (?), Arabic, Ethiopic, Armenian, and Hebrew; in the West it exists in Latin, French, Italian, German, English, Spanish, Bohemian, and Polish. As early as 1204, a king of Norway translated it into Icelandic, and at a later time it was translated by a Jesuit missionary into Tagala, the classical language of the Philippine Islands. But this is not all, Barlaam and Josaphat have actually risen to the rank of saints, both in the Eastern and in the Western churches. In the Eastern church, August 26 is the saints' day of Barlaam and Josaphat; in the Roman Martyrologium, November 27 is assigned to them.

There have been from time to time misgivings about the historical character of these two saints. Leo Allatius, in his *Prolegomena,* ventured to ask the question, whether the story of *Barlaam and Josaphat* was more real than the *Cyropaedia* of Xenophon, or the *Utopia* of Thomas More; but, *en bon Catholique,* he replied, that as Barlaam and Josaphat were mentioned, not only in the Menæa of the Greek, but also in the Martyrologium of the Roman Church, he could not bring himself to believe that their history was imaginary. Billius thought that to doubt the conclud- ing words of the author, who says that he received the story of *Barlaam and Josaphat* from men incapable of falsehood, would be to trust more in one's own suspicions than in Christian charity, which believeth all things. Bellarminus thought he could prove the truth of the story by the fact that, at the end of it, the author himself invokes the two saints Barlaam and Josaphat! Leo Allatius admitted, indeed, that some of the speeches and conversations occurring in the story might be the work of Joannes Damascenus, because Josaphat, having but recently been converted, could not have quoted so many passages from the Bible. But he implies that even this could be explained, because the Holy Ghost might have taught St. Josaphat what to say. At all events, Leo has no mercy for those "quibus omnia sub sanctorum nomine prodita male olent, quemadmodum de sanctis Georgio, Christophoro, Hippolyto, Catarina, aliisque nusquam eos in rerum natura extitisse impudentissime nugantur." The Bishop of Avranches had likewise his doubts but he calmed them by saying: "Non pas que je veuille soustenir que tout en soit supposé: il y auroit de la témerité à desavouer qu'il y ait jamais eû de Barlaam ni de Josaphat. Le témoignage du Martyrologe, qui les met au nombre des Saints, et leur intercession que Saint Jean Damascene reclame à la fin de cette histoire ne permettent pas d'en douter."[46]

With us the question as to the historical or purely imaginary character of Josaphat has assumed a new and totally different aspect. We willingly accept the statement of Joannes Damascenus that the story of *Barlaam and Josaphat* was told him by men who came from India. We know that in India a story was current of a prince who lived in the sixth century B.C., a prince of whom it was predicted that he would resign the throne, and devote his life to meditation in order to rise to the rank of a Buddha. The story tells us that his father did everything to prevent this; that he kept him in a palace secluded from the world, surrounded by all that makes life enjoyable; and that he tried to keep him in ignorance of sickness, old age, and death. We know from the same story that at last the young prince obtained permission to drive into the country, and that, by meeting an old man, a sick man, and a corpse, his eyes were opened to the unreality of life, and the vanity of this life's pleasures; that he escaped from his palace, and, after defeating the assaults of all adversaries, became the founder of a new religion. This is the story, it may be the legendary story, but at all events the recognized story of Gautama Sâkyamuni, best known to us under the name of Buddha.

If, then, Joannes Damascenus tells the same story, only putting the name of Joasaph or Josaphat, i.e., Bodhisattva, in the place of Buddha; if all that is human and personal in the life of St. Josaphat is taken from the *Lalita Vistara,* what follows? It follows that, in the same sense in which La Fontaine's Perrette is the Brahman of the Pañkatantra, St. Josaphat is the Buddha of the Buddhist canon. It follows that Buddha has become a saint in the Roman Church; it follows that, though under a different name, the sage of Kapilavastu, the founder of a religion which, whatever we may think of its dogma, is, in the purity of its morals, nearer to Christianity than any other religion, and which counts even now, after an existence of 2,400 years, 455 million believers, has received the highest honors that the Christian Church can bestow. And whatever we may think of the sanctity of saints, let those who doubt the right of Buddha to a place among them read the story of his life as it is told in the Buddhist canon. If he lived the life which is there described, few saints have a better claim to the title than Buddha; and no one either in the Greek or in the Roman Church need be ashamed of having paid to Buddha's memory the honor that was intended for St. Josaphat, the prince, the hermit, and the saint.

History, here as elsewhere, is stranger than fiction; and a kind fairy, whom men call Chance, has here, as elsewhere, remedied the ingratitude and injustice of the world.

Notes

1. La Fontaine, *Fables,* livre vii., fable 10.
2. Phædon, 61, 5.
3. Robert, *Fables Inédites,* des XIIe, XIIIe, et XIVe Siècles; Paris, 1825, vol. 1, p. ccxxvii.
4. *Pantschatantrum sive Quinquepartitum,* edidit I.G.L. Kosegarten. Bonnæ, 1848; *Pantschatantra, Fünf Bücher indischer Fablen, aus dem Sanskrit übersetzt,* von Th. Benfey. Leipzig, 1859; *Hitopadesa,* with interlinear translation, grammatical analysis, and English translation, in Max Müller's *Handbooks for the Study of Sanskrit.* London, 1864; *Hitopadesa, eine alte indische Fabelsammlung aus dem Sanskrit zum ersten Mal in das Deutsche übersetzt,* von Max Müller. Leipzig, 1844.
5. *Pañkatantra,* v. 10.
6. *Hitopadesa,* ed. Max Müller, p. 120; German translation, p. 159.
7. *Hottentot Fables and Tales,* by Dr. W.H.I. Bleek, London, 1894, p. 19.
8. *Academy,* vol. 5, p. 548.
9. *Die Märchen des Siddhi-kür, or Tales of an Enchanted Corpse,* translated from Kalmuk into German by B. Jülg, 1866. (This is based on the Vetâlapañkavimsati.) *Die Geschichte des Ardschi-Bordschi Chan,* translated from Mongolian by Dr. B. Jülg, 1868. (This in based on the Simhâsanadvâtrimsati.) A Mongolian translation of the *Kalila and Dimnah* is ascribed

to Mélik Said Iftikhar eddin Mohammed ben Abou Nasr, who died A.D. 1280. See Barbier de Meynard, "Description de la Ville de Kazvin," *Journal Asiatique*, 1857, p. 284; Lancereau, Pantchatantra, p. xxv.

10. Plato's expression, "As I have put on the lion's skin" (*Kratylos*, 411), seems to show that he knew the fable of an animal or a man having assumed the lion's skin without the lion's courage. The proverb ὄνος παρὰ Κυμαίους seems to be applied to men boasting before people who have no means of judging. It presupposes the story of a donkey appearing in a lion's skin.

A similar idea is expressed in a fable of the Pañkatantra (IV. 8) where a dyer, not being rich enough to feed his donkey, puts a tiger's skin on him. In this disguise the donkey is allowed to roam through all the cornfields without being molested, till one day he see a female donkey, and begins to bray. Thereupon the owners of the field kill him.

In the *Hitopadesa* (III. 3) the same fable occurs, only that there it is the keeper of the field who on purpose disguises himself as a she-donkey and when he hears the tiger bray, kills him.

In the Chinese Avadânas, translated by Stanislas Julien (vol. 2, p. 59), the donkey takes a lion's skin and frightens everybody, till he begins to bray, and is recognized as a donkey.

In this case it is again quite clear that the Greeks did not borrow their fable and proverb from the Pañkatantra; but it is not so easy to determine positively whether the fable was carried from the Greeks to the East, or whether it arose independently in two places.

11. *Calilah et Dimna, ou, Fables de Bidpai, en Arabe, précédées d'un Mémoire sur l'origine de ce livre.* Par Sylvestre de Sacy. Paris, 1816.

12. Loiseleur Deslongchamps, *Essai sur les Fables Indiennes, et sur leur Introduction en Europe.* Paris, 1838.

13. Pantschatantra, *Fünf Bücher indischer Fabeln, Märchen und Erzählungen, mit Einleitung.* Von. Th. Benfey. Leipzig, 1859.

14. See Weil, *Geschichte der Chalifen*, vol. 2, p. 84.

15. Benfey, p. 60.

16. Cf., *Barlaam et Joasaph*, ed. Boissonade, p. 37.

17. *Kalila and Dimna; or, the Fables of Bidpai, translated from the Arabic*, by the Rev. Wyndham Knatchbull, A.M. Oxford, 1819.

18. *Specimen Sapientiæ Indorum Veterum, id est Liber Ethico-Politicus pervetustus, dictus Arabice Kalîlah ve Dimnah, Græce Stephanites et Ichnelates, nunc primum Græce ex MS. Cod. Holsteiniano prodit cum versione Latina, opera S.G. Starkii.* Berolini, 1697.

19. This expression, a four-winged house, occurs also in the Pañkatantra. As it does not occur in the Arabic text, published by De Sacy, it is clear that Symeon must have followed another Arabic text in which this adjective, belonging to the Sanskrit and no doubt to the Pehlevi text also, had been preserved.

20. Benfey, *Orient und Occident*, vol. 1, p. 138.

21. Ibid., vol. 1, p. 501. Its title is: *Exemplario contra los Engaños y Peligros del Mundo*, ibid., pp. 167, 168.

22. *Discorsi degli animali, di Messer Agnolo Firenzuola, in prose di M. A. F.* Fiorenza, 1548.

23. *La Moral Filosophia del Doni, tratta da gli antichi scrittori.* Vinegia, 1552; *Trattati Diversi di Sendebar Indiano, filosopho morale.* Vinegia, 1552.

P. 65, *Trattato Quarto*: A woman tells her husband to wait till her son is born, and says:

Stava uno Romito domestico ne i monti di Brianza a far penitenza e teneva alcune cassette d' api per suo spasso, e di quelle a suoi tempi ne cavava il *Mele*, e di quello ne vendeva alcuna parte tal volta per i suoi besogni. Avenne che un' anno ne fu una gran carestia, e egli attendeva a conservarlo, e ogni giorno lo guardava mille volte, e gli pareva cent' anni ogni hora, che e gli indugiava a empierlo di Mele,' etc.

24. *Le Plaisant et Facétieux Discours des Animaux, novellement traduict de Tuscan en François.* Lyon, 1556, par Gabriel Cottier; and *Deux Livres de Filosofie Fabuleuse, le Premier Pris des Discours de M. Ange Firenzuola, le Second Extraict des Traictez de Sandebar Indien, par Pierre de La Rivey.* Lyon, 1579. The second book is a translation of the second part of Doni's *Filosofia Morale.*

25. *The Anvar-i Suhaili, or the Lights of Canopus, being the Persian version of the Fables of Pilpay, or the Book, Kalilah and Damnah, rendered into Persian by Husain Vâ'iz U'l-Káshifi, literally translated by E.B. Eastwick.* Hertford, 1854.

26. Translation: "I have a great fear that this entire enterprise will resemble the farce of the *pitcher of milk* out of which a shoemaker makes himself rich in his dreams, and then, after the pitcher breaks, didn't have anything for dinner."—Sondra Bacharach.

27. *Dialogues of Creatures moralysed,* sm. 4to, circ. 1517. It is generally attributed to the press of John Rastell, but the opinion of Mr. Haslewood, in his preface to the reprint of 1816, that the book was printed on the continent, is perhaps the correct one (*Quaritch's Catalogue,* July, 1870).

28. The Latin text is more simple.... *Dialogus Creaturarum optime moralizatus* (ascribed to Nicolaus Pergaminus, supposed to have lived in the thirteenth century). He quotes Elynandus, in *Gestis Romanorum.* First edition, "per Gerardum leeu [?] in oppido Goudensi inceptum; munere Dei finitus est, Anno Domini, 1480."

29. My learned German translator, Dr. Felix Liebrecht, says in a note: "Other books in which our story appears before La Fontaine are *Esopus,* by Burkhard Waldis, ed. H. Kurz, Leipzig, 1862, ii. 177; note to *Des Bettlers Kaufmannschaft;* and Oesterley, in Kirchoff's *Wendunmuth,* v. 44, note to i. 171, *Vergebene Anschleg reich zuwerden* (Bibl. des liter. Vereins zu Stuttg. No. 99)."

30. *Ricerche intorno al Libro di Sindibad.* Milano, 1869.

31. The Greek text was first published in 1832 by Boissonade, in his *Anecdota Græca,* vol. 4.... Joannes Monachus occurs as the name of the author in other works of Joannes Damascenus. See Leo Allatius, Prolegomena, p. L, in *Damasceni Opera Omnia.* Ed. Lequien, 1748. Venice.... See also Wiener, *Jahrbücher,* vol. 63, pp. 44–83; vol. 72, pp. 274–288; vol. 73, pp. 176–202.

32. Littré, *Journal des Savants,* 1865, p. 337.

33. The *Martyrologium Romanum,* whatever its authority may be, states distinctly that the acts of Barlaam and Josaphat were written by Sanctus Joannes Damascenus. "Apud Indos Persis finitimos sanctorum Barlaam et Josaphat, quorum actus mirandos sanctus Joannes Damascenus conscripsit." See Leonis Allatii Prolegomena, in *Joannis Damasceni Opera,* ed. Lequien, vol. 1, p. xxvi....

34. [Incidentally], the story of the caskets, well known from the *Merchant of Venice,* occurs in *Barlaam and Josaphat,* though it is used there for a different purpose.

35. Cf., Benfey, *Pantschatantra,* vol. 1, p. 80; vol. 2, p. 528; *Les Avadanas, Contes et Apologues indiens,* par Stanislas Julien, vol. 1, pp. 132, 191; *Gesta Romanorum,* cap. 168; *Homáyun Nameh,* cap. iv; Grimm, *Deutsche Mythologie,* pp. 758, 759; Liebrecht. *Jahrbücher für Rom. und Engl. Literatur,* 1860.

36. *Lalita Vistara,* ed. Calcutt., p. 126.

37. Ibid., p. 225.

38. See M. M.'s *Chips from a German Workshop* (Amer. ed.), vol. 1, p. 207.

39. Minayeff, *Mélanges Asiatiques,* vi. 5, p. 584, remarks: "According to a legend in the *Mahâvastu* of Yasas or Yasoda (in a less complete form to be found in Schiefner, *Eine tibetische Lebensbeschreibung Sâkyamunis,* p. 247; Hardy, *Manual of Buddhism,* p. 187; Bigandet, *The Life or Legend of Gaudama,* p. 113), a merchant appears in Yosoda's [*sic*]

house, the night before he has the dream which induces him to leave his paternal house, and proclaims to him the true doctrine."

40. *Journal of the American Oriental Society,* vol. 3, p. 21.

41. In some places one might almost believe that Joannes Damascenus did not only hear the story of Buddha, as he says, from the mouth of people who had brought it to him from India, but that he had before him the very text of the *Lalita Vistara.* Thus in the account of the three or four drives we find indeed that the Buddhist canon represents Buddha as seeing on three successive drives, first an old, then a sick, and at last a dying man, while Joannes makes Joasaph meet two men on his first drive, one maimed, the other blind, and an old man, who is nearly dying, on his second drive. So far there is a difference which might best be explained by admitting the account given by Joannes Damascenus himself, viz., that the story was brought from India, and that it was simply told him by worthy and truthful men. But, if it was so, we have here another instance of the tenacity with which oral tradition is able to preserve the most minute points of the story. The old man is described by a long string of adjectives both in Greek and in Sanskrit, and many of them are strangely alike. The Greek γέρων, old, corresponds to the Sanskrit *gîrna;* πεπαλαιώμενος, aged, is Sanskrit *vriddha;* ἐρρικνώμενος τὸ πρόσωπον, shriveled in his face, is *balînikitakâya,* the body covered with wrinkles; παρείμενος τὰς κνήμας, weak in his knees, is *pravedhayamânah sarvângapratyangaih,* trembling in all his limbs; συγκεκυφώς, bent, is *kubga;* πεπολιώμενος, gray, is *palitakesa;* ἐστερήμενος τοὺς ὀδόντας, toothless, is *khandadanta;* ἐγκεκομένα λαλῶν, stammering, is *khurakhurâvasaktakantha.*

42. *Zeitschrift der Deutschen Morgenländischen Gesellschaft,* vol. 24, p. 480.

43. *Débats,* 1859, Juillet 21 and 26.

44. *Die Quellen des Barlaam und Josaphat,* in *Jahrbuch für roman. und engl. Literatur,* vol. 2, p. 314, 1860.

45. *Travels of Fah-hian and Sung-yun, Buddhist Pilgrims from China to India* (A.D. 400 and A.D. 518). Translated from the Chinese by Samuel Beal. London, Trübner & Co., 1869.

46. Littré, *Journal des Savants.* 1865, p. 337; Translation: "Not that I want to maintain that everything is assumed; it would be rash to disavow that there was neither Barlaam nor Josaphat. The testimony of the Martyrologium, which places them with the Saints and their intercession that Saint Jean Damascene demands at the end of this history, do not allow us to doubt it."—Sondra Bacharach.

Chapter 8

On the Philosophy of Mythology

A Lecture Delivered at the Royal Institution in 1871

What can be in our days the interest of mythology? What is it to us that Kronos was the son of Uranos and Gaia, and that he swallowed his children, Hestia, Demeter, Hera, Pluton, and Poseidon, as soon as they were born? What have we to do with the stories of Rhea, the wife of Kronos, who, in order to save her youngest son from being swallowed by his father, gave her husband a stone to swallow instead? And why should we be asked to admire the exploits of this youngest son, who, when he had grown up, made his father drink a draught, and thus helped to deliver the stone and his five brothers and sisters from their paternal prison? What shall we think if we read in the most admired of classic poets that these escaped prisoners became afterwards the great gods of Greece, gods believed in by Homer, worshipped by Sokrates, immortalized by Pheidias? Why should we listen to such horrors as that Tantalos killed his own son, boiled him, and placed him before the gods to eat?; or that the gods collected his limbs, threw them into a cauldron, and thus restored Pelops to life, *minus,* however, his shoulder, which Demeter had eaten in a fit of absence, and which had therefore to be replaced by a shoulder made of ivory?

Can we imagine anything more silly, more savage, more senseless, anything more unworthy to engage our thoughts, even for a single moment? We may pity our children that, in order to know how to construe and understand the master-works of Homer and Virgil, they have to fill their memory with such idle tales; but we might justly suppose that men who have serious work to do in this world would banish such subjects forever from their thoughts.

And yet, how strange, from the very childhood of philosophy, from the first faintly-whispered Why? to our own time of matured thought and fearless inquiry, mythology has been the ever-recurrent subject of anxious wonder and careful study. The ancient philosophers, who could pass by the petrified shells on mountain-tops and the fossil trees buried in their quarries without ever asking the question how they came to be there, or what they signified, were ever ready with doubts and surmises when they came to listen to ancient stories of their gods and heroes. And, more curious still, even modern philosophers cannot resist the attraction of these ancient problems. That stream of philosophic thought which, springing from Descartes (1596–1650), rolled on through the seventeenth and eighteenth centuries in two beds—the *idealistic,* marked by the names of Malebranche (1638–1715), Spinoza (1632–1677), and Leibniz (1646–1716); and the *sensualistic,* marked by the names of Locke (1632–1704), David Hume (1711–1776), and Condillac (1715–1780), till the two arms united again in Kant (1724–1804), and the full stream was carried on by Schelling (1775–1854) and Hegel (1770–1831),—this stream of modern philosophic thought has ended where ancient philosophy began—in a Philosophy of Mythology, which, as you know, forms the most important part of Schelling's final system, of what he called himself his *Positive Philosophy*, given to the world after the death of that great thinker and poet, in the year 1854.

I do not mean to say that Schelling and Aristotle looked upon mythology in the same light, or that they found in it exactly the same problems; yet there is this common feature in all who have thought or written on mythology, that they look upon it as something which, whatever it may mean, does certainly not mean what it seems to mean; as something that requires an explanation, whether it be a system of religion, or a phase in the development of the human mind, or an inevitable catastrophe in the life of language.

According to some, mythology is history changed into fable; according to others, fable changed into history. Some discover in it the precepts of moral philosophy enunciated in the poetical language of antiquity; others see in it a picture of the great forms and forces of nature, particularly the sun, the moon, and the stars, the changes of day and night, the succession of the seasons, the return of the years—all this reflected by the vivid imagination of ancient poets and sages.

Epicharmos, for instance, the pupil of Pythagoras, declared that the gods of Greece were not what, from the poems of Homer, we might suppose them to be—personal beings, endowed with superhuman powers, but liable to many of the passions and frailties of human nature. He maintained that these gods were really the Wind, the Water, the Earth, the Sun, the Fire, and the Stars. Not long after his time, another philosopher, Empedokles, holding that the whole of nature consisted in the mixture and separation of the four elements, declared that Zeus was the element of Fire, Here [*sic*] the element of Air, Aidoneus or Pluton the element of Earth, and Nestis the element of Water. In fact, whatever the free thinkers of Greece discovered successively as the first principles of Being and Thought, whether the air of Anaximenes, or the fire of Herakleitos, or the Nous or Mind of Anaxagoras, was readily identified with Zeus and the other divine persons of Olympian mythology. Metrodoros, the contemporary of Anaxagoras, went even farther. While Anaxagoras would have been satisfied with looking upon Zeus as but another name of his Nous, the highest intellect, the mover, the disposer, the governor of all things, Metrodoros resolved not only the persons of Zeus, Here [*sic*], and Athene, but likewise those of human kings and heroes—such as Agamemnon, Achilles, and Hektor—into various combinations and physical agencies, and treated the adventures ascribed to them as natural facts hidden under a thin veil of allegory.

Sokrates, it is well known, looked upon such attempts at explaining all fables allegorically as too arduous and unprofitable; yet he, too, as well as Plato, pointed frequently to what they called the *hypónoia*, the under-current, or, if I may say so, the under-meaning of ancient mythology.

Aristotle speaks more explicitly: "It has been handed down," he says,

by early and very ancient people, and left to those who came after, in the form of myths, that these (the first principles of the world) are the gods, and that the divine embraces the whole of nature. The rest has been added mythically, in order to persuade the many, and in order to be used in support of laws and other interests. Thus they say that the gods have a human form, and that they are like to some of the other living beings, and other things consequent on this, and similar to what has been said. If one separated out of these fables, and took only that first point, namely, that they believed the first essences to be gods, one would think that it had been divinely said, and that while every art and every philosophy was probably invented ever so many times and lost again, these opinions had, like fragments of them, been preserved until now. So far only is the opinion of our fathers, and that received from our first ancestors, clear to us.

I have quoted the opinions of these Greek philosophers, to which many more might have been added, partly in order to show how many of the most distinguished minds of ancient Greece agreed in demanding an interpretation, whether physical or metaphysical, of Greek mythology, partly in order to satisfy those classical scholars, who, forgetful of their own classics, forgetful of their own Plato and Aristotle, seem to imagine that the idea of seeing in the gods and heroes of Greece anything beyond what they appear to be in the songs of Homer, was a mere fancy and invention of the students of Comparative Mythology.

There were, no doubt, Greeks, and eminent Greeks too, who took the legends of their gods and heroes in their literal sense. But what do these say of Homer and Hesiod? Xenophanes, the contemporary of Pythagoras, holds Homer and Hesiod responsible for the popular superstitions of Greece. In this he agrees with Herodotus, when he declares that these two poets made the theogony for the Greeks, and gave to the gods their names, and assigned to them their honors and their arts, and described their appearances. But he then continues in a very different strain from the pious historian.[1] "Homer," he says, "and Hesiod ascribed to the gods whatever is disgraceful and scandalous among men, yea, they declared that the gods had committed nearly all unlawful acts, such as theft, adultery, and fraud." "Men seem to have created their gods, and to have given to them their own mind, voice, and figure. The Ethiopians made their gods black and flat-nosed; the Thracians red-haired and blue-eyed."[2] This was spoken about 500 B.C. Herakleitos, about 460 B.C., one of the boldest thinkers of ancient Greece, declared that Homer deserved to be ejected from public assemblies and flogged; and a story is told that Pythagoras (about 540 B.C.) saw the soul of Homer in Hades, hanging on a tree and surrounded by serpents, as a punishment for what he had said of the gods. And what can be stronger than the condemnation passed on Homer by Plato? I shall read an extract from the *Republic,* from the excellent translation lately published by Professor Jowett:

'But what fault do you find with Homer and Hesiod, and the other great story-tellers of mankind?'
'A fault which is most serious,' I said: 'the fault of telling a lie, and a bad lie.'
'But when is this fault committed?'
'Whenever an erroneous representation is made of the nature of gods and heroes—like the drawing of a limner which has not the shadow of a likeness to the truth.'
'Yes,' he said, 'that sort of thing is certainly very blamable; but what are the stories which you mean?'
'First of all,' I said, 'there was that greatest of all lies in high places, which the poet told about Uranos, and which was an immoral lie too—I mean what Hesiod says that

Uranos did, and what Kronos did to him. The fact is that the doings of Kronos, and the sufferings which his son inflicted upon him, even if they were true, ought not to be lightly told to young and simple persons; if possible, they had better be buried in silence. But if there is an absolute necessity for their mention, a very few might hear them in a mystery, and then let them sacrifice not a common (Eleusinian) pig, but some huge and unprocurable victim; this would have the effect of very greatly reducing the number of the hearers.'

'Why, yes,' said he, 'these stories are certainly objectionable.'

'Yes, Adeimantos, they are stories not to be narrated in our state; the young man should not be told that in committing the worst of crimes he is far from doing anything outrageous, and that he may chastise his father when he does wrong in any manner that he likes, and in this will only be following the example of the first and greatest of the gods.'

'I quite agree with you,' he said; 'in my opinion those stories are *not fit to be repeated.*'

'Neither, if we mean our future guardians to regard the habit of quarrelling as dishonorable, should anything be said of the wars in heaven, and of the plots and fightings of the gods against one another, which are quite untrue. Far be it from us to tell them of the battles of the giants, and embroider them on garments; or of all the innumerable other quarrels of gods and heroes with their friends and relations. If they would only believe us, we would tell them that quarrelling is unholy, and that never up to this time has there been any quarrel between citizens; this is what old men and old women should begin by telling children, and the same when they grow up. And these are the sort of fictions which the poets should be required to compose. But the narrative of Hephaestos binding Here [sic] his mother, or how, on another occasion, Zeus sent him flying for taking her part when she was being beaten—such tales must not be admitted in our state, whether they are supposed to have an allegorical meaning or not. For the young man cannot judge what is allegorical and what is literal, and anything that he receives into his mind at that age is apt to become indelible and unalterable; and therefore the tales which they first hear should be models of virtuous thoughts.'

To those who look upon mythology as an ancient form of religion, such freedom of language as is here used by Xenophanes and Plato must seem startling. If the Iliad were really the Bible of the Greeks, as it has not unfrequently [sic] been called, such violent invectives would have been impossible. For let us bear in mind that Xenophanes, though he boldly denied the existence of all the mythological deities, and declared his belief in One God, "neither in form nor in thought like unto mortals,"[3] was not therefore considered a heretic. He never suffered for uttering his honest convictions; on the contrary, as far as we know, he was honored by the people among whom he lived and taught. Nor was Plato ever punished on account of his unbelief, and though he, as well as his master, Sokrates, became obnoxious to the dominant party at Athens, this was due to political far more than to theological motives. At all events, Plato, the pupil, the friend, the apologist of Sokrates, was allowed to teach at Athens to the end of his life, and few men commanded greater respect in the best ranks of Greek society.

But, although mythology was not religion in our sense of the word, and although the Iliad certainly never enjoyed among Greeks the authority either of the Bible, or even of the Veda among the Brahmans, or the Zend-Avesta among the Parsis, yet I would not deny altogether that in a certain sense the mythology of the Greeks belonged to their religion. We must only be on our guard, here as everywhere else, against the misleading influence of words. The word Religion has, like most words, had its history; it has grown and changed with each century, and it cannot, therefore, have meant with the Greeks and Brahmans what it means with us. Religions have sometimes been divided into *national* or *traditional,* as distinguished from *individual* or *statutable* religion. The former are, like languages, home-grown, autochthonic, without an historical beginning, generally without any recognized founder, or even an authorized code; the latter have been founded by historical persons, generally in antagonism to traditional systems, and they always rest on the authority of a written code. I do not consider this division as very useful for a scientific study of religion,[4] because in many cases it is extremely difficult, and sometimes impossible, to draw a sharp line of demarcation, and to determine whether a given religion should be considered as the work of one man, or as the combined work of those who came before him, who lived with him, nay, even of those who came after him. For our present purpose, however, for showing at once the salient difference between what the Greeks and what we ourselves should mean by Religion, this division is very serviceable. The Greek religion was clearly a national and traditional religion, and, as such, it shared both the advantages and disadvantages of this form of religious belief; the Christian religion is an historical and, to a great extent, an individual religion, and it possesses the advantage of an authorized code and of a settled system of faith. Let it not be supposed, however, that between traditional and individual religions the advantages are all on one, the disadvantages on the other side. As long as the immemorial religions of the different branches of the human race remained in their natural state, and were not pressed into the service of political parties or an ambitious priesthood, they allowed great freedom of thought and a healthy growth of real piety, and they were seldom disgraced by an intolerant or persecuting spirit. They were generally either honestly believed, or, as we have just seen, honestly attacked, and a high tone of intellectual morality was preserved, untainted by hypocrisy, equivocation, or unreasoning dogmatism.

The marvellous development of philosophy in Greece, particularly in ancient Greece, was chiefly due, I believe, to the absence of an established religion and an influential priesthood; and it is impossible to overrate the blessing which the fresh, pure, invigorating, and elevating air of that ancient Greek philosophy has conferred on all ages, not excepting our own. I shudder at the thought of what the world would have been without Plato and Aristotle, and I tremble at the idea that the youth of the future should ever be deprived of the teaching and the example of these true prophets of the absolute freedom of thought. Unfortunately, we know but little of the earliest fathers of Greek philosophy; we have but fragments, and those not always trustworthy, nor easily intelligible, of what they taught on the highest questions that can stir the heart of man. We have been accustomed to call the oracular sayings of men like Thales, Pythagoras, Xenophanes, or Herakleitos, philosophy, but there was in them as much of religion as in the songs of Homer and Hesiod. Homer and Hesiod were great powers, but their poems were not the only feeders of the religious life of Greece. The stream of ancient wisdom and philosophy flowed parallel with the stream of legend and poetry; and both were meant to support the religious cravings of the soul. We have only to attend without prejudice to the utterances of these ancient prophets, such as Xenophanes and Herakleitos, in order to convince ourselves that these men spoke with authority to the people,[5] that they considered themselves the equals of Homer and Hesiod, nay, their betters, and in no way fettered by the popular legends about gods and goddesses. While modern religions assume in general a hostile attitude towards philosophy, ancient religions have either included philosophy as an integral part, or they have at least tolerated its growth in the very precincts of their temples.

After we have thus seen what limitations we must place on the meaning of the word Religion, if we call mythology the religion of the ancient world, we may now advance another step.

We have glanced at the principal interpretations which have been proposed by the ancients themselves of the original purpose and meaning of mythology. But there is one question which none, either of the ancient or of the modern interpreters of mythology, has answered, or even asked, and on which, nevertheless, the whole problem of mythology seems to turn. If mythology is history changed into fable, why was it so changed? If it is fable represented as history, why were such fables invented? If it contains precepts of moral philosophy, whence their immoral disguise? If it is a picture of the great forms and forces of nature, the same question still returns, why were these forms and forces represented as heroes and heroines, as nymphs and shepherds, as gods and goddesses? It is easy enough to call the sun a god, or the dawn a goddess, after these predicates have once been framed. But how were these predicates framed? How did people come to know of gods and goddesses, heroes and nymphs, and what meaning did they originally connect with these terms? In fact, the real question which a philosophy of mythology has to answer is this: Is the whole of mythology an invention, the fanciful poetry of a Homer or Hesiod, or is it a growth? Or, to speak more definitely: Was mythology a mere accident, or was it inevitable? Was it only a false step, or was it a step that could not have been left out in the historical progress of the human mind?

The study of the history of language, which is only a part of the study of the history of thought, has enabled us to give a decisive answer to this question. Mythology is inevitable, it is natural, it is an inherent necessity of language, if we recognize in language the outward form and manifestation of thought; it is, in fact, the dark shadow which language throws on thought, and which can never disappear till language becomes altogether commensurate with thought, which it never will. Mythology, no doubt, breaks out more fiercely during the early periods of the history of human thought, but it never disappears altogether. Depend upon it, there is mythology now as there was in the time of Homer, only we do not perceive it, because we ourselves live in the very shadow of it, and because we all shrink from the full meridian light of truth. We are ready enough to see that if the ancients called their kings and heroes Διογενεῖς [Diogeneis], sprung of Zeus, that expression, intended originally to convey the highest praise which man can bestow on man, was apt to lapse into mythology. We easily perceive how such a conception, compatible in its origin with the highest reverence for the gods, led almost inevitably to the growth of fables, which transferred to divine beings the incidents of human paternity and sonship. But we are not so ready to see that it is our fate, too, to move in allegories which illustrate things intellectual by visions exhibited to the fancy. In our religion, too, the conceptions of paternity and sonship have not always been free from all that is human, nor are we always aware that nearly every note that belongs to human paternity and sonship must be taken out of these terms before they can be pronounced safe against mythological infection. Papal decisions on immaculate conception are of no avail against that mythology. The mind must become immaculate and rise superior to itself; or it must close its eyes and shut its lips in the presence of the Divine.

If then we want to understand mythology, in the ordinary and restricted sense of the word, we must discover the larger circle of mental phenomena to which it belongs. Greek mythology is but a small segment of mythology; the religious mythologies of all the races of mankind are again but a small segment of mythology. Mythology, in the highest sense, is the power exercised by language on thought in every possible sphere of mental activity; and I do not hesitate to call the whole history of philosophy, from Thales down to Hegel, an uninterrupted battle against mythology, a constant protest of thought against language. This will require some explanation.

Ever since the time of Wilhelm von Humboldt, all who have seriously grappled with the highest problems of the Science of Language have come to the conviction that thought and language are inseparable, that language is as impossible without thought as thought is without language; that they stand to each other somewhat like soul and body, like power and function, like substance and form. The objections which have been raised against this view arise generally from a mere misunderstanding. If we speak of language as the outward realization of thought, we do not mean language as deposited in a dictionary, or sketched in a grammar; we mean language as an act, language as being spoken, language as living and dying with every word that is uttered. We might perhaps call this speech, as distinguished from language.

Secondly, though if we speak of language, we mean chiefly phonetic articulate language, we do not exclude the less perfect symbols of thought, such as gestures, signs, or pictures. They, too, are language in a certain sense, and they must be included in language before we are justified in saying that discursive thought can be realized in language only. One instance will make this clear. We hold that we cannot think without language. But can we not count without language? We certainly can. We can form the conception of *three* without any spoken word, by simply holding up three fingers. In the same manner, the hand might stand for five, both hands for ten, hands and feet for twenty.[6] This is how people who possessed no organs of speech would speak; this is how the deaf and dumb *do* speak. Three fingers are as good as three strokes, three strokes are as good as three clicks of the tongue, three clicks of the tongue are as good as the sound *three,* or *trois,* or *drei,* or *shalosh* in Hebrew, or *san* in Chinese. All these are signs, more or less perfect, but being signs, they fall under the category of language; and all we maintain is, that without some kind of sign, discursive thought is impossible, and that in that sense, language, or λόγος, is the only possible realization of human thought.

Another very common misunderstanding is this: people imagine that, if it be impossible to think except in language, language and thought must be one and the same thing. But a true philosophy of language leads to the very opposite result. Every philosopher would say that matter cannot exist without form, nor form without matter, but no philosopher would say that therefore it is impossible to distinguish between form and matter. In the same way, though we maintain that thought cannot exist without language nor language without thought, we do distinguish between thought and language, between the inward and the outward λόγος, between the substance and the form. Nay, we go a step beyond. We admit that language necessarily reacts on thought, and we see in this reaction, in this refraction of the rays of language, the real solution of the old riddle of mythology.

You will now see why these somewhat abstruse disquisitions were necessary for our immediate purpose, and I can promise those who have hitherto followed me on this rather barren and rugged track, that they will now be able to rest, and command, from the point of view which we have reached, the whole panorama of the mythology of the human mind.

We saw just now that the names of numbers may most easily be replaced by signs. Numbers are simple analytical conceptions, and for that very reason they are not liable to mythology: name and conception being here commensurate, no misunderstanding is possible. But as soon as we leave this department of thought, mythology begins. I shall try by at least one example to show how mythology not only pervades the sphere of religion or religious tradition, but infects more or less the whole realm of thought.

When man wished for the first time to grasp and express a distinction between the body and something else within him distinct from the body, an easy name that suggested itself was *breath*. The breath seemed something immaterial and almost invisible, and it was connected with the life that pervaded the body, for as soon as the breath ceased, the life of the body became extinct. Hence the Greek name ψυχή,[7] which originally meant breath, was chosen to express at first the principle of life, as distinguished from the decaying body, afterwards the incorporeal, the immaterial, the undecaying, the immortal part of man—his soul, his mind, his Self. All this was very natural. When a person dies, we too say that he has given up the ghost, and ghost, too, meant originally spirit, and spirit meant breath.

A very instructive analogous case is quoted by Mr. E. B. Tylor from a compendium of the theology of the Indians of Nicaragua, the record of question and answer in an inquest held by Father Francisco de Bobadilla in the early days of the Spanish conquest. Asked, among other things, concerning death, the Indians said: "Those who die in their houses go underground, but those who are killed in war go to serve the gods (*teotes*). When men die, there comes forth from their mouth something which resembles a person, and is called *julio* (Aztec *yuli*, 'to live'). This being is like a person, but does not die, and the corpse remains here." The Spanish ecclesiastics inquired whether those who go on high keep the same body, features, and limbs as here below; to which the Indians answered, "No, there is only the heart." "But," said the Spaniards, "as the hearts are torn out" (they meant in the case of warriors who fell into the hands of the enemy), "what happens then?" Hereupon the Indians replied: "It is not precisely the heart, but that which is in them, and makes them live, and which quits the body when they die"; and again they said, "It is not their heart which goes up on high, but that which makes them live, that is, the breath coming out from their mouth, which is called *julio*." "Then," asked the Spaniards, "does this heart, *julio*, or soul, die with the body?" "When the deceased has lived well," replied the Indians, "the *julio* goes up on high with our gods; but when he has lived ill, the *julio* perishes with the body, and there is an end of it."

The Greeks expressed the same idea by saying that the ψυχή had left the body,[8] had fled through the mouth, or even through a bleeding wound,[9] and had gone into

Hades, which meant literally no more than the place of the Invisible ('Άίδης). That the breath had become invisible was matter of fact; that it had gone to the house of Hades was mythology springing spontaneously from the fertile soil of language.

The primitive mythology was by no means necessarily religious. In the very case which we have chosen, philosophical mythology sprang up by the side of religious mythology. The religious mythology consisted in speaking of the spirits of the departed as ghosts, as mere breath and air, as fluttering about the gates of Hades, or ferried across the Styx in the boat of Charon.[10]

The philosophical mythology, however, that sprang from this name was much more important. We saw that *Psyche*, meaning originally the breathing of the body, was gradually used in the sense of vital breath, and as something independent of the body; and that at last, when it had assumed the meaning of the immortal part of man, it retained that character of something independent of the body, thus giving rise to the conception of a soul, not only as a being without a body, but in its very nature opposed to body. As soon as that opposition had been established in language and thought, philosophy began its work in order to explain how two such heterogeneous powers could act on each other—how the soul could influence the body, and how the body could determine the soul. Spiritualistic and materialistic systems of philosophy arose, and all this in order to remove a self-created difficulty, in order to join together again what language had severed, the living body and the living soul. The question whether there is a soul or spirit, whether there is in man something different from the mere body, is not at all affected by this mythological phraseology. We certainly can distinguish between body and soul, but as long as we keep within the limits of human knowledge, we have no right to speak of the living soul as a breath, or of spirits and ghosts as fluttering about like birds or fairies. The poet of the nineteenth century says:

> The spirit does but mean the breath,
> I know no more.

And the same thought was expressed by Cicero two thousand years ago: "Whether the soul is air or fire, I do not know." As men, we only know of embodied spirits, however ethereal their bodies may be conceived to be, but of spirits, separate from body, without form or frame, we know as little as we know of thought without language, or of the Dawn as a goddess, or of the Night as the mother of the Day.

Though breath, or spirit, or ghost are the most common names that were assigned through the metaphorical nature of language to the vital, and afterwards to the intellectual, principle in man, they were by no means the only possible names. We speak, for instance, of the *shades* of the departed, which meant originally their shadows. Those who first introduced this expression—and we find it in the most distant parts of the world[11]—evidently took the shadow as the nearest approach to what they wished to express; something that should be incorporeal, yet closely connected with the body. The Greek εἰδώλον, too, is not much more than the shadow, while the Latin *manes* meant probably in the beginning no more than the Little Ones, the Small Folk.[12] But the curious part, as showing again the influence of language on thought, an influence more powerful even than the evidence of the senses, is this, that people who speak of the life or soul as the shadow of the body, have brought themselves to believe that a dead body casts no shadow, because the shadow has departed from it; that it becomes, in fact, a kind of Peter Schlemihl.[13]

Let us now return to mythology in the narrower sense of the word. One of the earliest objects that would strike and stir the mind of man, and for which a sign or a name would soon be wanted, is surely the Sun. It is very hard for us to realize the feelings with which the first dwellers on the earth looked upon the sun, or to understand fully what they meant by a morning prayer, or a morning sacrifice. Perhaps there are few people here present who have watched a sunrise more than once or twice in their lives; few people who have ever known the true meaning of a morning prayer, or a morning sacrifice. But think of man at the very dawn of time: forget for a moment, if you can, after having read the fascinating pages of Mr. Darwin, forget what man is supposed to have been before he was man; forget it, because it does not concern us here whether his bodily form and frame were developed once for all in the mind of a Creator, or gradually in the creation itself, which from the first monad or protoplasm to the last of the primates, or man, is not, I suppose, to be looked on as altogether causeless, meaningless, purposeless; think of him only as man (and man means the thinker), with his mind yet lying fallow, though full of germs—germs of which I hold as strongly as ever no trace has ever, no trace will ever, be discovered anywhere but in man; think of the Sun awakening the eyes of man from sleep, and his mind from slumber! Was not the Sunrise to him the first wonder, the first beginning of all reflection, all thought, all philosophy? Was it not to him the first revelation, the first beginning of all trust, of all religion? To us that wonder of wonders has ceased to exist, and few men now would even venture to speak of the sun as Sir John Herschel has spoken, calling him "the Almoner of the Almighty, the delegated dispenser to us of light and warmth, as well as the centre of attraction, and as such, the immediate source of all our comforts, and, indeed, of the very possibility of our existence on earth."[14]

Man is a creature of habit, and wherever we can watch him, we find that before a few generations have passed he has lost the power of admiring what is regular, and that he can see signs and wonders only in what is irregular. Few nations only have preserved in their ancient poetry some remnants of the natural awe with which the earliest dwellers on the earth saw that brilliant being slowly rising from out the darkness of the night, raising itself by its own might higher and higher, till it stood triumphant on the arch of heaven, and then descended and sank down in its fiery glory into the dark abyss of the heaving and hissing sea. In the hymns of the Veda the poet still wonders whether the sun will rise again; he asks how he can climb the vault of heaven?, why he does not fall back?, why there is no dust on his path? And when the rays of the morning rouse him from sleep and call him back to new life; when he sees the sun, as he says, stretching out his golden arms to bless the world and rescue it from the terrors of darkness, he exclaims, "Arise, our life, our spirit has come back!; the darkness is gone, the light approaches!"

For so prominent an object in the primeval picture-gallery of the human mind, a sign or a name must have been wanted at a very early period. But how was this to be achieved? As a mere sign, a circle would have been sufficient, such as we find in the hieroglyphics of Egypt, in the graphic system of China, or even in our own astronomical tables. If such a sign was fixed upon, we have a beginning of language in the widest sense of the word, for we have brought the Sun under the general concept of roundness, and we have found a sign for this concept which is made up of a large number of single sensuous impressions. With such definite signs mythology has little chance; yet the mere fact that the sun was represented as a circle would favor the idea that the sun was round; or, as ancient people, who had no adjective as yet for round or *rotundus*,[15] would say, that the sun was a wheel, a *rota*. If, on the contrary, the round sign reminded the people of an eye, then the sign of the sun would soon become the eye of heaven, and germs of mythology would spring up even from the barren soil of such hieroglyphic language.

But now, suppose that a real name was wanted for the sun, how could that be achieved?

We know that all words are derived from roots, that these roots express general concepts, and that, with few exceptions, every name is founded on a general concept under which the object that has to be named can be ranged. How these roots came to be, is a question into which we need not enter at present. Their origin and growth form a problem of psychology rather than of philology, and each science must keep within its proper bounds. If a name was wanted for snow, the early framers of language singled out one of the general predicates of snow, its whiteness, its coldness, or its liquidity, and called the snow the white, the cold, or the liquid, by means of roots conveying the general idea of whiteness, coldness, or liquidity. Not only Nix, nivis, but Niobe too, was a name of the snow, and meant the melting;[16] the death of her beautiful children by the arrows of Apollon and Artemis represents the destruction of winter by the rays of the sun. If the sun itself was to be named, it might be called the brilliant, the awakener, the runner, the ruler, the father, the giver of warmth, of fertility, of life, the scorcher, the destroyer, the messenger of death, and many other names; but there was no possibility of naming it, except by laying hold of one of its characteristic features, and expressing that feature by means of one of the conceptual or predicative roots.

Let us trace the history of at least one of these names. Before the Aryan nations separated, before there was a Latin, a Greek, or a Sanskrit language, there existed a root svar or sval, which meant to beam, to glitter, to warm. It exists in Greek, σέλας, splendor; σελήνη, moon; in Anglo-Saxon, as swélan, to burn, to sweal; in modern German, schwül, oppressively hot. From it we have in Sanskrit the noun svar, meaning sometimes the sky, sometimes the sun; and exactly the same word has been preserved in Latin, as sol; in Gothic as sauil; in Anglo-Saxon, as sol. A secondary form of svar is the Sanskrit sûrya for svârya, the sun, which is the same word as the Greek ἥλιος.

All these names were originally mere predicates; they meant bright, brilliant, warm. But as soon as the name svar or sûrya was formed, it became, through the irresistible influence of language, the name, not only of a living, but of a male being. Every noun in Sanskrit must be either a masculine or a feminine (for the neuter gender was originally confined to the nominative case), and as sûryas had been formed as a masculine, language stamped it once for all as the sign of a male being, as much as if it had been the name of a warrior or a king. In other languages where the name for sun is a feminine, and the sun is accordingly conceived as a woman, as a queen, as the bride of the moon, the whole mythology of the love-making of the heavenly bodies is changed.

You may say that all this shows, not so much the influence of language on thought, as of thought on language; and that the sexual character of all words reflects only the peculiarities of a child's mind, which can conceive of nothing except as living, as male or female. If a child hurts itself against a chair, it beats and scolds the chair. The chair is looked upon not as *it,* but as *he;* it is the naughty chair, quite as much as a boy is a naughty boy. There is some truth in this, but it only serves to confirm the right view of the influence of language on thought; for this tendency, though in its origin intentional, and therefore the result of thought, became soon a mere rule of tradition in language, and it then reacted on the mind with irresistible power. As soon, in fact, as *sûryas* or ἥλιος appears as a masculine, we are in the very thick of mythology. We have not yet arrived at Helios as a god—that is a much later stage of thought, which we might describe almost in the words of Plato at the beginning of the seventh book of the *Republic:* "And after this, he will reason that the sun is he who gives the seasons and the years, and is the guardian of all that is in the visible world, and in a certain way the cause of all things which he and his fellows have been accustomed to behold." We have not yet advanced so far, but we have reached at least the first germs of a myth. In the Homeric hymn to Helios, Helios is not yet called an immortal, but only ἐπιείκελος ἀθανάτοισι, like unto immortals, yet he is called the child of Euryphaessa, the son of Hyperion, the grandson of Uranos and Gæa.[17]

All this is mythology; it is ancient language going beyond its first intention.

Nor is there much difficulty in interpreting this myth. Helios, the sun, is called the son of Hyperion, sometimes Hyperion himself. This name Hyperion is derived from the preposition ὑπέρ, the Latin *super,* which means above. It is derived by means of the suffix -ίων, which originally was not a patronymic, but simply expressed belonging to. So if Helios was called Hyperion, this simply meant he who dwells on high, and corresponds to Latin *Summanus* or *Superior,* or *Excelsior.* If, on the contrary, Helios is called Hyperionides, this, too, which meant originally no more than he who comes from, or belongs to those who dwell on high,[18] led to the myth that he was the descendant of Hyperion; so that in this case, as in the case of Zeus Kronion, the son really led to the conception of his father. Zeus Kronion meant originally no more than Zeus the eternal, the god of ages, the ancient of days; but -ίων becoming usual as a patronymic suffix, Kronion was supposed to mean the son of Kronos. Kronos, the father, was created in order to account for the existence of the name Kronion. If Hyperion is called the son of Euryphaessa, the wide-shining, this requires no commentary; for even at present a poet might say that the sun is born of the wide-shining dawn. You see the spontaneous generation of mythology with every new name that is formed. As not only the sun, but also the moon and the dawn could be called dwellers on high, they, too, took the name of Hyperionis or Hyperionides; and hence Homer called Selene, the Moon, and Eos, the Dawn, sisters of Helios, and daughters of Hyperion and Euryphaessa, the Dawn doing service twice, both as mother, Euryphaessa, and as daughter, Eos. Nay, according to Homer, Euryphaessa, the Dawn, is not only the wife, but also the sister of Helios. All this is perfectly intelligible, if we watch the growth of language and mythology; but it leads, of course, to the most tragic catastrophes as soon as it is all taken in a literal sense.

Helios is called ἀκάμας, the never-tiring; πανδερκής, the all-seeing; φαέθων, the shining; and also φοῖβος, the brilliant. This last epithet φοῖβος has grown into an independent deity Phœbus, and it is particularly known as a name of Apollon, Phoibos Apollon; thus showing what is also known from other sources, that in Apollo, too, we have one of the many mythic disguises of the sun.

So far all is clear, because all the names which we have to deal with are intelligible, or, at all events, yield to the softest etymological pressure. But now if we hear the story of Phoibos Apollon falling in love with Daphne, and Daphne praying to her mother, the Earth, to save her from Phoibos; and if we read how either the earth received her in her lap, and then a laurel tree sprang up where she had disappeared, or how she herself was changed into a laurel tree, what shall we think of this? It is a mere story, it might be said, and why should there be any meaning in it? My answer is, because people do not tell such stories of their gods and heroes, unless there is some sense in them. Besides, if Phoibos means the sun, why should not Daphne have a meaning too? Before, therefore, we can decide whether the story of Phoibos and Daphne is a mere invention, we must try to find out what can have been the meaning of the word Daphne.

In Greek it means a laurel,[19] and this would explain the purely Greek legend that Daphne was changed into a laurel tree. But who was Daphne? In order to answer this question, we must have recourse to etymology, or, in other words, we must examine the history of the word. Etymology, as you know, is no longer what it used to be; and though there may still be a classical scholar here and there who crosses himself at the idea of a Greek word being explained by a reference to Sanskrit, we naturally look to Sanskrit as the master-key to many a lock which no Greek key will open. Now Daphne, as I have shown, can be traced back to Sanskrit *Ahanâ* and Ahanâ in Sanskrit means the dawn. As soon as we know this, everything becomes clear. The story of Phoibos and Daphne is no more than a description of what every one may see every day; first, the appearance of the Dawn in the eastern sky, then the rising of the Sun as if hurrying after his bride, then the gradual fading away of the bright Dawn at the touch of the fiery rays of the sun, and at last her death or disappearance in the lap of her mother, the Earth. All this seems to me as clear as daylight, and the only objection that could be raised against this reading of the ancient myth would be, if it could be proved, that Ahanâ does not mean Dawn, and that Daphne cannot be traced back to Ahanâ, or that *Helios* does not mean the Sun.

I know there is another objection, but it seems to me so groundless as hardly to deserve an answer. Why, it is asked, should the ancient nations have told these endless stories about the Sun and the Dawn, and why should they have preserved them in their mythology? We might as well ask why the ancient nations should have invented so many irregular verbs, and why they should have preserved them in their grammar. A fact does not cease to be a fact because we cannot at once explain it. As far as our knowledge goes at present, we are justified in stating that the Aryan nations preserved not only their grammatical structure, and a large portion of their dictionary, from the time which preceded their separation, but that they likewise retained the names of some of their deities, some legends about their gods, some popular sayings and proverbs, and in these, it may be, the seeds of parables, as part of their common Aryan heirloom. Their mythological lore fills, in fact, a period in the history of Aryan thought, half-way between the period of language and the period of literature, and it is this discovery which gives to mythology its importance in the eyes of the student of the most ancient history and psychology of mankind.

And do not suppose that the Greeks, or the Hindus, or the Aryan nations in general, were the only people who possessed such tales. Wherever we look, in every part of the world, among uncivilized as well as a civilized people, we find the same kind of stories, the same traditions, the same myths.

I shall give one story from the extreme North, another from the extreme South. Among the Esquimaux of Repulse Bay, on the west side of Hudson's Bay, on the Arctic Circle, Mr. John Rae picked up the following story:

> Many years ago, a great Esquimaux Conqueror gained so much power that he was able to rise unto the heavens, taking with him on one occasion a sister, a very beautiful girl, and some fire. He added much fuel to the fire, and thus formed the Sun. For some time he and his sister lived in great harmony, but after a time he became very cruel, and ill-treated his sister in many ways. She bore it at first with great patience, until at last he threw fire at her, and scorched one side of her face. This spoiling of her beauty was beyond endurance; she therefore ran away from him, and formed the Moon. Her brother then began, and still continues to chase her; but although he sometimes got near, he has not yet overtaken her, nor ever will.
>
> When it is New Moon, the burnt side of the face is towards us; at Full Moon it is the reverse.

There are dialectic varieties in the mythology of the Esquimaux as of the Greeks and Hindus, and, with a change of gender between Sun and Moon, the same story occurs among other tribes in the following form:

> There was a girl at a party, and some one told his love for her by shaking her shoulders, after the manner of the country. She could not see who it was in the dark hut, so she smeared her hands with soot, and when he came back she blackened his cheek with her hand. When a light was brought she saw that it was her brother and fled. He ran after her, followed her, and as she came to the end of the earth, he sprang out into the sky. Then she became the sun, and he the moon, and this is why the moon is always chasing the sun through the heavens, and why the moon is sometimes dark as he turns his blackened cheek towards the earth.[20]

We now turn to the South, and here, among the lowest of the low, among the Hottentots, who are despised even by their black neighbors, the Zulus, we find the following gem of a fable, beaming with mingled rays of religion and philosophy:

> The Moon, it is said, sent once an insect to men, saying, 'Go thou to men, and tell them, "As I die, and dying live, so ye shall also die, and dying live."' The insect started with the message, but whilst on his way was overtaken by the hare, who asked: 'On what errand art thou bound?' The insect answered, 'I am sent by the Moon to men, to tell them that as she dies and dying lives, they also shall die and dying live.' The hare said, 'As thou art an awkward runner, let me go' (to take the message). With these words he ran off, and when he reached men, he said, 'I am sent by the Moon to tell you, "As I die, and dying perish, in the same manner ye also shall die and come wholly to an end."' Then the hare returned to the Moon, and told her what he had said to men. The

Moon reproached him angrily, saying, 'Darest thou tell the people a thing which I have not said?' With these words she took up a piece of wood, and struck him on the nose. Since that day the hare's nose is slit.

Of this story, too, there are various versions and in one of them the end is as follows:

The hare, having returned to the Moon, was questioned as to the message delivered, and the Moon, having heard the true state of the case, became so enraged with him that she took up a hatchet to split his head; falling short, however, of that, the hatchet fell upon the upper lip of the hare, and cut it severely. Hence it is that we see the 'hare-lip.' The hare, being duly incensed at having received such treatment, raised his claws, and scratched the Moon's face; and the dark parts which we now see on the surface of the Moon are the scars which she received on that occasion.[21]

The Finns, Lapps, and Esthonians do not seem a very poetical race, yet there is poetry even in their smoky huts, poetry surrounded with all the splendor of an arctic night, and fragrant with the perfume of moss and wild flowers. Here is one of their legends:

Wanna Issi had two servants, Koit and Ämmarik, and he gave them a torch which Koit should light every morning, and Ämmarik should extinguish in the evening. In order to reward their faithful services, Wanna Issi told them they might be man and wife, but they asked Wanna Issi that he would allow them to remain forever bride and bridegroom. Wanna Issi assented, and henceforth Koit handed the torch every evening to Ämmarik, and Ämmarik took it and extinguished it. Only during four weeks in summer they remain together at midnight; Koit hands the dying torch to Ämmarik, but Ämmarik does not let it die, but lights it again with her breath. Then their hands are stretched out, and their lips meet, and the blush of the face of Ämmarik colors the midnight sky.

This myth requires hardly any commentary; yet as long as it is impossible to explain the names, Wanna Issi, Koit, and Ämmarik, it might be said that the story was but a love story, invented by an idle Lapp, or Finn, or Esthonian. But what if Wanna Issi in Esthonian means the Old Father, and if Koit means the Dawn? Can we then doubt any longer that Ämmarik[22] must be the Gloaming and that their meeting in the summer reflects those summer evenings when, particularly in the North, the torch of the sun seems never to die, and when the Gloaming is seen kissing the Dawn?

I wish I could tell you some more of these stories which have been gathered from all parts of the world, and which, though they may be pronounced childish and tedious by some critics, seem to me to glitter with the brightest dew of nature's own poetry, and to contain those very touches that make us feel akin, not only with Homer or Shakespeare, but even with Lapps, and Finns, and Kaffirs.

I cannot resist, however, the temptation of inserting here a poetical rendering of the story of Koit and Ämmarik, sent to me from the New World, remarking only that instead of Lapland, Esthonia is really the country that may claim the original story.

A Legend of Lapland
Two servants were in Wanna Issi's pay;
A blazing torch their care;

Each morning Koit must light it till its ray
Flamed through the air;
And every evening Ämmarik's fair hand
Must quench the waning light;
Then over all the weary, waiting land
Fell the still night.

So passed the time; then Wanna Issi said,
'For faithful service done,
Lo, here reward! To-morrow shall ye wed,
And so be one.'

'Not so,' said Koit; 'for sweeter far to me
The joy that neareth still;
Then grant us ever fast betrothed to be.'
They had their will.

And now the blazing lustre to transfer
Himself, is all his claim;
Warm from her lover's hand it comes to her,
To quench the flame.

Only for four times seven lengthening days,
At midnight, do they stand
Together, while Koit gives the dying blaze
To Ämmarik's hand.

O wonder then! She lets it not expire,
But lights it with her breath —
The breath of love, that, warm with quickening fire,
Wakes life from death.

Then hands stretch out, and touch, and clasp on high,
Then lip to lip is pressed,
And Ämmarik's blushes tinge the midnight sky
From east to west.
                                        —Anna C. Brackett

If people cannot bring themselves to believe in solar and celestial myths among the Hindus and Greeks, let them study the folk-lore of the Semitic and Turanian races. I know there is, on the part of some of our most distinguished scholars, the same objection against comparing Aryan to non-Aryan myths, as there is against any attempt to explain the features of Sanskrit or Greek by a reference to Finnish or Basque. In one sense that objection is well founded, for nothing would create greater confusion than to ignore the genealogical principle as the only safe one in a scientific classification of languages, of myths, and even of customs. We must first classify our myths and legends, as we classify our languages and dialects. We must first of all endeavor to explain what wants explanation in one member of a family by a reference to other members of the same family, before we allow ourselves to glance beyond. But there is in a comparative study of languages and myths not only a philological, but also a philosophical, and, more particularly, a psychological interest, and though even in this more general study of mankind the frontiers of language and race ought never to disappear, yet they can no longer be allowed to narrow or intercept our view. How much the student of Aryan mythology and ethnology may gain for his own progress by allowing himself a wider survey over the traditions and customs of the whole human race, is best known to those who have studied the works of Klemm, Waitz, Bastian, Sir John Lubbock, Mr. Tylor, and Dr. Callaway. What is prehistoric in language among the Aryan nations, is frequently found as still historic among Turanian races. The same applies with regard to religions, myths, legends, and customs. Among Finns and Lapps, among Zulus and Maoris, among Khonds and Karens, we sometimes find the most startling analogies to Aryan traditions, and we certainly learn, again and again, this one important lesson, that as in language, so in mythology, there is nothing which had not originally a meaning, that every name of the gods and heroes had a beginning, a purpose, and a history.

Jupiter was no more called Jupiter by accident, than the Polynesian *Maui,* the Samoyede *Num,* or the Chinese *Tien.*[23] If we can discover the original meaning of these names, we have reached the first ground of their later growth. I do not say that, if we can explain the first purpose of the mythological names, we have solved the whole riddle of mythology, but I maintain that we have gained firm ground. I maintain that every true etymology gives us an historical fact, because the first giving of a name was an historical fact, and an historical fact of the greatest importance for the later development of ancient ideas. Think only of this one fact, which no one would now venture to doubt, that the supreme deity of the Greeks, the Romans, the Germans, is called by the same name as the supreme deity of the earliest Aryan settlers in India. Does not this one fact draw away the curtain from the dark ages of antiquity, and open before our eyes an horizon which we can hardly measure by years? The Greek *Zeus* is the same word as the Latin *Ju* in *Jupiter,* as the German *Tiu;* and all these were merely dialectic varieties of the Vedic *Dyaus.*[24] Now *dyaus* in Sanskrit is the name of the sky, if used as a feminine; if used as a masculine as it is still in the Veda, it is the sky as a man or as a god—it is Zeus, the father of gods and men. You know, of course, that the whole language of ancient India is but a sister dialect of Greek, Latin, of German, Keltic, and Slavonic, and that if the Greek says *es-ti,* he is, if the Roman says *est,* the German *ist,* the Slave [*sic*] *yesté,* the Hindu, three thousand years ago, said *as-ti,* he is. This *as-ti* is a compound of a root *as,* to be, and the pronoun *ti.* The root meant originally *to breathe,* and dwindled down after a time to the meaning of *to be.* All this must have happened before a single Greek or German reached the shores of Europe, and before a single Brahman descended into the plains of India. At that distant time we must place the gradual growth of language and ideas, of a language which we are still speaking, of ideas which we are still thinking; and at the same time only can we explain the framing of those names which were the first attempts at grasping supernatural powers, which became in time the names of the deities of the ancient world, the heroes of mythology, the chief actors in many a legend, nay, some of which have survived in the nursery tales of our own time.[25]

My time, I see, is nearly over, but before I finish, I feel that I have a duty to perform from which I ought not to shrink. Some of those who have honored me with their presence to-night may recollect that about a year ago a lecture was delivered in this very room by Professor Blackie, in which he tried to throw discredit on the scientific method of the interpretation of popular myths, or on what I call Comparative Mythology. Had he confined his remarks to the subject itself, I should have felt most grateful for his criticisms, little minding the manner in which they were conveyed—for a student of language knows what words are made of. Nor, had his personal reflections concerned myself alone, should I have felt called upon to reply to them thus publicly, for it has always seemed to me that unless we protest against unmerited praise, we have no right to protest against unmerited abuse. I believe I can appeal to all here present, that during the many years I have had the honor to lecture in this institution, I have *not once* allowed myself to indulge in any personal remarks, or attacked those who, being absent, cannot defend themselves. Even when I had to answer objections, or to refute false theories, I have always most carefully avoided mentioning the names of living writers. But as Professor Blackie has directed his random blows, not against myself, but against a friend of mine, Mr. Cox, the author of a work on Aryan Mythology, I feel that I must for once try to get angry, and return blow for blow. Professor Blackie speaks of Mr. Cox as if he had done nothing beyond repeating what I had said before. Nothing can be more unfair. My own work in Comparative Mythology has consisted chiefly in laying down some of the general principles of that science, and in the etymological interpretation of some of the ancient names of gods, goddesses, and heroes. In fact, I have made it a rule never to interpret or to compare the legends of India, Greece, Italy, or Germany, except in cases where it was possible, first of all, to show an identity or similarity in the Sanskrit, Greek, Latin, or German names of the principal actors. Mr. Cox, having convinced himself that the method which I have followed in mythology rests on sound and truly scientific principles, has adopted most, though by no means all, of my etymological interpretations. Professor Blackie, on the contrary, without attempting any explanation of the identity of mythological names in Greek and Sanskrit which must be either disproved or explained, thunders forth the following sentence of condemnation: "Even under the scientific guidance of a Bopp, a Bott, a Grimm, and a Müller, a sober man may sometimes, even in the full blaze of the new sun of comparative philology, allow himself to drink deep draughts, if not of *maundering madness,* at least of *manifest hallucination.*"

If such words are thrown at my head, I pick them up chiefly as etymological curiosities, and as striking illustrations of what Mr. Tylor calls "survivals in culture," showing how the most primitive implements of warfare, rude stones and unpolished flints, which an ethnologist would suppose to be confined to prehistoric races, to the red Indians of America or the wild Picts of Caledonia, turn up again most unexpectedly at the present day in the very centre of civilized life. All I can say is, that if, as a student of Comparative Mythology, I have been drinking deep draughts of maundering madness, I have been drinking in good company. In this respect Mr. Cox has certainly given me far more credit than I deserve. I am but one out of many laborers in this rich field of scientific research, and he ought to have given far greater prominence to the labors of Grimm, Burnouf, Bopp, and, before all, of my learned friend, Professor Kuhn.

But while, with regard to etymology, Mr. Cox contents himself with reporting the results of other scholars, he stands quite independent in his own treatment of Comparative Mythology. Of this Professor Blackie seems to have no suspicion whatever. The plan which Mr. Cox follows is to collect the coincidences in the legends themselves, and to show how in different myths the same story with slight variations is told again and again of different gods and heroes. In this respect his work is entirely original and very useful; for although these coincidences may be explained in different ways, and do not afford a proof of a common historical origin of the mythologies of India, Greece, Italy, and Germany, they are all the more interesting from a purely psychological point of view, and supply important material for further researches. Mr. Tylor has lately worked with great success in the same rich mine, extending the limits of mythological research far beyond the precincts of the Aryan world, and showing that there are solar myths wherever the sun shines. I differ from Mr. Cox on many points, as he differs from me. I shall certainly keep to my own method of never attempting an interpretation or a comparison, except where the ground has first been cleared of all uncertainty by etymological research, and where the names of different gods and heroes have been traced back to a common source. I call this the *nominalistic* as opposed to the *realistic* method of Comparative Mythology, and it is the former only that concerns the student of the Science of Language. I gratefully acknowledge, however, the help which I have received from Mr. Cox's work, particularly as suggesting new clusters of myths that might be disentangled by etymological analysis.

But not only has Professor Blackie failed to perceive the real character of Mr. Cox's researches, but he has actually charged him with holding opinions which both Mr. Cox and myself have repeatedly disavowed, and most strenuously opposed. Again and again have we warned the students of Comparative Mythology that they must not expect to be able to explain everything. Again and again have we pointed out that there are irrational elements in mythology, and that we must be prepared to find grains of local history on which, as I said,[26] the sharpest tools of Comparative Mythology must bend or break. Again and again have we shown that historical persons[27]—not only Cyrus and Charlemagne, but Frederick Barbarossa and even Frederick the Great—have been drawn into the vortex of popular mythology. Yet these are the words of Professor Blackie: "The cool way in which Max Müller and his English disciple, Mr. Cox, assume that there are no human figures and historical characters in the whole gallery of heroes and demi-gods in the Greek Mythology, is something very remarkable."

I readily admit that some of the etymologies which I have proposed of mythological names are open to criticism; and if, like other scholars, Professor Blackie had pointed out to me any cases where I might seem to him to have offended against Grimm's law or other phonetic rules, I should have felt most grateful; but if he tells me that the Greek Erinys should not be derived from the Sanskrit Saranyû, but from the Greek verb ἐρινύειν, to be angry, he might as well derive *critic* from *to criticise;*[28] and if he maintains that a name may have two or three legitimate etymologies, I can only answer that we might as well say that a child could have two or three legitimate mothers.

I have most reluctantly entered upon these somewhat personal explanations, and I should not have done so if I alone had been concerned in Professor Blackie's onslaught. I hope, however, that I have avoided anything that could give just offence to Professor Blackie, even if he should be present here to-night. Though he abuses me as a German, and laughs at the instinctive aversion to external facts and the extravagant passion for self-evolved ideas as national failings of all Germans (I only wonder that the story of the camel and the inner consciousness did not come in), yet I know that for many years German poetry and German scholarship have had few more ardent admirers, and German scholars few more trusty friends, than Professor Blackie. Nationality, it seems to me, has as little to do with scholarship as with logic. On the contrary, in every nation he that will work hard and reason honestly may be sure to discover some grains of truth. National jealousies and animosities have no place in the republic of letters, which is, and I trust always will be, the true international republic of all friends of work, of order, and of truth.

Notes

1. Herodotus ii. 53.
2. Sext. Emp., *adv. Math.*, 1289; ix, 193; Clem. Alex. *Strom.*, v, p. 601, c and vii, p. 711; and B. *Historia Philosophiæ*, ed. Ritter et Preller, cap. iii.
3. Clem. Alex., *Strom.*, v, p. 601, c.
4. See *Introduction to the Science of Religion,* p. 139.
5. Empedokles, *Carmina*, v. 411 (*Fragm. Philos. Græc.*, vol. 1, p. 12).
6. *Daily Life and Origin of the Tasmanians,* by J. Bonwick, 1870, p. 143.
7. The word ψυχή is clearly connected in Greek with ψύχω, which meant originally blowing, and was used either in the sense of cooling by blowing, or breathing by blowing. In the former acceptation it produced ψῦχος, coldness; ψυχρός, cold; ψυχάω, I cool; in the latter ψυχή, breath, then life, then soul. So far the purely Greek growth of words derived from ψύχω is clear. But ψύχω itself is difficult. It seems to point to a root *spu,* meaning to blow out, to spit; Lat. *spuo,* and *spuma,* foam; Goth. *speivan;* Gr. πτύω, supposed to stand for σπίύω. Hesychius mentions ψύττει = πτύει, ψυττόν = πτύελον. (Pott, *Etym. Forsch.,* No. 355) Curtius connects this root with Gr. φυ, in φῦσα, blowing, bellows; φυσάω, to blow; φυσιάω, to snort; ποι-φύσσω, to blow; and with Lat. *spirare* (i.e., *spoisare*). See E. B. Tylor, "The Religion of Savages," *Fortnightly Review,* 1866, p. 73.

   Stahl, who rejected the division of life and mind adopted by Bacon, and returned to the Aristotelian doctrine, falls back on Plato's etymology of ψυχή as φυσέχη from φύσιν ἔχειν or ὀχεῖν, *Crat.,* 400 B; see Stahl, *Theoria Medica Vera* (Halæ, 1708), p. 44.
8. *Iliad,* ix. 408.
9. Ibid., xiv. 517.
10. Virg., *Æneid,* ii. 792.
11. See E. B. Tylor, *Fortnightly Review,* 1866, p. 74.
12. *Im-manis,* originally "not small," came to mean enormous or monstrous. See Preller, *Römische Mythologie,* p. 72 seq.
13. *Unkulunkulu; or the Tradition of Creation as existing among the Amazulu and other Tribes of South Africa,* by the Rev. J. Callaway, M.D., Natal, 1868. Part I, p. 91.
14. See J. Samuelson, *Views of the Deity, Traditional and Scientific,* p. 144, Williams & Norgate, 1871.
15. "It has already been implied that the Aborigines of Tasmania had acquired very limited powers of abstraction or generalization. They possessed no words representing abstract ideas; for each variety of gum-tree and wattle-tree, etc., etc., they had a name, but they had no equivalent for the expression, 'a tree'; neither could they express abstract qualities, such as hard, soft, warm, cold, long, short, round, etc.; for 'hard' they would say 'like a stone;' for 'tall' they would say 'long legs,' etc.; for 'round' they said 'like a ball,' 'like the moon,' and so on, usually suiting the action to the word, and confirming by some sign the meaning to be understood." Milligan, *Vocabulary of the Dialects of some of the Aboriginal Tribes of Tasmania,* Hobart Town, 1866, p. 34.

16. If Signor Ascoli blames me for deriving *Niobe* with other names for snow from the root *snu*, instead of from the root *snigh*, this can only be due to an oversight. I am responsible for the derivation of Niobe, and for the admission of a secondary root *snyu* or *nyu*, and so far I may be either right or wrong. But Signor Ascoli ought to have known that the derivation of Gothic *snáiv-s*, Old High-German *snêo*, or *snê*, gen. *snêwe-s*, Lithuanian *snéga-s*, Slav. *snjeg*, Hib. *sneachd*, from the root *snu*, rests on the authority of Bopp (*Glossarium*, 1847, s.v. *snu*; see also Grimm, *Deutsche Grammatik*, ii, p. 700). He ought likewise to have known that in 1852 Professor Schweizer-Siedler, in his review of Bötticher's *Arica* (Kuhn's *Zeitschrift*, i, p. 479), had pointed out that *snigh* may be considered as a secondary root by the side of *snu* and *snâ* (cf., σμάω, σμήχω; ψάω, ψήχω; νάω, νήχω). The real relation of *snu* to *snigh* had been explained as early as 1842 by Benfey, *Wurzellexicon*, ii, p. 54; and Signor Ascoli was no doubt aware of what Professor Curtius had written on the relation of *snigh* to *snu* (*Grundzüge der Griechischen Etymologie*, p. 297). Signor Ascoli has certainly shown with greater minuteness than his predecessors that not only Zend *snizh* and Lithuanian *snéga-s*, but likewise Gothic *snaiv-s*, Greek νίφει, Latin *nix*, *niv-is*, and *ninguis*, may be derived from *snigh*; but if from *snigh*, a secondary development of the root *snu*, we can arrive at νίφ-α, and at νίβα, the other steps that lead on to Niobe will remain just the same.

17. At the end of the hymn, the poet … seem[s] to imply that [he] looked upon Helios as a half-god, almost as a hero, who had once lived on earth.

18. Corssen, *Über Steigerungsendungen;* Kuhn's *Zeitschrift*, iii, p. 299.

19. See *Selected Essays*, vol. 1, p. 399.

20. E. Clodd, *The Childhood of the World*, p. 62.

21. *Reynard the Fox in South Africa, or Hottentot Fables and Tales*, by W. H. I. Bleek, 1864, p. 69. Dr. Theophilus Hahn, *Die Sprache der Nama*, 1870, p. 59. As a curious coincidence, it may be mentioned that in Sanskrit, too, the Moon is called *sasanka*, i.e., "having the marks of a hare," the black marks in the moon being taken for the likeness of the hare. Another coincidence is that the Namaqua Hottentots will not touch hare's flesh (see Sir James E. Alexander's *Expedition of Discovery into the Interior of Africa*, vol. 1, p. 269), because the hare deceived men, while the Jews abstain from it, because the hare is supposed to chew the cud (Lev. 11.6).

   A similar tradition on the meaning of death occurs among the Zulus but as they do not know of the Moon as a deity, the message that men are not to die, or that they are to die, is sent there by Unkulunkulu, the ancestor of the human race, and thus the whole story loses its point. See Dr. Callaway, *Unkulunkulu*, p. 4; and Gray, *Polynesian Mythology*, pp. 16–58.

22. According to a letter just received from an Esthonian lady, *ämmarik* does mean the gloaming in the language of the common people of Esthonia. Bertram (*Ilmatar*, Dorpat, 1870, p. 265) remarks that Koit is the dawn, *Koido täht*, the morning-star, also called *eha täht*. *Ämarik*, the ordinary name for the dawn, is used as the name for the evening twilight, or the gloaming in the well-known story, published by Fählmann (*Verhandlungen der gelehrten Estnischen Gesellschaft zu Dorpat*, vol. 1). In Finnish *hämära* is twilight in general.

23. See *Lectures on the Science of Religion*, pp. 194, 200.

24. See my *Lectures on the Science of Language* (10th ed.), vol. 2, p. 468.

25. See a most interesting essay, *Le Petit Poucet* (Tom Thumb), by Gaston Paris.

26. *Selected Essays*, vol. 1, p. 478: "Here then we see that mythology does not always create its own heroes, but that it lays hold of real history, and coils itself round it so closely that it is difficult, nay, almost impossible, to separate the ivy from the oak, the lichen from the granite to which it clings. And here is a lesson which comparative mythologists ought not to neglect. They are naturally bent on explaining everything that can be explained; but they should bear in mind that there may be elements in every mythological riddle which resist etymological analysis, for the simple reason that their origin was not etymological, but historical."

27. *Lectures on the Science of Language,* vol. 2, p. 581.
28. Professor Blackie quotes Pausanias in support of this etymology. He says: "The account of Pausanias (viii. 25, 26), according to which the terrible impersonation of conscience, or the violated moral law, is derived from ἐρινύειν, an old Greek verb originally signifying to be angry, has sufficient probability, not to mention the obvious analogy of Ἀραί, another name sometimes given to the awful maids (σεμναί), from ἀρά, an imprecation." If Professor Blackie will refer to Pausanias himself, he will find that the Arcadians assigned a very different cause to the anger of Demeter, which is supposed to have led to the formation of her new name Erinys.

CHAPTER 9

THE PERCEPTION OF THE INFINITE

from *Lectures on the Origin and Growth of Religion* (1878)

The Problem of the Origin of Religion

How is it that we have a religion? This is a question which has not been asked for the first time in these latter days, but it is, nevertheless, a question which sounds startling even to ears that have been hardened by the din of many battles, fought for the conquest of truth. How it is that we exist, how it is that we perceive, how it is that we form concepts, how it is that we compare percepts and concepts, add and subtract, multiply and divide them—all these are problems with which everybody is more or less familiar, from the days in which he first opened the pages of Plato or Aristotle, of Hume or Kant. Sensation, perception, imagination, reasoning, everything in fact which exists in our own consciousness, has had to defend the right and reason of its existence; but the question, Why we believe, why we are, or imagine we are conscious of things which we can neither perceive with our senses, nor conceive with our reason—a question, it would seem more natural to ask than any other—has but seldom received, even from the greatest philosophers, that attention which it so fully deserves.

David Strauss: Have We Still Any Religion?

What can be less satisfactory than the manner in which this problem has lately been pushed into the foreground of popular controversy? Strauss, in many respects a most acute reasoner, puts before us in his last work, *The Old and the New Faith,* the question, "Have we still any religion?" To a challenge put in this form the only answer that could be given would be an appeal to statistics; and here we should soon be told that, out of a hundred thousand people, there is hardly one who professes to be either without a religion or without religion. If another answer was wanted, the question ought to have been put in a different form. Strauss ought before all things to have told us clearly what he himself understands by religion. He ought to have defined religion both in its psychological and historical development. But what does he do instead? He simply takes the old definition which Schleiermacher gave of religion, viz., that it consists in a feeling of absolute dependence, and he supplements it by a definition of Feuerbach's, that the essence of all religion is covetousness, which manifests itself in prayer, sacrifice, and faith. He then concludes, because there is less of prayer, crossing, and attending mass in our days than in the Middle Ages, that therefore there is little left of real piety and religion. I have used, as much as possible, Strauss's own words.

But where has Strauss or anybody else proved that true religion manifests itself in prayer, crossing, and attending mass only, and that all who do not pray, who do not cross themselves, and who do not attend mass, have no longer any religion at all, and no belief in God? If we read on, we are almost tempted to admit that M. Renan was right in saying that those poor Germans try very hard to be irreligious and atheistical, but never succeed. Strauss says: "The world is to us the workshop of the Rational and the Good. That on which we feel ourselves absolutely dependent is by no means a brute power before which we must bow in silent resignation. It is order and law, reason and goodness, to which we surrender ourselves with loving confidence. In our inmost nature we feel a kinship between ourselves and that on which we depend. In our dependence we are free, and pride and humility, joy and resignation, are mingled together in our feeling for all that exists."

If that is not religion, what is it to be called? The whole argument of Strauss amounts, in fact, to this. He retains religion as the feeling of dependence, in the full sense assigned to it by Schleiermacher, but he rejects the element added by Feuerbach, namely, the motive of covetousness, as both untrue, and unworthy of religion. Strauss himself is so completely in the dark as to the true essence of religion that when, at the end of the second chapter of his book, he asks himself whether he still has a religion, he can only answer, "Yes, or No, according as you understand it."

Yes, but this is the very point which ought to have been determined first, namely, what we ought to understand by religion. And here I answer that in order to understand what religion is, we must first of all see what it has been, and how it has come to be what it is.

Antiquity of Religion

Religion is not a new invention. It is, if not as old as the world, at least as old as the world we know. As soon almost as we know anything of the thoughts and feelings of man, we find him in possession of religion, or rather possessed by religion. The oldest literary documents are almost everywhere religious. "Our earth," as Herder says, "owes the seeds of all higher culture to a religious tradition, whether literary or oral."[1] Even if we go beyond the age of literature, if we explore the deepest levels of human thought, we can discover, in the crude ore which was made to supply the earliest coins or counters of the human mind, the presence of religious ingredients. Before the Aryan languages separated—and who is to tell how many thousand years before the first hymn of the Veda or the first line of Homer that ethnic schism may have happened?—there existed in them an expression for light, and from it, from the root *div*, to shine, the adjective *deva* had been formed, meaning originally "bright." Afterwards this word *deva* was applied, as a comprehensive designation, to all the bright powers of the morning and the spring, as opposed to all the dark powers of the night and the winter; but when we meet with it for the first time in the oldest literary documents, it is already so far removed from this its primitive etymological meaning, that in the Veda there are but few passages where we can with certainty translate it still by "bright." The bright dawn is addressed in the Veda as *devî ushas,* but it must remain doubtful whether the old poets still felt in that address the etymological meaning of brightness, or, whether we ought not to translate *deva* in the Veda, as *deus* in Latin, by God, however difficult we may find it to connect any definite meaning with such a translation. Still, what we know for certain is that *deva* came to mean "god," because it originally meant "bright," and we cannot doubt that something beyond the meaning of brightness had attached itself to the word *deva*, before the ancestors of the Indians and Italians broke up from their common home.

Thus, whether we descend to the lowest roots of our own intellectual growth, or ascend to the loftiest heights of modern speculation, everywhere we find religion as a power that conquers, and conquers even those who think that they have conquered it.

Science of Religion

Such a power did not escape the keen-eyed philosophers of ancient Greece. They, to whom the world of thought seems to have been as serene and transparent as the air which revealed the sea, and shore, and the sky of Athens, were startled at a very early time by the presence of religion, as by the appearance of a phantom which they could not explain. Here was the beginning of the science of religion, which is not, as has often been said, a science of to-day or of yesterday. The theory on the origin of religion put forward by Feuerbach in his work *On the Essence of Christianity*, which sounds to us like the last note of modern despair, was anticipated more than two thousand years ago by the philosophers of Greece. With Feuerbach religion is a radical evil, inherent in mankind—the sick heart of man is the source of all religion, and of all misery. With Herakleitos, in the sixth century B.C., religion is a disease, though a sacred disease.[2] Such a saying, whatever we may think of its truth, shows, at all events, that religion and the origin of religious ideas had formed the subject of deep and anxious thought at the very beginning of what we call the history of philosophy.

I doubt, however, whether there was in the sayings of Herakleitos the same hostile spirit against all religion as that which pervades the writings of Feuerbach. The idea that to believe is meritorious, was not an ancient Greek idea, and therefore to doubt was not yet regarded as a crime, except where it interfered with public institutions. There was, no doubt, an orthodox party in Greece, but we can hardly say that it was fanatical;[3] nay, it is extremely difficult to understand at what time it acquired its power and whence it took its coherence.[4]

Herakleitos certainly blames those who follow singers (ἀοιδοί),[5] and whose teacher is the crowd, who pray to idols, as if they were to gossip with the walls of houses, not knowing what gods and heroes really are. Epikouros does the same. But, unlike Epikouros, Herakleitos nowhere denies the existence of invisible Gods or of the One Divine. Only when he saw people believing in what the singers, such as Homer and Hesiod, told them about Zeus and Hera, about Hermes and Aphrodite, he seems to have marvelled; and the only explanation which he could find of so strange a phenomenon was that it arose from an affection of the mind, which the physician might try to heal, whensoever it showed itself, but which he could never hope to stamp out altogether.

In a certain sense, therefore, the science of religion is as little a modern invention as religion itself. Wherever there is human life, there is religion, and wherever there is religion, the question whence it came cannot be long suppressed. When children once begin to ask questions, they ask the why and the wherefore of everything, religion not excepted; nay, I believe that the first problems of what we call philosophy were suggested by religion.

It has sometimes been asked why Thales should be called a philosopher, and should keep his place on the first page of every history of philosophy. Many a schoolboy may have wondered why to say that water was the beginning of all things, should be called philosophy. And yet, childish as that saying may sound to us, it was anything but childish at the time of Thales. It was the first bold denial that the gods had made the world; it was the first open protest against the religion of the crowd—a protest that had to be repeated again and again before the Greeks could be convinced that such thinkers as Herakleitos[6] and Xenophanes had at least as good a right to speak of the gods or of God as Homer and other itinerant singers.

No doubt, at that early time, what was alone important was to show that what was believed by the crowd was purely fanciful. To ask how those fanciful opinions of the crowd had arisen, was a problem belonging to a later age. Still, even that problem was not entirely absent from the minds of the earliest thinkers of Greece; for no one could have given the answer ascribed to Herakleitos, who had not asked himself the question which we ask ourselves to-day: What, then, is the origin of religion?; or, to put it into more modern language, How is it that we believe, that we accept what, as we are told by enemy and friend, cannot be either supplied to us by our senses or established by our reason?

Difference between Ancient and Modern Belief

It may be said that, when Herakleitos pondered on οἴησις, or belief, he meant something very different from what we mean by religion. No doubt he did; for if there is a word that has changed from century to century, and has a different aspect in every country in which it is used—nay, which conveys peculiar shades of meaning, as it is used by every man, woman, or child—it is religion. In our ordinary language we use religion in at least three different senses: first, as the object of belief; secondly, as the power of belief; thirdly, as the manifestation of belief, whether in acts of worship or in acts of real piety.

The same uncertainty prevails in other languages. It would be difficult to translate our word religion into Greek or Sanskrit; nay, even in Latin, *religio* does by no means cover all that religion comprehends in English. We need not be surprised, therefore, at the frequent misunderstandings, and consequent wranglings, between those who write on religion, without at least having made so much clear to themselves and others, whether by religion they mean religious dogma, religious faith, or religious acts.

I have dwelt on this point in order to show you that it is not from mere pedantry if, at the very outset of these lectures, I insist on the necessity of giving a definition of religion, before we attempt another step in our journey that is to lead us as near as possible to the hidden sources of our faith.

Definitions of Religion

It was, I think, a very good old custom never to enter upon the discussion of any scientific problem without giving beforehand definitions of the principal terms that had to be employed. A book on logic or grammar generally opened with the question, What is logic? What is grammar? No one would write on minerals without first explaining what he meant by a mineral, or on art without defining, as well as he might, his idea of art. No doubt it was often as troublesome for the author to give such preliminary definitions, as it seemed useless to the reader, who was generally quite incapable of appreciating in the beginning their full value. Thus it happened that the rule of giving verbal definitions came to be looked upon after a time as useless and obsolete. Some authors actually took credit for no longer giving these verbal definitions, and it soon became the fashion to say that the only true and complete definition of what was meant by logic or grammar, by law or religion, was contained in the books themselves which treated of these subjects.

But what has been the result? Endless misunderstandings and controversies, which might have been avoided in many cases, if both sides had clearly defined what they did, and what they did not understand by certain words.

With regard to religion, it is no doubt extremely difficult to give a definition. The word rose to the surface thousands of years ago; it was retained while what was meant by it went on changing from century to century, and it is now often applied to the very opposite of what it was originally intended to signify.

Etymological Meaning of *Religio*

It is useless with words of this kind to appeal to their etymological meaning. The etymological meaning of a word is always extremely important, both psychologically and historically, because it indicates the exact point from which certain ideas started. But to know the small source of a river is very different from knowing the whole course of it; and to know the etymology of a word is very different from being able to trace it through all the eddies and cataracts through which it has been tossed and tumbled, before it became what it is now.

Besides, as with rivers, so with words, it is by no means easy to put our finger on the exact spot from whence they bubble forth. The Romans themselves felt doubtful as to the original meaning of *religio*. Cicero, as is well known, derived it from *re-legere*, to gather up again, to take up, to consider, to ponder—opposed to *nec-ligere*, to neglect; while others derived it from *re-ligare*, to fasten, to hold back. I believe myself that Cicero's etymology is the right one; but if *religio*[7] meant originally attention, regard, reverence, it is quite clear that it did not continue long to retain that simple meaning.

Historical Aspect of Religion

It must be clear that when we have to use words which have had a long history of their own, we can neither use them in their primitive etymological meaning, nor can we use them at one and the same time in all the senses through which they have passed. It is utterly useless to say, for instance, that religion meant this, and did not mean that; that it meant faith or worship, or morality or ecstatic vision, and that it did not mean fear or hope, or surmise, or reverence of the gods. Religion may mean all this; perhaps at one time or other the name was used in every one of these meanings; but who has a right to say that religion shall at present or in future have one of these meanings, and one only? The mere savage may not even have a name for religion; still when the Papua squats before his *karwar*, clasping his hands over his forehead, and asking himself whether what he is going to do is right or wrong, that is to him religion. Among several savage tribes, where there was no sign of a knowledge of divine beings, missionaries have discovered in the worship paid to the spirits of the departed the first faint beginnings of religion; nor should we hesitate to recognise the last glimmerings of religion when we see a recent philosopher, after declaring both God and gods obsolete, falling down before a beloved memory, and dedicating all his powers to the service of humanity. When the publican, standing afar off, would not lift up so much as his eyes unto heaven, but smote upon his breast, saying, "God be merciful to me a sinner," that was to him religion. When Thales declared that all things were full of the gods, and when Buddha denied that there were any *devas* or gods at all, both were stating their religious convictions.

When the young Brahman lights the fire on his simple altar at the rising of the sun, and prays, in the oldest prayer of the world, "May the Sun quicken our minds"; or when, later in life, he discards all prayer and sacrifice as useless, nay, as hurtful, and silently buries his own self in the Eternal Self—all this is Religion. Schiller declared that he professed no religion; and when asked why?, he answered, From religion. How, then, shall we find a definition of religion sufficiently wide to comprehend all these phases of thought?

Definitions of Religion by Kant and Fichte

It may be useful, however, to examine at least a few of the more recent definitions of religion, if only to see that almost every one is met by another, which takes the very opposite view of what religion is or ought to be. According to Kant, religion is morality. When we look upon all our moral duties as divine commands, that, he thinks, constitutes religion.[8] And we must not forget that Kant does not consider that duties are moral duties because they rest on a divine command (that would be, according to Kant, merely revealed religion); on the contrary, he tells us that because we are directly conscious of them as duties, therefore we look upon them as divine commands. Any outward divine authority is, in the eyes of a Kantian philosopher, something purely phenomenal, or, as we should say, a mere concession to human weakness. An established religion[9] or the faith of a Church, though it cannot at first dispense with statutory laws which go beyond pure morality, must, he thinks, contain in itself a principle which in time will make the religion of good moral conduct its real goal, and enable us in the end to surrender the preliminary faith of the Church.

Fichte, Kant's immediate successor, takes the very opposite view. Religion, he says, is never practical, and was never intended to influence our life. Pure morality suffices for that, and it is only a corrupt society that has to use religion as an impulse to moral action. Religion is knowledge. It gives to a man a clear insight into himself, answers the highest questions, and thus imparts to us a complete harmony with ourselves, and a thorough sanctification to our mind.

Now Kant may be perfectly right in saying that religion *ought* to be morality, or Fichte may be perfectly right in saying that it *ought* to be knowledge. What I protest against is that either the one or the other should be taken as a satisfactory definition of what is or was universally meant by the word religion.

Religion, with or without Worship

There is another view according to which religion consists in the worship of divine beings, and it has been held by many writers to be impossible that a religion could exist without some outward forms, without what is called a *cultus*. A religious reformer has a perfect right to say so, but the historian of religion could easily point out that religions have existed, and do exist still, without any signs of external worship.

In the last number of the *Journal of the Anthropological Society* (February 1878), Mr. C. H. E. Carmichael draws our attention to a very interesting account of a mission established by Benedictine monks in New Nursia in Western Australia, north of the Swan River, in the diocese assigned to the Roman Catholic Bishop of Perth in 1845.[10] These Benedictine monks took great pains to ascertain the religious sentiments of the natives, and for a long time they seem to have been unable to discover even the faintest traces of anything that could be called religion. After three years of mission life, Monsignor Salvado declares that the natives do not adore any deity, whether true or false. Yet he proceeds to tell us that they believe in an Omnipotent Being, creator of heaven and earth, whom they call *Motogon,* and whom they imagine as a very tall, powerful, and wise man of their own country and complexion. His mode of creation was by breathing. To create the earth, he said, "Earth, come forth!," and he breathed, and the earth was created. So with the sun, the trees, the kangaroo, &c., *Motogon,* the author of good, is confronted by *Cienga,* the author of evil. This latter being is the unchainer of the whirlwind and the storm, and the invisible author of the death of their children, wherefore the natives fear him exceedingly. Moreover, as *Motogon* has long since been dead and decrepit, they no longer pay him any worship. Nor is *Cienga,* although the natives believe that he afflicts them with calamities, propitiated by any service. "Never," the bishop concludes, "did I observe any act of external worship, nor did any indication suggest to me that they practised any internal worship."

If from one savage race we turn to another, we find among the Hidatsa or Grosventre Indians of the Missouri the very opposite state. Mr. Matthews, who has given us an excellent account of this tribe, says: "If we use the term worship in its most extended sense, it may be said that, besides 'the Old Man Immortal' or 'the Great Spirit,' 'the Great Mystery,' they worship everything in nature. Not man alone, but the sun, the moon, the stars, all the lower animals, all trees and plants, rivers and lakes, many boulders and other separated rocks, even some hills and buttes which stand alone—in short, everything not made by human hands, which has an independent being, or can be individualized, possesses a spirit, or, more properly, a shade. To these shades some respect or consideration is due, but not equally to all. ... The sun is held in great veneration, and many valuable sacrifices are made to it."[11]

Here then among the very lowest of human beings we see how some worship everything, while others worship nothing, and who shall say which of the two is the more truly religious?

Let us now look at the conception of religion, such as we find it among the most cultivated races of Europe, and we shall find among them the same divergence. Kant declares that to attempt to please the Deity by acts which have no moral value, by mere *cultus,* i.e., by external worship, is not religion, but simply superstition.[12] I need not quote authorities on the other side who declare that a silent religion of the heart, or even an active religion in common life, is nothing without an external worship, without a priesthood, without ritual.

We might examine many more definitions of religion, and we should always find that they contain what certain persons thought that religion ought to be; but they are hardly ever wide enough to embrace all that has been called religion at different periods in the history of the world. That being so, the next step has generally been to declare that whatever is outside the pale of any one of these definitions, does not deserve to be called religion, but should be called superstition, or idolatry, or morality, or philosophy, or any other more or less offensive name. Kant would call much of what other people call religion, hallucination; Fichte would call Kant's own religion mere legality. Many people would qualify all brilliant services, whether carried on in Chinese temples or Roman Catholic cathedrals, as mere superstition; while the faith of the silent Australians, and the half-uttered convictions of Kant, would by others be classed together as not very far removed from atheism.

Definition of Schleiermacher (Dependence), and of Hegel (Freedom)

I shall mention one more definition of religion, which in modern times has been rendered memorable and popular by Schleiermacher. According to him religion consists in our consciousness of absolute dependence on something which, though it determines us, we cannot determine in turn.[13] But here again another class of philosophers steps in, declaring that feeling of dependence the very opposite of religion. There is a famous, though not very wise, saying of Hegel, that if the consciousness of dependence constituted religion, the dog would possess most religion. On the contrary, religion, according to Hegel, is or ought to be perfect freedom; for it is, according to him, neither more nor less than the Divine Spirit becoming conscious of himself through the finite spirit.

Comte and Feuerbach

From this point it required but another step, and that step was soon taken by Feuerbach in Germany, and by Comte in France, to make man himself, not only the subject, but also the object of religion and religious worship. We are told that man cannot know anything higher than man; that man therefore is the only true object of religious knowledge and worship, only not man as an individual, but man as a class. The generic concept of man, or the genius of humanity, is to be substantiated, and then humanity becomes at once both the priest and the deity.

Nothing can be more eloquent, and in some passages really more solemn and sublime, than the religion of humanity, as preached by Comte and his disciples. Feuerbach, however, dissipates the last mystic halo which Comte had still left. "Self-love," he says, "is a necessary, indestructible, universal law and principle, inseparable from every kind of love. Religion must and does confirm this on every page of its history. Wherever man tries to resist that human egoism, in the sense in which we explained it, whether in religion, philosophy, or politics, he sinks into pure nonsense and insanity; for the sense which forms the foundation of all human instincts, desires, and actions is the satisfaction of the human being, the satisfaction of human egoism."[14]

Difficulty of Defining Religion

Thus we see that each definition of religion, as soon as it is started, seems at once to provoke another which meets it by a flat denial. There seem to be almost as many definitions of religion as there are religions in the world, and there is almost the same hostility between those who maintain these different definitions of religion as there is between the believers in different religions. What, then, is to be done? Is it really impossible to give a definition of religion, that should be applicable to all that has ever been called religion, or by some similar name? I believe it is, and you will yourselves have perceived the reason why it is so. Religion is something which has passed, and is still passing through an historical evolution, and all we can do is to follow it up to its origin, and then to try to comprehend it in its later historical developments.

Specific Characteristic of Religion

But though an adequate definition, or even an exhaustive description, of all that has ever been called religion is impossible, what is possible is to give some specific characteristic which distinguishes the objects of religious consciousness from all other objects, and at the same time distinguishes our consciousness, as applied to religious objects, from our consciousness when dealing with other objects supplied to it by sense and reason.

Let it not be supposed, however, that there is a separate consciousness for religion. There is but oneself and one consciousness, although that consciousness varies according to the objects to which it is applied. We distinguish between sense and reason, though even these two are in the highest sense different functions only of the same conscious self. In the same manner, when we speak of faith as a religious faculty in man, all that we can mean is our ordinary consciousness, so developed and modified as to enable us to take cognisance of religious objects. This is not meant as a new sense, by the side of the other senses, or as a new reason by the side of our ordinary reason— a new soul within the soul. It is simply the old consciousness applied to new objects, and reacted upon by them. To admit faith as a separate religious faculty, or a theistic instinct, in order to explain religion as a fact, such as we find it everywhere, would be like admitting a vital force in order to explain life; it would be a mere playing with words, a trifling with truth. Such explanations may have answered formerly, but at present the battle has advanced too far for any peace to be concluded on such terms.

Religion as a Subjective Faculty for the Apprehension of the Infinite

In a course of introductory lectures on the Science of Religion, delivered at the Royal Institution in 1873, I tried to define the subjective side of religion, or what is commonly called faith, in the following words:

> Religion is a mental faculty or disposition which, independent of, nay, in spite of sense and reason, enables man to apprehend the Infinite under different names and under varying disguises. Without that faculty, no religion, not even the lowest worship of idols and fetishes, would be possible; and if we will but listen attentively, we can hear in all religions a groaning of the spirit, a struggle to conceive the inconceivable, to utter the unutterable, a longing after the Infinite, a love of God.[15]

I do not quote these words because I altogether approve of them now. I very seldom approve altogether of what I have written myself some years ago. I fully admit the force of many objections that have been raised against that definition of religion, but I still think that the kernel of it is sound. I should not call it now an exhaustive definition of religion, but I believe it supplies such characteristics as will enable us to distinguish between religious consciousness on one side, and sensuous and rational consciousness on the other.

What has been chiefly objected to in my definition of religion, was that I spoke of it as a mental faculty. "Faculty" is a word that rouses the anger of certain philosophers, and to some extent I share their objections. It seems to be imagined that faculty must signify something substantial, a spring as it were, setting a machine in motion; a seed or a pip that can be handled, and will spring up when planted in proper soil. How faculty could be used in such a sense, I have never been able to comprehend, though I cannot deny that it has often been thus used. Faculty signifies a mode of action, never a substantial something. Faculties are neither gods nor ghosts, neither powers nor principalities. Faculties are inherent in substances, quite as much as forces or powers are. We generally speak of the faculties of conscious, of the forces of unconscious substances. Now we know that there is no force without substance, and no substance without force. To speak of gravity, for instance, as a thing by itself, would be sheer mythology. If the law of gravity had been discovered at Rome, there would have been a temple built to the goddess of gravity. We no longer build temples, but the way in which some natural philosophers speak of gravity is hardly less mythological. The same danger exists, I fully admit, with regard to the manner in which certain philosophers speak of our mental faculties, and we know that one faculty at least, that of Reason, had actually an altar erected to her not very long ago. If, therefore, faculty is an ambiguous and dangerous, or if it is an unpopular word, let us by all means discard it. I am perfectly willing to say "potential energy" instead, and therefore to define the subjective side of religion as the potential energy which enables man to apprehend the infinite. If the English language allowed it, I should even propose to replace "faculty" by the *Not-yet*, and to speak of the *Not-yet* of language and religion, instead of their faculties or potential energies.[16] Professor Pfleiderer, to whom we owe some excellent contributions to the science of religion, finds fault with my definition because it admits, not only a *facultas,* but a *facultas occulta.* All depends here again on the sense which we attach to *facultas occulta.* If it means no more than that there is in men, both individually and generally (ontogenetically and phylogenetically), something that develops into perception, conception, and faith, using the last word as meaning the apprehension of the infinite, then I fully admit a *facultas occulta*. Everything that develops may from one point of view be called occult. This, however, applies not only to the faculty of faith, but likewise to the faculties of sense and reason.

The Three Functions of Sense, Reason, and Faith

Secondly, it has been objected that there is something mysterious in this view of religion. As to myself, I cannot see that in admitting, besides the sensuous and rational, a third function of the conscious self, for apprehending the infinite, we introduce a mysterious element into psychology. One of the essential elements of all religious knowledge is the admission of beings which can neither be apprehended by sense nor comprehended by reason. Sense and reason, therefore, in the ordinary acceptation of these terms, would not be sufficient to account for the facts before us. If, then, we openly admit a third function of our consciousness for the apprehension of what is infinite, that function need not be more mysterious than those of sense and reason. Nothing is in reality more mysterious than sensuous perception. It is the real mystery of all mysteries. Yet we have accustomed ourselves to regard it as the most natural of all things. Next comes reason which, to a being restricted to sensuous perception, might certainly appear very mysterious, and which even by certain philosophers has been represented as altogether incomprehensible. Yet we know that reason is only a development of sensuous perception, possible under certain conditions. These conditions correspond to what we call the potential energy or faculty of reason. They belong to one and the same conscious self, and though reason is active in a different manner, yet, if kept under proper control, reason works in perfect harmony with sense. The same applies to religion, in its subjective sense of faith. It is, as I shall try to show, simply another development of sensuous perception, quite as much as reason is. It is possible under certain conditions, and these conditions correspond to what we call the potential energy of faith. Without this third potential energy, the facts which are before us in religion, both subjectively and objectively, seem to me inexplicable. If they can be explained by a mere appeal to sense and reason, in the ordinary meaning of these words, let it be done. We shall then have a rational religion, or an intuitional faith. None of my critics, however, has done that yet; few, I believe, would like to do it.

When I say that our apprehension of the Infinite takes place independent of, nay, in spite of sense and reason, I use these two words in their ordinary acceptation. If it is true that sense supplies us with finite objects only, and if reason has nothing to work on except those finite objects, then our assumed apprehension of anything infinite must surely be independent of, nay, in spite of sense or reason. Whether the premises [*sic*] are right is another question, which we shall have to discuss presently.

The Meaning of Infinite

Let us now see whether we can agree on some general characteristic of all that forms the object of our religious consciousness. I chose "infinite" for that purpose, as it seemed best to comprehend all that transcends our senses and our reason, taking these terms in their ordinary meaning. All sensuous knowledge, whatever else it may be, is universally admitted to be finite, finite in space and time, finite also in quantity and quality, and as our conceptual knowledge is based entirely on our sensuous knowledge, that also can deal with finite objects only. Finite being then the most general predicate of all our so-called positive knowledge, I thought infinite the least objectionable term for all that transcends our senses and our reason, always taking these words in their ordinary meaning. I thought it preferable to indefinite, invisible, supersensuous, supernatural, absolute or divine, as the characteristic qualification of the objects of that large class of knowledge which constitutes what we call religion. All these terms are meant for the same thing. They all express different aspects of the same object. I have no predilection for infinite, except that it seems to me the widest term, the highest generalization. But if any other term seems preferable, again I say, let us adopt it by all means.

Only let us now clearly understand what we mean by infinite, or any other of these terms that may seem preferable.

If the infinite were, as certain philosophers suppose, simply a negative abstraction (*ein negativer Abstractions-begriff*), then, no doubt, reason would suffice to explain how we came to be possessed of it. But abstraction will never give us more than that from which we abstract. From a given number of perceptions we can abstract the concept of a given multitude. Infinite, however, is not contained in finite, and by no effort whatever shall we be able to abstract the infinite from the finite. To say, as many do, that the infinite is a negative abstract concept, is a mere playing with words. We may form a negative abstract concept, when we have to deal with serial or correlative concepts, but not otherwise. Let us take a serial concept, such as blue, then not-blue means green, yellow, red, any colour, in fact, except blue. Not-blue means simply the whole concept of colour, *minus* blue. We might of course comprehend sweet, or heavy, or crooked by the negative concept of not-blue—but our logic, or our language, does not admit of such proceedings.

If we take correlative concepts, such as crooked and straight, then not-straight may by logicians be called a negative concept, but it is in reality quite as positive as crooked, not-straight being crooked, not-crooked being straight.

Now let us apply this to finite. Finite, we are told, comprehends everything that can be perceived by the senses, or counted by reason. Therefore, if we do not only form a word at random, by adding the ordinary negative particle to finite, but try to form a really negative concept, then that concept of infinite would be outside the concept of finite, and as, according to a premiss [*sic*] generally granted, there is nothing known to us outside the concept of the finite, the concept of the infinite would simply comprise nothing. Infinite therefore cannot be treated simply as a negative concept. If it were no more than that, it would be a word formed by false analogy, and signify nothing.

Can the Finite Apprehend the Infinite?

All the objections which we have hitherto examined proceed from friendly writers. They are amendments of my own definition of religion, they do not amount to a moving of the previous question. But it is well known that that previous question also has been moved. There is a large class, not only of philosophers by profession, but of independent thinkers in all classes of society, who look upon any attempt at defining religion as perfectly useless, who would not listen even to a discussion whether one religion was false or another true, but who simply deny the possibility of any religion whatsoever, on the ground that men cannot apprehend what is infinite, while all religions, however they may differ on other points, agree in this, that their objects transcend, either partially or entirely, the apprehensive and comprehensive powers of our senses and of our reason. This is the ground on which what is now called positive philosophy takes its stand, denying the possibility of religion, and challenging all who admit any source of knowledge except sense and reason, to produce their credentials.

This is not a new challenge, nor is the ground on which the battle has to be fought new ground. It is the old battle-field measured out long ago by Kant, only that the one opening which was still left in his time, viz., the absolute certainty of moral truth, and through it the certainty of the existence of a God, is now closed up. There is no escape in that direction.[17] The battle between those who believe in something which transcends our senses and our reason, who claim for man the possession of a faculty or potential energy for apprehending the infinite, and those who deny it on purely psychological grounds, must end in the victory of one, and the surrender of the other party.

Conditions Accepted on Both Sides

Before we commit ourselves to this struggle for life or death, let us inspect once more the battle-field, as it is measured out for us, and survey what is the common ground on which both parties have agreed to stand or to fall. What is granted to us is that all consciousness begins with sensuous perception, with what we feel, and hear, and see. This gives us sensuous knowledge. What is likewise granted is that out of this we construct what may be called conceptual knowledge, consisting of collective and abstract concepts. What we call thinking consists simply in addition and subtraction of percepts and concepts. Conceptual knowledge differs from sensuous knowledge, not in substance, but in form only. As far as the material is concerned, nothing exists in the intellect except what existed before in the senses. The organ of knowledge is throughout the same, only that it is more highly developed in animals that have five senses, than in animals that have but one sense, and again more highly developed in man who counts and forms concepts, than in all other animals who do not.

On this ground and with these weapons we are to fight. With them, we are told, all knowledge has been gained, the whole world has been conquered. If with them we can force our way to a world beyond, well and good; if not, we are asked to confess that all that goes by the name of religion, from the lowest fetishism to the most spiritual and exalted faith, is a delusion, and that to have recognised this delusion is the greatest triumph of our age.

I accept these terms, and I maintain that religion, so far from being impossible, is inevitable, if only we are left in possession of our senses, such as we really find them, not such as they have been defined for us. Thus the issue is plain. We claim no special faculty, no special revelation. The only faculty we claim is perception, the only revelation we claim is history, or, as it is now called, historical evolution.

For let it not be supposed that we find the idea of the infinite ready made in the human mind from the very beginning of our history. There are even now millions of human beings to whom the very word would be unintelligible. All we maintain is that the germ or the possibility, the *Not-yet* of that idea, lies hidden in the earliest sensuous perceptions, and that as reason is evolved from what is finite, so faith is evolved from what, from the very beginning, is infinite in the perceptions of our senses.

Positive philosophy imagines that all that is supplied to us through the senses is by its very nature finite, that whatever transcends the finite is a mere delusion, that the very word "infinite" is a mere jingle, produced by an outward joining of the negative particle with the adjective "finite," a particle which is rightly used with serial, or correlative concepts, but which is utterly out of place with an absolute or exclusive concept, such as finite. If the senses tell us that *all* is finite, and if reason draws all her capital from the senses, who has a right, they say, to speak of the infinite? It may be true that an essential element of all religious knowledge is the admission of beings which can neither be apprehended by sense, nor comprehended by reason, which are in fact infinite, and not finite. But instead of admitting a third faculty or potential energy in order to account for these facts of religion, positive philosophers would invert the argument, and prove that, for that very reason, religion has no real roots in our consciousness, that it is a mere mirage in the desert, alluring the weary traveller with bright visions, and leaving him to despair, when he has come near enough to where the springs of living water seemed to flow.

Some philosophers have thought that a mere appeal to history would be a sufficient answer to this despairing view. No doubt, it is important that, so long as we know man in possession of sense and reason, we also find him in possession of religion. But not even the eloquence of Cicero has been able to raise this fact to the dignity of an invulnerable argument. That all men have a longing for the gods is an important truth, but not even the genius of Homer could place that truth beyond the reach of doubt. Who has not wondered at those simple words of Homer (*Odyssey,* iii. 48), πάντες δὲ θεῶν χατέουσ' ἄνθρωποι, "All men crave for the gods"; or, as we might render it still more literally and truthfully, "as young birds open their mouth for food, all men crave for the gods" For χατεῖν, as connected with χαίνειν, meant originally to gape, to open the mouth, then to crave, to desire. But even that simple statement is met with an equally simple denial. Some men, we are told, in very ancient times, and some in very modern times, know of no such cravings. It is not enough therefore to show that man has always transcended the limits which sense and reason seem to trace for him. It is not enough to show that, even in the lowest fetish worship, the fetish is not only what we can see, or hear, or touch, but something else, which we cannot see, or hear, or touch. It is not enough to show that in the worship paid to the objects of nature, the mountains, trees, and rivers are not simply what we can see, but something else which we cannot see; and that when the sky and the heavenly bodies are invoked, it is not the sun or the moon and the stars, such as they appear to the bodily eye, but again something else which cannot be seen, that forms the object of religious belief. The rain is visible; he who sends the rain is not. The thunder is heard, the storm is felt; but he who thunders and rides on the whirlwind is never seen by human eye. Even if the gods of the Greeks are sometimes seen, the Father of gods and men is not; and he who in the oldest Aryan speech was called Heaven-Father (Dyaus Pitar), in Greek Ζεὺς πατήρ, in Latin Jupiter, was no more an object of sensuous perception than He whom we call our Father in Heaven.

All this is true, and it will be the object of these lectures to watch this important development of religious thought from its very beginning to its very end, though in one stream only, namely, in the ancient religion of India. But before we can do this, we have to answer the preliminary and more abstract question: Whence comes that something else, which, as we are told, neither sense nor reason can supply? Where is the rock for him to stand on, who declines to rest on anything but what is called the evidence of the senses, or to trust in anything but the legitimate deductions derived from it by reason, and who nevertheless maintains his belief in something which transcends both sense and reason?

Apprehension of the Infinite

We have granted that all our knowledge begins with the senses, and that out of the material, supplied by the senses, reason builds up her marvellous structure. If therefore all the materials which the senses supply are finite, whence, we ask, comes the concept of the infinite?

Hobbes calls the idea of the Infinite an absurd speech.

"Whatsoever we imagine," he writes, "is *finite*. Therefore there is no idea, or conception of anything we call *infinite*. No man can have in his mind an image of infinite magnitude; nor conceive infinite swiftness, infinite time, or infinite force, or infinite power. When we say anything is infinite, we signify only, that we are not able to conceive the ends and bounds of the things named; having no conception of the thing, but of our own inability. And therefore the name of God is used, not to make us conceive him, for he is incomprehensible; but that we may honour him. Also because, whatsoever, as I said before, we conceive, has been perceived first by sense, either all at once, or by parts; not that anything is all in this place, and all in another place at the same time; nor that two, or more things can be in one, and the same place at once: for none of these things ever have, nor can be incident to sense; but are absurd speeches, taken upon credit, without any signification at all, from deceived philosophers, and deceived, or deceiving schoolmen."[18]

Condillac thinks that we might have avoided all difficulties if, instead of Infinite, we had used the word Indefinite. "If we had called the Infinite," he writes, "the Indefinite, by this small change of a word, we should have avoided the error of imagining that we have a positive idea of infinity, from whence so many false reasonings have been carried on, not only by metaphysicians, but even by geometricians."

That a useful distinction might be made between *infinite* and *indefinite* has been proved by Kant in his *Critique of Pure Reason*, p. 448.

What I want to prove in this course of lectures is that indefinite and infinite are in reality two names of the same thing, the former expressing its phenomenal, the latter its real character; that the history of religion is a history of all human efforts to render the Infinite less and less indefinite, that, in spite of all these efforts, the Infinite must always remain *to us* the Indefinite.

1. The Infinitely Great

The first point that has to be settled—and on that point all the rest of our argument turns—is this: "Are all the materials which the senses supply finite, and finite only?"

Is it true that all we can see, and feel, and hear has a beginning and an end, and is it only by apprehending these beginnings and ends that we gain sensuous knowledge? We perceive a body by perceiving its outline; we perceive green in large intervals between blue and yellow; we hear the musical note D between where C ends and E begins; and so with all other perceptions of the senses. This is true—true at least for all practical purposes. But let us look more carefully. When our eye has apprehended the furthest distance which it can reach, with or without instruments, the limit to which it clings is always fixed on the one side by the finite, but on the other side by what to the eye is not finite, and what may be called indefinite or infinite. Let us remember that we have accepted the terms of our opponents, and that therefore we look upon man as simply endowed with sense. To most philosophers it would appear much more natural, and, I doubt not, much more convincing, to derive the idea of the infinite from a necessity of our human reason. Whenever we try to fix a point in space or time, they say, we are utterly unable to fix it so as to exclude the possibility of a point beyond. In fact, our very idea of limit implies the idea of a beyond, and thus forces the idea of the infinite upon us, whether we like it or not.

This is perfectly true, but we must think not of our friends, but of our opponents, and it is well known that our opponents do not accept that argument. If on one side, they say, our idea of a limit implies a beyond and leads us to postulate an infinite, on the other, our idea of a whole excludes a beyond, and thus leads us to postulate a finite. These antinomies of human reason have been fully discussed by Kant, and later philosophers have naturally appealed to them to show that what we call necessities, may be after all but weaknesses of human reason, and that, like all other ideas, those of finite and infinite also, if they are to be admitted at all, must be shown to be the result not of speculation, but of experience, and, as all experience is at first sensuous, the result of sensuous experience. This is the argument we have to deal with, and here neither Sir W. Hamilton nor Lucretius can help us.

We have accepted the primitive savage with nothing but his five senses. These five senses supply him with a knowledge of finite things; our problem is how such a being ever comes to think or speak of anything not finite, but infinite.

I answer, without any fear of contradiction, that it is his senses which give him the first impression of infinite things, and force him to the admission of the infinite. Everything of which his senses cannot perceive a limit is to a primitive savage, or to any man in an early stage of intellectual activity, unlimited or infinite. Man sees, he sees to a certain point; and there his eyesight breaks down. But exactly where his sight breaks down, there presses upon him, whether he likes it or not, the perception of the unlimited or the infinite. It may be said that this is not perception, in the ordinary sense of the word. No more it is, but still less is it mere reasoning. In perceiving the infinite, we neither count, nor measure, nor compare, nor name. We know not what it is, but we know that it is, and we know it, because we actually feel it and are brought in contact with it. If it seems too bold to say that man actually sees the invisible, let us say that he suffers from the invisible, and this invisible is only a special name for the infinite.

Therefore, as far as mere distance or extension is concerned, it would seem difficult to deny that the eye, by the very same act by which it apprehends the finite, apprehends also the infinite. The more we advance, the wider no doubt grows our horizon; but there never is and never can be to our senses a horizon unless as standing between the visible and finite on one side, and the invisible and infinite on the other. The infinite, therefore, instead of being merely a late abstraction, is really implied in the earliest manifestations of our sensuous knowledge. Theology begins with anthropology. We must begin with men living on high mountains, or in a vast plain, or on a coral island without hills and streams, surrounded on all sides by the endless expanse of the ocean, and screened above by the unfathomable blue of the sky; and we shall then understand how, from the images thrown upon them by the senses, some idea of the infinite would arise in their minds earlier even than the concept of the finite, and would form the omnipresent background of the faintly dotted picture of their monotonous life.

2. The Infinitely Small

But that is not all. We apprehend the infinite not only as beyond, but also as within the finite; not only as beyond all measure great, but also as beyond all measure small. However much our senses may contract the points of their tentacles, they can never touch the smallest objects. There is always a beyond, always a something smaller still. We may, if we like, postulate an atom in its original sense, as something that cannot be cut asunder; our senses—and we speak of them only, for we have been restricted to them by our opponents—admit of no real atoms, nor of imponderable substances, or of what Robert Mayer called the last gods of Greece, "immaterial matter." In apprehending the smallest extension, they apprehend a smaller extension still. Between the centre and the circumference, which every object must have in order to become visible, there is always a *radius;* and that omnipresent and never entirely vanishing radius gives us again and again the sensuous impression of the infinite— of the infinitely small, as opposed to the infinitely great.

And what applies to space, applies equally to time, applies equally to quality and quantity.

When we speak of colours or sounds, we seem for all practical purposes to move entirely within the finite. This is red, we say, this is green, this is violet. This is C, this is D, this is E. What can apparently be more finite, more definite? But let us look more closely. Let us take the seven colours of the rainbow; and where is the edge of an eye sharp enough to fix itself on the point where blue ends and green begins, or where green ends and yellow begins? We might as well attempt to put our clumsy fingers on the point where one millimetre ends and another begins. We divide colour by seven rough degrees, and speak of the seven colours of the rainbow. Even those seven rough degrees are of late date in the evolution of our sensuous knowledge. Xenophanes says that what people call Iris is a cloud, purple (πορφύρεον), red (φοινίκεον), and yellow (χλωρόν). Even Aristotle[19] still speaks of the tricoloured rainbow, red (φοινικῆ), yellow (ξανθή), and green (πρασίνη), and in the Edda the rainbow is called a three-coloured bridge. Blue, which seems to us so definite a colour, was worked out of the infinity of colours at a comparatively late time. There is hardly a book now in which we do not read of the blue sky. But in the ancient hymns of the Veda,[20] so full of the dawn, the sun, and the sky, the blue sky is never mentioned; in the Zend-Avesta the blue sky is never mentioned; in Homer the blue sky is never mentioned; in the Old, and even in the New Testament, the blue sky is never mentioned.

It has been asked whether we should recognise in this a physiological development of our senses, or a gradual increase of words capable of expressing finer distinctions of light. No one is likely to contend that the irritations of our organs of sense, which produce sensation, as distinguished from perception, were different thousands of years ago from what they are now. They are the same for all men, the same even for certain animals, for we know that there are insects which react very strongly against differences of colour. No, we only learn here again, in a very clear manner, that conscious perception is impossible without language. Who would contend that savages, unable, as we are told, to count beyond three—that is to say, not in possession of special numerals beyond three—do not receive the sensuous impression of the four legs of a cow, and know four legs as different from three or two? No, in this evolution of consciousness of colour we see once more how perception, as different from sensation, goes hand in hand with the evolution of language, and how by a very slow process every definite concept is gained out of an infinitude of indistinct perceptions. Demokritos knew of four colours, viz., black and white, which he treated as colours, red and yellow. Are we to say that he did not see the blue sky because he never called it blue, but either dark or bright? In China the number of colours was originally five.[21] That number was increased with the increase of their power of distinguishing and of expressing their distinctions in words. In common Arabic, as Palgrave tells us, the names for green, black, and brown are constantly confounded to the present day. It is well known that among savage nations we seldom find distinct words for blue and black,[22] but in our language too we shall find a similar indefiniteness of expression when we inquire into its historical antecedents. Though *blue* now does no longer mean black, we see in such expressions as "to beat black and blue" the closeness of the two colours. In Old Norse too, *blár, blá, blátt* now means blue, as distinct from *blakkr,* black. But in O.N. *bláman,* the livid colour of a bruise, we see the indefiniteness of meaning between black and blue, and in *blá-madr,* a black man, a negro, *blá* means distinctly black. The etymology of these words is very obscure. Grimm derives blue, O.H.G. *plâo, plawes,* Med. Lat. *blavus* and *blavius,* It. *biavo,* Fr. *bleu,* from Goth. *bliggvan,* to strike, so that it would originally have conveyed the black and blue colour of a bruise. He appeals in support of his derivation to Latin *lividus,* which he derives from **fligvidus* and *fligere;* nay even to *flavus,* which he proposes to derive from **flagvus* and **flagere. Caesius* also is quoted as an analogy, supposing it is derived from *caedere.* All this is extremely doubtful, and the whole subject of the names of colour requires to be treated most accurately, and yet in the most comprehensive way before any certain results can be expected in the place of ingenious guesses. Most likely the root *bhrag* and *bhrâg,* with *r* changed to *l,* will be found as a fertile source of names of colour. To that root *bleak,* A.S. *blâc, blæc,* O.N. *bleikr,* O.H.G. *pleih,* has been referred, meaning originally bright, then pale; and to the same family, though the vowel is different, *black* also will probably have to be traced back, A.S. *blac,* O.N. *blakkr.*

As languages advance, more and more distinctions are introduced, but the variety of colours always stands before us as a real infinite, to be measured, it may be, by millions of ethereal vibrations in one second, but beyond that immeasurable and indivisible even to the keenest eye.

What applies to colour applies to sounds. Our ear begins to apprehend tone when there are thirty vibrations in one second; it ceases to apprehend tone when there are four thousand vibrations in one second. It is the weakness of our ears which determines these limits; but as there is beyond the violet, which we can perceive, an ultra-violet which to our eye is utter darkness, while it is revealed in hundreds of lines through the spectroscope, so there may be to people with more perfect powers of hearing, music where to us there is but noise. Though we can distinguish tones and semitones, there are many smaller divisions which baffle our perception, and make us feel, as many other things, the limited power of our senses before the unlimited wealth of the universe, which we try slowly to divide, to fix, to comprehend, and to name.

Growth of the Idea of the Infinite

I hope I shall not be misunderstood—or, I ought rather to say, I fear I shall be—as if I held the opinion that the religion of the lowest savages begins with the barren idea of the infinite, and with nothing else. As no concept is possible without a name, I shall probably be asked to produce from the dictionaries of Veddas and Papúas any word to express the infinite; and the absence of such a word, even among more highly civilised races, will be considered a sufficient answer to my theory.

Let me, therefore, say once more that I entirely reject such an opinion. I am acting at present on the defensive only; I am simply dealing with the preliminary objections of those philosophers who look upon religion as outside the pale of philosophy, and who maintain that they have proved once for all that the infinite can never become a legitimate object of our consciousness, because our senses, which form the only avenue to the whole domain of our human consciousness, never come in contact with the infinite. It is in answer to that powerful school of philosophy, which on that one point has made converts even amongst the most orthodox defenders of the faith, that I felt it was necessary to point out, at the very outset, that their facts are no facts, but that the infinite was present from the very beginning in all finite perceptions, just as the blue colour was there, though we find no name for it in the dictionaries of Veddas and Papúas. The sky was blue in the days of the Vedic poets, of the Zoroastrian worshippers, of the Hebrew prophet[s], of the Homeric singers, but though they saw it, they knew it not: they had no name for that which is the sky's own peculiar tint, the sky-blue. We know it, for we have a name for it. We know it, at least to a certain extent, because we can count the millions of vibrations that make up what we now call the blue of the sky. We know it quantitatively, but not qualitatively. Nay, to most of us it is, and it always will be, nothing but visible darkness, half veiling and half revealing the infinite brightness beyond.

It is the same with the infinite. It was there from the very first, but it was not yet defined or named. If the infinite had not from the very first been present in our sensuous perceptions, such a word as "infinite" would be a sound, and nothing else. For that reason I felt it incumbent upon me to show how the presentiment of the infinite rests on the sentiment of the finite, and has its real roots in the real, though not yet fully apprehended presence of the infinite in all our sensuous perceptions of the finite. This presentiment or incipient apprehension of the infinite passes through endless phases and assumes endless names. I might have traced it in the wonderment with which the Polynesian sailor dwells on the endless expanse of the sea, in the jubilant outburst with which the Aryan shepherd greets the effulgence of the dawn, or in the breathless silence of the solitary traveller in the desert when the last ray of the sun departs, fascinating his weary eyes, and drawing his dreamy thoughts to another world. Through all these sentiments and presentiments there vibrates the same chord in a thousand tensions, and if we will but listen attentively we can still perceive its old familiar ring even in such high harmonics as Wordsworth's:

> Obstinate questionings
> Of sense and outward things,
> Fallings from us, vanishings;
> Blank misgivings of a Creature
> Moving about in worlds not realized.

No Finite without an Infinite

What I hold is that with every finite perception there is a concomitant perception, or, if that word should seem too strong, a concomitant sentiment or presentiment of the infinite; that from the very first act of touch, or hearing, or sight, we are brought in contact, not only with a visible, but also at the same time with an invisible universe. Those therefore who deny the possibility or the legitimacy of the idea of the infinite in our human consciousness, must meet us here on their own ground. All our knowledge, they say, must begin with the senses. Yes, we say, and it is the senses which give us the first intimation of the infinite. What grows afterwards out of this intimation supplies materials both to the psychologist and to the historian of religion, and to both of them this indisputable sentiment of the infinite is the first pre-historic impulse to all religion. I do not say that in the first dark pressure of the infinite upon us, we have all at once the full and lucid consciousness of that highest of all concepts. I mean the very opposite. I simply say we have in it a germ, and a living germ, we have in it that without which no religion would have been possible, we have in that perception of the infinite the root of the whole historical development of human faith.

And let it not be supposed that in insisting on an actual perception of the infinite, I indulge in poetical language only, though I am the last to deny that poetical language may sometimes convey much truth, nay often more than is to be found in the confused webs of argumentative prose. I shall quote at least one of these poetical pleadings in favour of the reality of the infinite: "Et qu'on ne dise pas que l'infini et l'éternel sont inintelligibles; c'est le fini et le passager qu'on serait souvent tenté de prendre pour un rêve; car la pensée ne peut voir de terme à rien, et l'être ne saurait concevoir le néant. On ne peut approfondir les sciences exactes elles-mêmes, sans y rencontrer l'infini et l'éternel; et les choses les plus positives appartiennent autant, sous de certains rapports, à cet infini et à cet éternel, que le sentiment et l'imagination."[23]

I fully admit that there is much truth in these impassioned utterances, but we must look for the deepest foundation of that truth, otherwise we shall be accused of using poetical or mystic assertions, where only the most careful logical argument can do real good. In postulating, or rather in laying my finger on the point where the actual contact with the infinite takes place, I neither ignore nor do I contravene any one of the stringent rules of Kant's *Critik der reinen Vernunft*. Nothing, I hold, can be more perfect than Kant's analysis of human knowledge. "Sensuous objects cannot be known except such as they appear to us, never such as they are in themselves; supersensuous objects are not to us objects of theoretic knowledge." All this I fully accept. But though there is no theoretic knowledge of the supersensuous, is there no knowledge of it at all? Is it no knowledge, if we know that a thing is, though we do not know what it is? What would Kant say, if we were to maintain that because we do not know what the *Ding an sich* is, therefore we do not know that it is. He carefully guards against such a misunderstanding, which would change his whole philosophy into pure idealism. "Nevertheless," he says, "it should be observed that we must be able, if not to know, at all events to be conscious of the same objects, also as *Dinge an sich*. Otherwise we should arrive at the irrational conclusion that there is appearance without something that appears."[24]

If I differ from Kant, it is only in going a step beyond him. With him the supersensuous or the infinite would be a mere *Nooumenon*, not a *Phainomenon*. I maintain that before it becomes a *Nooumenon*, it is an *Aistheton*, though not a *Phainomenon;* it is felt, though not yet represented. I maintain that we, as sentient beings, are in constant contact with the infinite, and that this constant contact is the only legitimate basis on which the infinite can and does exist for us afterwards, whether as a *Nooumenon* or as a *Pisteuomenon*. I maintain that, here as elsewhere, no legitimate concept is possible without a previous percept, and that that previous percept is as clear as daylight to all who are not blinded by traditional terminologies.

We have been told again and again that a finite mind cannot approach the infinite, and that therefore we ought to take our Bible and our Prayer-book, and rest there and be thankful. This would indeed be taking a despairing view both of ourselves and of our Bible and Prayer-book. No, let us only see and judge for ourselves, and we shall find that, from the first dawn of history, and from the first dawn of our own individual consciousness, we have always been face to face with the infinite. Whether we shall ever be able to gain more than this sentiment of the real presence of the infinite, whether we shall ever be able, not only to apprehend, but to comprehend it, that is a question which belongs to the end, not to the beginning of our subject. At present we are concerned with history only, in order to learn from its sacred annals how the finite mind has tried to pierce further and further into the infinite, to gain new aspects of it, and to raise the dark perception of it into more lucid intuitions and more definite names. There may be much error in all the names that man has given to the infinite, but even the history of error is full of useful lessons. After we have seen how it is possible for man to gain a presentiment of something beyond the finite, we shall watch him looking for the infinite in mountains, trees, and rivers, in the storm and lightning, in the moon and the sun, in the sky and what is beyond the sky, trying name after name to comprehend it, calling it thunderer, bringer of light, wielder of the thunderbolt, giver of rain, bestower of food and life;

and, after a time, speaking of it as maker, ruler, and preserver, king and father, lord of lords, god of gods, cause of causes, the Eternal, the Unknown, the Unknowable. All this we shall see in at least one great evolution of religious thought, preserved to us in the ancient literature of India.

There are many other historical evolutions, in other countries, each leading to its own goal. Nothing can be more different than the evolution of the consciousness of the Infinite or the Divine among Aryan, Semitic, and Turanian races. To some the infinite first revealed itself, as to the Vedic poets, in certain visions of nature. Others were startled by its presence in the abyss of their own hearts. There were whole tribes to whom the earliest intimation of the infinite came from the birth of a child, or from the death of a friend; and whose idea of beings more than human was derived from the memory of those whom they had loved or feared in life. The sense of duty, which in ancient times had always a religious character, seems in some cases to have sprung from that feeling of burning shame which was none the less real because it could not be accounted for; while other tribes became conscious of law by witnessing the order in nature, which even the gods could not transgress. And love, without which no true religion can live, while in some hearts it bursts forth as a sudden warmth kindled by the glances of the morning light, was roused in others by that deep sympathy of nature—that suffering in common—which, whether we like it or not, makes our nerves quiver at the sight of a suffering child; or was called into life by that sense of loneliness and finiteness which makes us long for something beyond our own narrow, finite self, whether we find it in other human selves, or in that infinite Self in which alone we have our being, and in which alone we find in the end our own true self.

Each religion had its own growth, each nation followed its own path through the wilderness. If these lectures continue, as I hope they may, other and better analysts of the human mind will hereafter disentangle and lay before you the manifold fibres that enter into the web of the earliest religious thoughts of man; other and more experienced guides will hereafter lead you through the valleys and deserts which were crossed by the great nations of antiquity, the Egyptians, the Babylonians, the Jews, the Chinese, it may be, or the Greeks and Romans, the Celts, the Slavs, and Germans, nay, by savage and as yet hardly human races, in their search after the infinite, that infinite which surrounded them, as it surrounds us, on every side, and which they tried, and tried in vain, to grasp and comprehend.

I shall confine myself to one race only, the ancient Aryans of India, in many respects the most wonderful race that ever lived on earth. The growth of their religion is very different from the growth of other religions; but though each religion has its own peculiar growth, the seed from which they spring is everywhere the same. That seed is the perception of the infinite, from which no one can escape, who does not wilfully [sic] shut his eyes. From the first flutter of human consciousness, that perception underlies all the other perceptions of our senses, all our imaginings, all our concepts, and every argument of our reason. It may be buried for a time beneath the fragments of our finite knowledge, but it is always there, and, if we dig but deep enough, we shall always find that buried seed, supplying the living sap to the fibres and feeders of all true faith.

For many reasons I could have wished that some English student, who in so many respects would have been far better qualified than I am, should have been chosen to inaugurate these lectures. There was no dearth of them, there was rather, I should say, an *embarras de richesse*. How ably would a psychological analysis of religion have been treated by the experienced hands of Dr. Martineau or Principal Caird! If for the first course of these Hibbert Lectures you had chosen Egypt and its ancient religion, you had such men as Birch, or Le Page Renouf; for Babylon and Nineveh, you had Rawlinson or Sayce; for Palestine, Stanley or Cheyne; for China, Legge or Douglas; for Greece, Gladstone, or Jowett, or Mahaffy; for Rome, Munro or Seeley; for the Celtic races, Rhŷs; for the Slavonic races, Morfill or Ralston; for the Teutonic races, Skeat or Sweet; for savage tribes in general, Tylor or Lubbock. If after considerable hesitation I decided to accept the invitation to deliver the first course of these lectures, it was because I felt convinced that the ancient literature of India, which has been preserved to us as by a miracle, gives us opportunities for a study of the origin and growth of religion such as we find nowhere else;[25] and, I may add, because I know from past experience, how great indulgence is shown by an English audience to one who, however badly he may say it, says all he has to say, without fear, without favour, and, as much as may be, without offence.

Notes

1. Herder, *Ideen zur Geschichte der Menschheit*, 9. Buch, p. 130 (ed. Brockhaus).
2. See *Heracliti Ephesii Reliquiæ*, ed. Bywater, p. 57, l.18, from *Vita Heracliti e Diogene Laertio*, ix. 1. Mr. Bywater places the saying τὴν τε οἴησιν ἱερὰν νόσον ἔλεγε among the Spuria, p. 51. It seems to me to have the full, massive, and noble ring of Herakleitos. It is true that οἴησις means rather opinion and prejudice in general than religious belief, but to the philosophical mind of Herakleitos the latter is a subdivision only of the former. Opinion in general might be called a disease, but hardly a sacred disease, nor can sacred disease be taken here either in the sense of great and fearful disease, or in the technical sense of epilepsy. If I am wrong, I share my error with one of the best Greek scholars and mythologists, for Welcker takes the words of Herakleitos in the same sense in which I have taken them. They are sometimes ascribed to Epikouros; anyhow they belong to the oldest wisdom of Greece.
3. Lange, *Geschichte des Materialismus*, i, 4.
4. See E. Curtius, "Über die Bedeutung von Delphi für die Griechische Cultur," Festrede am 22 Februar, 1878.
5. *Heracliti Reliquiæ*, cxi, cxxvi.
6. Ibid., xx.
7. *Rĕligio*, if it was derived from *rĕ-legere*, would have meant originally gathering again, taking up again, considering carefully. Thus *dī-ligo* meant originally to gather, to take up from among other things: then to esteem, to love. *Negligo* (nec-lego) meant not to take up, to leave unnoticed, to neglect. *Intelligo* meant to gather together with other things, to connect together, to arrange, classify, understand.
   Rĕlego occurs in the sense of taking back, gathering up (Ovid, *Met.*, 8.173): Janua, difficilis filo est inventa relecto, "The difficult door was found by the thread [of Ariadne], which was gathered up again." It is frequently used in the sense of travelling over the same ground: Egressi relegunt campos (Val. Fl., 8.121). In this meaning Cicero thinks that it was used, when applied to religion: Qui omnia quæ ad cultum deorum pertinerent diligenter retractarent et tamquam relegerent, sunt dicti religiosi ex relegendo, ut eleganter ex eligendo, tamquam a diligendo diligenter, ex intelligendo intelligenter: his enim in verbis omnibus inest vis legendi eadem quæ in religioso (Cic., *de Nat. Deor.*, 2, 28, 72), "People were called religious from relegere, because they went over again, as it were, and reconsidered carefully whatever referred to the worship of gods."

Relegere would therefore have meant originally much the same as *respicere, revereri,* which, from meaning to look back, came to mean to respect.

An ancient author quoted by Gellius (4.9) makes a distinction between *religiosus,* which he uses in the sense of superstitious, and *religens.* "Religentem esse oportet," he says, "religiosum nefas": it is right to be reverent, wrong to be religious, i.e., superstitious. The difficulty that rēligio has retained its long ē, being also written sometimes relligio (from red-ligio), is not even mentioned by Cicero. Lucretius uses both rēduco and rēlatum with a long ē.

Religio, used subjectively, meant conscientiousness, reverence, awe, and was not originally restricted to reverence for the gods. Thus we read: "Religione jurisjurandi ac metu deorum in testimoniis dicendis commoveri," "to be moved in giving evidence by the reverence for an oath, and by the fear of the gods" (C. Font., 9.20). Very soon, however, it became more and more restricted to reverence for the gods and divine things. People began to speak of a man's religion, meaning his piety, his faith in the gods, his observance of ceremonies, till at last an entire system of faith was called *religiones* or *religio.*

The other derivation of *religio* is supported by high authorities, such as Servius, Lactantius, St. Augustine, who derive it from *religare,* to bind up, to fasten, to moor. From this point of view *religio* would have meant originally what binds us, holds us back. I doubt whether with Pott (*Etym. Forsch.,* i, p. 201) we can say that such a derivation is impossible. No doubt, a noun like *religio* cannot be derived direct from a verb of the first conjugation, such as *religare.* That would give us *religatio,* just as *obligare* gives us *obligatio.* But verbs of the first conjugation are themselves derivatives, and many of them exist by the side of words derived from their more simple roots. Thus by the side of *opinari,* we have *opinio* and *necopinus*; by the side of *rebellare, rebellis* and *rebellio.* Ebel (Kuhn's *Zeitschrift,* iv, p. 144) points out that by the side of *ligare,* we have *lictor,* originally a binder, and that, therefore, *religio* from *religare* could be defended, at all events, grammatically. I believe that is so. Still there is no trace of *religare* having been used by the Romans themselves in the sense of restraining, still less of revering or fearing, and these after all are the original meanings in which *religio* first appears in Latin. Ebel thinks that *lex, leg-is,* is likewise derived from *ligare,* like *jus,* from Sanskrit *yu,* to join. The Oscan *lig-ud, lêge,* might seem to confirm this. But Lottner's comparison of *lex,* with the Old N. *lög,* Eng. *law,* what is laid down, and is settled (*Gesetz* in German), deserves consideration (see Curtius, *Griech. Etymologie,* i, p. 367), though it must be borne in mind that the transition of *h* and χ into *g* is irregular.

8. "Religion ist (subjectiv betrachtet) das Erkenntniss aller unserer Pflichten als göttlicher Gebote."—Religion innerhalb der Grenzen der blossen Vernunft, iv, 1; *Werke* (ed. Rosenkranz), p. 183.

9. See Kant, loc. cit., p. 183.

10. *Memorie Storiche dell' Australia, particolarmente della Missione Benedettina di Nuova Norcia, e degli usi e costumi degli Australiani,* per Mgr. D. Rudesindo Salvado, O.S.B., Vescovo di Porto Vittoria. Roma, Tip. S. Cong. de Prop. Fide, 1851.

11. *Ethnography and Philology of the Hidatsa Indians.* By Washington Matthews. Washington, 1877, p. 48.

12. Loc. cit., iv. 2, pp. 205, 208.

13. This is, of course, a very imperfect account of Schleiermacher's view of religion, which became more and more perfect as he advanced in life. See on this point the excellent *Life of Schleiermacher,* by W. Dilthey, 1870.

14. Feuerbach, *Wesen der Religion,* p. 100.

15. *Introduction to the Science of Religion,* Lecture I, 1882, p. 13.

16. Instead of slaying the slain over again, I quote the following words of Locke, *On the Understanding,* Book ii. c. 21.17: "For if it be reasonable to suppose and talk of faculties as distinct beings, that can act (as we do, when we say the will orders, and the will is free), it is fit that we should make a speaking faculty, and a walking faculty, and a dancing

faculty, by which those actions are produced, which are but several modes of motion; as well as we make the will and understanding to be faculties by which the actions of choosing and perceiving are produced, which are but several modes of thinking; and we may as properly say, that it is the singing faculty sings, and the dancing faculty dances, as that the will chooses, or that the understanding conceives; or, as is usual, that the will directs the understanding, or the understanding obeys, or obeys not, the will; it being altogether as proper and intelligible to say, that the power of speaking directs the power of singing, or the power of singing obeys, or disobeys the power of speaking. This way of talking, nevertheless, has prevailed, and, as I guess, produced great confusion."

"In einem Dialog sollte einmal recht persiflirt werden, wie die Leute von einzelnen Seelenvermögen reden, z. B., Kant: die reine Vernunft schmeichelt sich."—*Schleiermacher,* von Dilthey, vol. 1, p. 122.

17. One of the first who pointed out the uncertainty of the foundation on which Kant attempted to reconstruct religion, in the widest sense of the word, was Wyttenbach, *Opusc.* ii, p. 190. See Prantl, *Sitzungsberichte der philos. philolog. und historischen Classe der K. B. Akademie der Wissenschaften,* 1877, p. 284.

18. Hobbes, *Leviathan,* i. 3.

19. *Meteor.* iii. 2.5.

20. See a very remarkable paper, "Über den Farbensinn der Urzeit und seine Entwickelung," by L. Geiger in his *Vorträge zur Entwickelungsgeschichte der Menschheit,* 1871, p. 45. The same subject is treated again in his *Ursprung und Entwickelung der menschlichen Sprache und Vernunft,* Zweiter Band, p. 304 seq.

21. See Victor von Strauss, "Bezeichnung der Farben Blau und Grün im Chinesischen Alterthum," in the *Zeitschrift der D.M.G.,* 1879, p. 502.

22. See Meyer, "Über die Mafoor'sche und einige andern Papúa-Sprachen," p. 52: "Blau, peisim, wird nicht von schwarz unterschieden." *Lectures on the Science of Language,* ii, p. 343.

23. Translation: "And do not say that the infinite and the eternal are unintelligible; it is the end and the passing passenger whom we are often tempted to take for a dream; because thought cannot see an end to anything and the being could not conceive of nothingness. We cannot go further deeper into the natural sciences themselves, without encountering the infinite and the eternal; and the most positive things belong as much, under certain circumstances, to the infinite and to the eternal as they do to the emotions and the imagination"—Sondra Bacharach.

24. *Critik der reinen Vernunft,* 2te Auflage, Vorr; 2. 676. What Kant says in his *Critik,* 1te Auflage, pp. 288, 289, is less distinct and liable to be misunderstood.

25. "Die Inder bildeten ihre Religion zu einer Art von urweltlicher Classicität aus, welche sie für alle Zeiten zum Schlüssel des Götterglaubens der ganzen Menschheit macht." Geiger, *Über Ursprung und Entwickelung der menschlicher Sprache und Vernunft,* vol. 2, p. 339.

CHAPTER 10

IS FETISHISM A PRIMITIVE FORM OF RELIGION?

from *Lectures on the Origin and Growth of Religion* (1878)

The First Impulse to the Perception of the Infinite

In my first lecture I tried to lay open the foundations on which alone a religion can be built up. If man had not the power—I do not say, to comprehend, but to apprehend the infinite, in its most primitive and undeveloped form—he would indeed have no right to speak of a world beyond this finite world, of time beyond this finite time, or of a Being which, even though he shrinks from calling it Zeus, or Jupiter, or Dyaus-pitar, or Lord, Lord, he may still feel after, and revere, and even love, under the names of the Unknown, the Incomprehensible, the Infinite. If, on the contrary, an apprehension of the infinite is possible and legitimate, if I have succeeded in showing that this apprehension of the infinite underlies and pervades all our perceptions of finite things, and likewise all the reasonings that flow from them, then we have firm ground to stand on, whether we examine the various forms which that sentiment has assumed among the nations of antiquity, or whether we sound the foundations of our own faith to its lowest depth.

The arguments which I placed before you in my first lecture were, however, of a purely abstract nature. It was the possibility, not the reality, of the perception of the infinite which alone I wished to establish. Nothing could be further from my thoughts than to represent the perfect idea of the infinite as the first step in the historical evolution of religious ideas. Religion begins as little with the perfect idea of the infinite as astronomy begins with the law of gravity; nay, in its purest form, that idea is the last rather than the first step in the march of the human intellect.

Mana, a Melanesian Name for the Infinite

How the idea of the infinite, of the unseen, or as we call it afterwards, the Divine, may exist among the lowest tribes in a vague and hazy form we may see, for instance, in the *Mana* of the Melanesians. Mr. R. H. Codrington, an experienced missionary and a thoughtful theologian, says in a letter, dated July 7, 1877, from Norfolk Island: "The religion of the Melanesians consists, as far as belief goes, in the persuasion that there is a supernatural power about, belonging to the region of the unseen; and, as far as practice goes, in the use of means of getting this power turned to their own benefit. The notion of a Supreme Being is altogether foreign to them, or indeed of any Being occupying a very elevated place in their world" (p. 14).

And again:

> There is a belief in a force altogether distinct from physical power, which acts in all kinds of ways for good and evil, and which it is of the greatest advantage to possess or control. This is Mana. The word is common, I believe, to the whole Pacific, and people have tried very hard to describe what it is in different regions. I think I know what our people mean by it, and that meaning seems to me to cover all that I hear about it elsewhere. It is a power or influence, not physical, and, in a way, supernatural; but it shows itself in physical force, or in any kind of power or excellence which a man possesses. This Mana is not fixed in anything, and can be conveyed in almost anything; but spirits, whether disembodied souls or supernatural beings, have it, and can impart it; and it essentially belongs to personal beings to originate it, though it may act through the medium of water, or a stone, or a bone. All Melanesian religion, in fact, consists in getting this Mana for one's self, or getting it used for one's benefit—all religion, that is, as far as religious practices go, prayers and sacrifices.

This Mana is one of the early, helpless expressions of what the apprehension of the infinite would be in its incipient stages, though even the Melanesian Mana shows ample traces both of development and corruption.

My first lecture, therefore, was meant to be no more than a preliminary answer to a preliminary assertion. In reply to that numerous and powerful class of philosophers who wish to stop us on the very threshold of our inquiries, who tell us that here on earth there is no admission to the infinite, and that, if Kant has done anything, he has for ever closed our approaches to it, we had to make good our right by producing credentials of the infinite, both within and without the finite, which even the most positive of positivists has to recognise, viz., the evidence of our senses.

We have now to enter upon a new path: we have to show how men in different parts of the world worked their way in different directions, step by step, from the simplest perceptions of the world around them, to the highest concepts of religion and philosophy; how, in fact, the consciousness of the infinite, which lay hidden in every fold of man's earliest impressions, was unfolded in a thousand different ways, till it became freer and freer of its coarser ingredients, reaching at last that point of purity which we imagine is the highest that can be reached by human thought. The history of that development is neither more nor less than the history of religion, closely connected, as that history always has been and must be, with the history of philosophy. To that history we now turn, as containing the only trustworthy illustration of the evolution of the idea of the infinite from the lowest beginnings to a height which few can reach, but to which we may all look up from the nether part of the mount.

Fetishism, the Original Form of All Religion

If you consulted any of the books that have been written during the last hundred years on the history of religion, you would find in most of them a striking agreement on at least one point, viz., that the lowest form of what can be called religion is *fetishism,* that it is impossible to imagine anything lower that would still deserve that name, and that therefore fetishism may safely be considered as the very beginning of all religion. Whenever I find so flagrant an instance of agreement, the same ideas expressed in almost the same words, I confess I feel suspicious, and I always think it right to go back to the first sources, in order to see under what circumstances, and for what special purpose, a theory which commands such ready and general assent has first been started.

De Brosses, the Inventor of Fetishism

The word *fetishism* was never used before the year 1760. In that year appeared an anonymous book called *Du Culte des Dieux Fétiches, ou, Parallèle de l'ancienne Religion de l'Egypte avec la Religion actuelle de Nigritie.* It is known that this little book was written by Charles De Brosses (1709–1777), the well-known President De Brosses, the correspondent of Voltaire, one of the most remarkable men of the Voltairian period. It was at the instigation of his friend, the great Buffon, that De Brosses seems to have devoted himself to the study of savage tribes, or to the study of man in historic and prehistoric times. He did so by collecting the best descriptions which he could find in the books of old and recent travellers, sailors, missionaries, traders, and explorers of distant countries, and he published in 1756 his *Histoire des navigations aux terres Australes,* two large volumes in quarto. Though this book is now antiquated, it contains two names which, I believe, occur here for the first time, which were, it seems, coined by De Brosses himself, and which will probably survive when all his other achievements, even his theory of fetishism, have been forgotten, viz., the names *Australia* and *Polynesia.*

Another book by the same author, more often quoted than read, is his *Traité de la Formation mécanique des Langues,* published in 1765. This is a work which, though its theories are likewise antiquated, well deserves a careful perusal even in these heydays of comparative philology, and which, particularly in its treatment of phonetics, was certainly far in advance of its time.

Between his book on Eastern Voyages and his treatise on the Mechanical Formation of Language, lies his work on the Worship of the Fetish Deities, which may rightly be described as an essay on the mechanical formation of religion. De Brosses was dissatisfied with the current opinions on the origin of mythology and religion, and he thought that his study of the customs of the lowest savages, particularly those on the west coast of Africa, as described by Portuguese sailors, offered him the means of a more natural explanation of that old and difficult problem: "The confused mass of ancient mythology," he says,

> has been to us an undecipherable chaos, or a purely arbitrary riddle, so long as one employed for its solution the *figurism* of the last Platonic philosophers, who ascribed to ignorant and savage nations a knowledge of the most hidden causes of nature, and perceived in a heap of trivial practices of gross and stupid people intellectual ideas of the most abstract metaphysics. Nor have they fared better who tried, mostly by means of forced and ill-grounded comparisons, to find in the ancient mythology the detailed, though disfigured, history of the Hebrew nation, a nation that was unknown almost to all others, and made a point never to communicate its doctrines to strangers.... Allegory is an instrument which will do anything. The system of a figurative meaning once admitted, one soon sees everything in everything, as in the clouds. The matter is never embarrassing, all that is wanted is spirit and imagination. The field is large and fertile, whatever explications may be required.

"Some scholars," he continues,

more judicious, better instructed also in the history of the early nations whose colonies first discovered the West, and familiar with Oriental languages, have at last, after clearing mythology of the rubbish with which the Greeks had covered it, found the true key of it in the actual history of the early nations, their opinions and their rulers, in the false translations of a number of simple expressions, the meaning of which had been forgotten by those who nevertheless continued to use them; and in the homonymies which out of one object, designated by various epithets, have made so many different beings or persons.

But these keys which open so well the meaning of historical fables, do not always suffice to give a reason for the singularity of the dogmatic opinions, nor of the practical rites of the early nations. These two portions of heathen theology depend either on the worship of the celestial bodies, well known by the name of *Sabeism,* or on the probably not less ancient worship of certain terrestrial and material objects, called *fétiche,* among the African negroes (he meant to say, by those who visited the African negroes), and which for this reason I shall call *Fétichisme.* I ask permission to use this term habitually, and though in the proper signification it refers in particular to the religion of the negroes of Africa only, I give notice beforehand that I mean to use it with reference also to any other nation paying worship to animals, or to inanimate things which are changed into gods, even when these objects are less gods, in the proper sense of the word, than things endowed with a certain divine virtue, such as oracles, amulets, or protecting talismans. For it is certain that all these forms of thought have one and the same origin, which belongs to one general religion, formerly spread over the whole earth, which must be examined by itself, constituting, as it does, a separate class among the various religions of the heathen world.

De Brosses divides his book into three parts. In the first he collects all the information which was then accessible on fetishism, as still practised by barbarous tribes in Africa and other parts of the world. In the second he compares it with the religious practices of the principal nations of antiquity. In the third he tries to show that, as these practices are very like to one another in their outward appearance, we may conclude that their original intention among the negroes of to-day and among the Egyptians, the Greeks, and Romans, was the same.

All nations, he holds, had to begin with fetishism, to be followed afterwards by polytheism and monotheism.

One nation only forms with him an exception—the Jews, the chosen people of God. They, according to De Brosses, were never fetish-worshippers, while all other nations first received a primeval divine revelation, then forgot it, and then began again from the beginning—viz., with fetishism.

It is curious to observe the influence which the prevalent theological ideas of the time exercised even on De Brosses. If he had dared to look for traces of fetishism in the Old Testament with the same keenness which made him see fetishes in Egypt, in Greece, in Rome, and everywhere else, surely the Teraphim, the Urim and Thummim, or the ephod, to say nothing of golden calves and brazen serpents, might have supplied him with ample material (Gen. 28.18; Jer. 2.27).

But though on this and some other points those who have more recently adopted and defended the theory of De Brosses would differ from him, on the whole his view of fetishism has been maintained intact during the last hundred years. It sounded so easy, so natural, so plausible, that it soon found its way into manuals and school-books, and I believe we all of us have been brought up on it.[1] I myself certainly held it for a long time, and never doubted it, till I became more and more startled by the fact that, while in the earliest accessible documents of religious thought we look in vain for any very clear traces of fetishism, they become more and more frequent everywhere in the later stages of religious development, and are certainly more visible in the later corruptions of the Indian religion,[2] beginning with the Âtharvana, than in the earliest hymns of the Rig-veda.

Origin of the Name of Fetish

Why did the Portuguese navigators,—who were Christians, but Christians in that metamorphic state which marks the popular Roman Catholicism of the last century—why did they recognise at once what they saw among the Negroes of the Gold Coast as *feitiços?* The answer is clear. Because they themselves were perfectly familiar with a *feitiço,* an amulet or talisman; and probably all carried with them some beads, or crosses, or images, that had been blessed by their priests before they started for their voyage. They themselves were fetish-worshippers in a certain sense. What was more natural therefore for them, if they saw a native hugging some ornament, or unwilling to part with some glittering stone, or it may be prostrating himself and praying to some bones, carefully preserved in his hut, than to suppose that the Negroes did not only keep these things for luck, but that they were sacred relics, something in fact like what they themselves would call *feitiço?* As they discovered no other traces of any religious worship, they concluded very naturally that this outward show of regard for these *feitiços* constituted the whole of the Negro's religion.

Suppose these Negroes, after watching the proceedings of their white visitors, had asked on their part what the religion of those white men might be, what would they have said? They saw the Portuguese sailors handling their rosaries, burning incense to dauby images, bowing before altars, carrying gaudy flags, prostrating themselves before a wooden cross. They did not see them while saying their prayers, they never witnessed any sacrifices offered by them to their gods, nor was their moral conduct such as to give the natives the idea that they abstained from any crimes, because they feared the gods. What would have been more natural therefore for them than to say that their religion seemed to consist in a worship of *gru-grus,* their own name for what the Portuguese called *feitiço,* and that they had no idea of a supreme spirit or a king of heaven, or offered any worship to him?

With regard to the word, it is well known that the Portuguese *feitiço* corresponds to Latin *factitius. Factitius,* from meaning what is made by hand, came to mean artificial, then unnatural, magical, enchanted, and enchanting. A false key is called in Portuguese *chave feitiça,* while *feitiço* becomes the recognised name for amulets and similar half-sacred trinkets. The trade in such articles was perfectly recognised in Europe during the Middle Ages, as it is still among the Negroes of Africa. A manufacturer or seller of them was called *feitiçero,* a word which, however, was likewise used in the sense of a magician or conjurer. How common the word was in Portuguese we see from its being used in its diminutive form as a term of endearment, *meu feitiçinho* meaning my little fetish, or darling.

We see a similar transition of meaning in the Sanskrit *kriyâ*, the Italian *fattura*, incantation, which occurs in mediæval Latin as far back as 1311;[3] also in *charme*, which was originally no more than *carmen*; and in the Greek ἐπῳδή.

Wrong Extension of the Name Fetish

It will be clear from these considerations that the Portuguese sailors—for it is to them that we are indebted for the introduction of the word *fetish*—could have applied that term to certain tangible and inanimate objects only, and that it was an unwarrantable liberty taken with the word which enabled De Brosses to extend it to animals, and to such things as mountains, trees, and rivers. De Brosses imagined that the name *feitiço* was somehow related to *fatum,* and its modern derivative *fata* (nom. plur. of the neuter, used afterwards as a nom. sing. of the feminine), a *fée,* fairy; and this may have made it appear less incongruous to him to apply the name of fetish, not only to artificial and material objects, but also to trees, mountains, rivers, and even to animals. This was the first unfortunate step on the part of De Brosses, for he thus mixed up three totally distinct phases of religion, first, physiolatry, or the worship paid to natural objects which impress the mind of man with feelings of awe or gratitude, such as rivers, trees, or mountains; secondly, zoolatry, or the worship paid to animals, as for instance by the highly-cultivated inhabitants of ancient Egypt; and lastly, fetishism proper, or the superstitious veneration felt and testified for mere rubbish, apparently without any claim to such distinction.

But this was not all. De Brosses did not keep what he calls fetish-worship distinct even from idolatry, though there is a very important distinction between the two. A fetish, properly so called, is itself regarded as something supernatural; the idol, on the contrary, was originally meant as an image only, a similitude or a symbol of something else. No doubt an idol was apt to become a fetish; but in the beginning, fetish worship, in the proper sense of the word, springs from a source totally different from that which produces idolatry.

Let us hear how De Brosses explains his idea of a fetish: "These fetishes," he says,

> are anything which people like to select for adoration—a tree, a mountain, the sea, a piece of wood, the tail of a lion, a pebble, a shell, salt, a fish, a plant, a flower, certain animals, such as cows, goats, elephants, sheep, or anything like these. These are the gods of the negro, sacred objects, talismans. The negroes offer them worship, address their prayers to them, perform sacrifices, carry them about in procession, consult them on great occasions. They swear by them, and such oaths are never broken.
>
> There are fetishes belonging to a whole tribe, and others belonging to individuals. National fetishes have a kind of public sanctuary; private fetishes are kept in their own place in the houses of private individuals.
>
> If the negroes want rain, they place an empty jar before the fetish. When they go to battle, they deposit their weapons before it or him. If they are in want of fish or meat, bare bones are laid down before the fetish; while, if they wish for palm-wine, they indicate their desire by leaving with the fetish the scissors with which the incisions are made in the palm-trees.[4] If their prayers are heard, all is right. But if they are refused, they think that they have somehow incurred the anger of their fetish, and they try to appease him.

Such is a short abstract of what De Brosses meant by fetishism, what he believed the religion of the Negroes to be, and what he thought the religion of all the great nations of antiquity must have been before they reached the higher stages of poly-theism and monotheism.

Usefulness of the Study of Savage Tribes

The idea that, in order to understand what the so-called civilised people may have been before they reached their higher enlightenment, we ought to study savage tribes, such as we find them still at the present day, is perfectly just. It is the lesson which geology has taught us, applied to the stratification of the human race. But the danger of mistaking metamorphic for primary igneous rocks is much less in geology than in anthropology. Allow me to quote some excellent remarks on this point by Mr. Herbert Spencer. "To determine," he writes,

> what conceptions are truly primitive, would be easy if we had accounts of truly primitive men. But there are sundry reasons for suspecting that existing men of the lowest types, forming social groups of the simplest kinds, do not exemplify men as they originally were. Probably most of them, if not all, had ancestors in higher states; and among their beliefs remain some which were evolved during those higher states. While the degradation theory, as currently held, is untenable, the theory of progression, taken in its unqualified form, seems to me untenable also. If, on the one hand, the notion that savagery is caused by lapse from civilisation is irreconcilable with the evidence, there is, on the other hand, inadequate warrant for the notion that the lowest savagery has always been as low as it is now. It is quite possible, and, I believe, highly probable, that retrogression has been as frequent as progression.[5]

These words contain a most useful warning for those ethnologists who imagine that they have only to spend a few years among Papuas, Fuegians, or Andaman Islanders, in order to know what the primitive ancestors of the Greeks and Romans may have been. They speak of the savage of to-day as if he had only just been sent into the world, forgetting that, as a living species, he is probably not a day younger than we ourselves.[6] He may be a more stationary being, but he may also have passed through many ups and downs before he reached his present level. Anyhow, even if it could be proved that there has been a continuous progression in everything else, no one could maintain that the same applies to religion.

Frequent Retrogression in Religion

That religion is liable to corruption is surely seen again and again in the history of the world. In one sense the history of most religions might be called a slow corruption of their primitive purity. At all events, no one would venture to maintain that religion always keeps pace with general civilisation. Even admitting therefore that, with regard to their tools, their dress, their manners and customs, the Greeks and Romans, the Germans and Celts may have been before the first dawn of history in the same state in which we find some of the Negro races of Africa at present, nothing would justify the conclusion that their religion also must have been the same, that they must have worshipped fetishes, stocks and stones, and nothing else.

We see Abraham, a mere nomad, fully impressed with the necessity of the unity of the godhead, while Solomon, famous among the kings of the earth, built high places for Chemosh and Moloch. Ephesus, in the sixth century before Christ, was listening to one of the wisest men that Greece ever produced, Herakleitos; while a thousand years later, the same town resounded with the frivolous and futile wranglings of Cyrillus, and the council of Ephesus. The Hindus, who, thousands of years ago, had reached in the Upanishads the loftiest heights of philosophy, are now in some places sunk into a grovelling worship of cows and monkeys.

Difficulty of Studying the Religion of Savages

But there is another and even greater difficulty. If we feel inclined to ascribe to the ancestors of the Greeks and Romans the religion of the Negroes and of other savages of the present day, have we seriously asked ourselves what we really know of the religious opinions of these so-called savages?

A hundred years ago there might have been some excuse for people speaking in the most promiscuous manner of the religion of savages. Savages were then looked upon as mere curiosities, and almost anything related of them was readily believed. They were huddled and muddled together much in the same manner as I have heard Neander and Strauss quoted from the pulpit, as representatives of German neology; and hardly any attempt was made to distinguish between Negro and Negro, between savage and savage.

At present, all such general terms are carefully avoided by scientific ethnologists. In ordinary parlance we may still use the name of Negro for black people in general, but when we speak scientifically, Negro is mostly restricted to the races on the west coast of Africa between the Senegal and the Niger, extending inland to the lake of Tchad and beyond, we hardly know how far. When the Negro is spoken of as the lowest of the low, it generally is this Negro of the west coast that is intended, he from whom Europeans first took their idea of a fetish-worship.

It is not the place here to discuss the ethnography of Africa as it has been established by the latest travellers. The classification as given by Waitz will suffice to distinguish the Negro of the Senegal and Niger from his nearest neighbours:

First, the Berber and Copt tribes, inhabiting the north of Africa. For historical purposes they may be said to belong to Europe rather than to Africa. These races were conquered by the Mohammedan armies, and rapidly coalesced with their conquerors. They are sometimes called Moors, but never Negroes.

Secondly, the races which inhabit Eastern Africa, the country of the Nile to the equator. They are Abyssinian or Nubian, and in language distantly allied to the Semitic family.

Thirdly, the Fulahs, who are spread over the greater part of Central Africa, and feel themselves everywhere as distinct from the Negroes.

Fourthly, from the equator downward as far as the Hottentots, the Kaffer and Congo races, speaking their own well-defined languages, possessed of religious ideas of great sublimity, and physically also very different from what is commonly meant by a Negro.

Lastly, the Hottentots and the Bushmen, differing from the rest, both by their language and their physical appearance.

These are only the most general divisions of the races which now inhabit Africa. If we speak of all of them simply as Negroes, we do so in the same loose manner in which the Greeks spoke of Scythians, and the Romans, before Caesar, of Celts. For scientific purposes the term Negro should either be avoided altogether, or restricted to the races scattered over about twelve degrees of latitude, from the Senegal to the Niger, and extending inland to the as-yet undefined regions where they are bounded by Berber, Nubian, and Kaffer tribes.

But though the ethnologist no longer speaks of the inhabitants of Africa as Negroes or "niggers," it is much more difficult to convince the student of history that these races cannot be lumped together as savages, but that here, too, we must distinguish before we can compare. People who talk very freely of savages, whether in Africa, or America, or Australia, would find it extremely difficult to give any definition of that term, beyond this, that savages are different from ourselves. Savages with us are still very much what barbarians were to the Greeks. But as the Greeks had to learn that some of these so-called barbarians possessed virtues which they might have envied themselves, so we also shall have to confess that some of these savages have a religion and a philosophy of life which may well bear comparison with the religion and philosophy of what we call the civilised and civilising nations of antiquity. Anyhow, the common idea of a savage requires considerable modification and differentiation, and there is perhaps no branch of anthropology beset with so many difficulties as the study of these so-called savage races.

Language of Savages

Let us examine a few of the prejudices commonly entertained with regard to these so-called savages. Their languages are supposed to be inferior to our own. Now here the science of language has done some good work. It has shown, first of all, that no human beings are without language, and we know what that implies. All the stories of tribes without language, or with languages more like the twitterings of birds than the articulate sounds of human beings, belong to the chapter of ethnological fables.

What is more important still is that many of the so-called savage languages have been shown to possess a most perfect, in many cases too perfect, that is to say, too artificial a grammar, while their dictionary possesses a wealth of names which any poet might envy.[7] True, this wealth of grammatical forms[8] and this superabundance of names for special objects are, from one point of view, signs of logical weakness and of a want of powerful generalisation. Languages which have cases to express nearness to an object, movement alongside an object, approach towards an object, entrance into an object, but which have no purely objective case, no accusative, may be called rich, no doubt, but their richness is truly poverty. The same applies to their dictionary. It may contain names for every kind of animal; again for the same animal when it is young or old, male or female; it may have different words for the foot of a man, a horse, a lion, a hare; but it probably is without a name for animal in general, or even for such concepts as member or body. There is here, as elsewhere, loss and gain on both sides. But however imperfect a language may be in one point or other, every language, even that of Papuas and Veddas, is such a masterpiece of abstract thought that it would baffle the ingenuity of many philosophers to produce anything like it. In several cases the grammar of so-called savage dialects bears evidence of a far higher state of mental culture possessed by these people in former times. And it must not be forgotten that every language has capacities, if they are only called out, and that no language has yet been found into which it was not possible to translate the Lord's Prayer.

Numerals of Savages

For a long time it was considered as the strongest proof of the low mental capacity of certain savages that they were unable to count beyond three or four or five. Now, first of all we want a good scholar to vouch for such facts when they exist;[9] but when they have been proved to exist, then let us begin to distinguish. There may be tribes by whom everything beyond five, beyond the fingers of one hand, is lumped together as many, though I confess I have grave doubts whether, unless they are idiots, any human beings could be found unable to distinguish between five, six, and seven cows.

But let us read the accounts of the absence of numerals beyond two or three more accurately. It was said, for instance, that the Abipones have no numbers beyond *three*.[10] What do we really find? That they express *four* by *three* plus *one*. Now this, so far from showing any mental infirmity, proves in reality a far greater power of analysis than if four were expressed, say, by a word for hands and feet, or for eyes and ears. Savages who expressed *four* by *two-two*, would never be in danger of considering the proposition that two and two make four as a synthetic judgment *à priori*; they would know at once that in saying "two and two make two-two," they were simply enunciating an analytical judgment.

We must not be too eager to assert the mental superiority of the races to which we ourselves belong. Some very great scholars have derived the Aryan word for *four* (whether rightly or wrongly I do not ask), the Sanskrit *ka-tur*, the Latin *quatuor*, from three, *tar*, preceded by *ka*, the Latin *que*, so that *katur*, in Sanskrit too, would have been conceived originally as *one* plus *three*.[11] If some African tribes express *seven* either by *five* plus *two* or *six* plus *one*,[12] why should this stamp them as the lowest of the low, whereas no one blames the French, marching at the head of European civilisation, for expressing ninety by *quatre-vingt-dix*, fourscore ten, or the Romans for saying *undeviginti* for nineteen?[13]

No; here too we must learn to mete to others that measure which we wish to be measured to us again. We must try to understand, before we presume to judge.

No History among Savages

Another serious charge brought against the savage in general is that he has no history. He hardly counts the days of a year, still less the years of a life. Some Negro tribes consider it wrong to do so, as showing a want of trust in God. As they have no knowledge of writing, there is of course no trace of what we call history among them.[14] I do not deny that an utter carelessness about the past and the future would be a sign of a low stage of culture; but this can by no means be charged against all so-called savages. Many of them remember the names and deeds of their fathers and grandfathers, and the marvel is that, without the power of writing, they should have been able to preserve their traditions, sometimes for many generations.

The following remarks, from a paper by the Rev. S. J. Whitmee, throw some curious light on this subject:

The keepers of these national traditions (among the brown Polynesians) usually belonged to a few families, and it was their duty to retain intact, and transmit from generation to generation, the myths and songs entrusted to their custody. The honour of the families was involved in it. It was the hereditary duty of the elder sons of these families to acquire, retain, and transmit them with verbal accuracy. And it was not only a sacred duty, but the right of holding such myths and songs was jealously guarded as a valuable and honourable privilege. Hence the difficulty of having them secured by writing. Care was taken not to recite them too frequently or too fully at one time. Sometimes they have been purposely altered in order to lead the bearers astray. Missionaries and other foreign residents, who have manifested an interest in these myths, have often been deceived in this way. Only a person thoroughly familiar with the language, quite conversant with the habits of the people, and who had their confidence, could secure a trustworthy version. And this was usually secured only after a promise made to the keepers of these treasures not to make them public in the islands.

But notwithstanding these difficulties, some missionaries and others have succeeded in making large collections of choice myths and songs, and I am not without hope, that before very long we may succeed in collecting them together for the formation of a comparative mythology of Polynesia.

Most of these legends and songs contain archaic forms, both idioms and words, unknown to most of the present generation of the people.

The way in which verbal accuracy in the transmission of the legends and songs has been secured is worth mentioning. In some islands all the principal stories, indeed all which are of value, exist in two forms, in *prose* and in *poetry*. The prose form gives the story in simple language. The poetic gives it in rhythm, and usually in rhyme also. The poetic form is used as a check on the more simple and more easily changed prose form. As it is easy to alter and add to the prose account, that is never regarded as being genuine, unless each particular has its poetic tally. An omission or interpolation in the poetic form would, of course, be easily detected. Thus the people have recognised the fact that a poetic form is more easily remembered than a prose form, and that it is better adapted for securing the strict accuracy of historical myths.[15]

Our idea of history, however, is something totally different. To keep up the memory of the kings of Egypt and Babylon, to know by heart the dates of their battles, to be able to repeat the names of their ministers, their wives and concubines, is, no doubt, something very creditable in a competitive examination, but that it is a sign of true culture I cannot persuade myself to believe. Sokrates was not a savage, but I doubt whether he could have repeated the names and dates of his own archons, much less the dates of the kings of Egypt and Babylon.

And if we consider how history is made in our own time, we shall perhaps be better able to appreciate the feelings of those who did not consider that every massacre between hostile tribes, every palaver of diplomatists, every royal marriage-feast deserved to be recorded for the benefit of future generations. The more one sees of how history is made, the less one thinks that its value can be what it was once supposed to be. Suppose Lord Beaconsfield, Mr. Gladstone, and Prince Gortshakoff were to write the history of the last two years, what would future generations have to believe? What will future generations have to believe of those men themselves, when they find them represented by observers who had the best opportunity of judging them, either as high-minded patriots or as selfish partisans? Even mere facts, such as the atrocities committed in Bulgaria, cannot be described by two eyewitnesses in the same manner. Need we wonder, then, that a whole nation, I mean the old Hindus, simply despised history, in the ordinary sense of the word, and instead of burdening their memories with names and dates of kings, queens, and battles, cared more to remember the true sovereigns in the realm of thought, and the decisive battles for the conquest of truth?

No Morals among Savages

Lastly, all savages were supposed to be deficient in moral principles. I am not going to represent the savage as Rousseau imagined him, or deny that our social and political life is an advance on the hermit or nomadic existence of the tribes of Africa and America. But I maintain that each phase of life must be judged by itself. Savages have their own vices, but they also have their own virtues. If the Negro could write a black book against the white man, we should miss in it few of the crimes which we think peculiar to the savage. The truth is that the morality of the Negro and the white man cannot be compared, because their views of life are totally different. What we consider wrong, they do not consider wrong. We condemn, for instance, polygamy; Jews and Mohammedans tolerate it, savages look upon it as honourable, and I have no doubt that, in their state of society, they are right. Savages do not consider European colonists patterns of virtue, and they find it extremely difficult to enter into their views of life.

Nothing puzzles the mere savage more than our restlessness, our anxiety to acquire and to possess, rather than to rest and to enjoy. An Indian chief is reported to have said to a European: "Ah, brother, you will never know the blessings of doing nothing and thinking nothing; and yet, next to sleep, that is the most delicious. Thus we were before our birth, thus we shall be again after death."[16] The young girls in Tahiti, who were being taught weaving, very soon left the looms, and said, "Why should we toil? Have we not as many breadfruits and cocoa-nuts as we can eat? You who want ships and beautiful dresses must labour indeed, but we are content with what we have."[17]

Such sentiments are certainly very un-European, but they contain a philosophy of life which may be right or wrong, and which certainly cannot be disposed of by being simply called savage.

A most essential difference between many so-called savages and ourselves is the little store they set on life. Perhaps we need not wonder at it. There are few things that bind them to this life. To a woman or to a slave, in many parts of Africa or Australia, death must seem a happy escape, if only they could feel quite certain that the next life would not be a repetition of this. They are like children, to whom life and death are like travelling from one place to another; and as to the old people, who have more friends on the other side of the grave than on this, they are mostly quite ready to go; nay, they consider it even an act of filial duty that their children should kill them, when life becomes a burden to them. However unnatural this may seem to us, it becomes far less so if we consider that among nomads those who can travel no more must fall a prey to wild animals or starvation. Unless we take all this into account, we cannot form a right judgment of the morality and religion of savage tribes.

Religion Universal among Savages

At the time when De Brosses wrote, the wonder was that black people should possess anything that could be called morality or religion, even a worship of stocks and stones. We have learnt to judge differently, thanks chiefly to the labours of missionaries who have spent their lives among savages, have learnt their languages and gained their confidence, and who, though they have certain prejudices of their own, have generally done full justice to the good points in their character. We may safely say that, in spite of all researches, no human beings have been found anywhere who do not possess something which to them is religion; or, to put it in the most general form, a belief in something beyond what they can see with their eyes.

As I cannot go into the whole evidence for this statement, I may be allowed to quote the conclusions which another student of the science of religion, Professor Tiele, has arrived at on this subject, particularly as, on many points, his views differ widely from my own. "The statement," he says, "that there are nations or tribes which possess no religion rests either on inaccurate observations, or on a confusion of ideas. No tribe or nation has yet been met with destitute of belief in any higher beings, and travellers who asserted their existence have been afterwards refuted by facts. It is legitimate, therefore, to call religion, in its most general sense, an universal phenomenon of humanity."[18]

Study of the Religion of Literary Nations

When, however, these old prejudices had been removed, and when it had been perceived that the different races of Africa, America, and Australia could no longer be lumped together under the common name of savages, the real difficulties of studying these races began to be felt, more particularly with regard to their religious opinions. It is difficult enough to give an accurate and scholar-like account of the religion of the Jews, the Greeks, the Romans, the Hindus and Persians; but the difficulty of understanding and explaining the creeds and ceremonials of those illiterate races is infinitely greater. Any one who has worked at the history of religion knows how hard it is to gain a clear insight into the views of Greeks and Romans, of Hindus and Persians on any of the great problems of life. Yet we have here a whole literature before us, both sacred and profane, we can confront witnesses, and hear what may be said on the one side and the other. If we were asked, however, to say, whether the Greeks in general, or one race of Greeks in particular, and that race again at any particular time, believed in a future life, in a system of rewards and punishments after death, in the supremacy of the personal gods or of an impersonal fate, in the necessity of prayer and sacrifice, in the sacred character of priests and temples, in the inspiration of prophets and lawgivers, we should find it often extremely hard to give a definite answer. There is a whole literature on the theology of Homer, but there is anything but unanimity among the best scholars who have treated on that subject during the last two hundred years.

Still more is this the case when we have to form our opinions of the religion of the Hindus and Persians. We have their sacred books, we have their own recognised commentaries; but who does not know that the decision whether the ancient poets of the Rig-veda believed in the immortality of the soul, depends sometimes on the right interpretation of a single word, while the question whether the author of the Avesta admitted an original dualism, an equality between the principle of Good and Evil, has to be settled in some cases on purely grammatical grounds?[19]

Let me remind you of one instance only. In the hymn of the Rig-veda, which is to accompany the burning of a dead body, there occurs the following passage:

> May the eye go to the sun, the breath to the wind,
> Go to heaven and to the earth, as it is right;
> Or go to the waters, if that is meet for thee,
> Rest among the herbs with thy limbs.
>
> The unborn part—warm it with thy warmth,
> May thy glow warm it and thy flame!
> With what are thy kindest shapes, O Fire,
> Carry him away to the world of the Blessed! (X.16, 3).

This passage has often been discussed, and its right apprehension is certainly of great importance. *Aga* means unborn, a meaning which easily passes into that of imperishable, immortal, eternal. I translate *ago bhâgah* by the unborn, the eternal part, and then admit a stop, in order to find a proper construction of the verse. But it has been pointed out that *aga* means also goat, and others have translated, "The goat is thy portion." They also must admit the same kind of aposiopesis, which no doubt is not very frequent in Sanskrit. It is perfectly true, as may be seen in the Kalpa-Sûtras, that sometimes an animal of the female sex was led after the corpse to the pile, and was burnt with the dead body. It was therefore called the *Anustaranî*, the covering. But, first of all, this custom is not general, as it probably would be, if it could be shown to be founded on a passage of the Veda. Secondly, there is actually a Sûtra that disapproves of this custom, because, as Kâtyâyana says, if the corpse and the animal are burnt together, one might in collecting the ashes confound the bones of the dead man and of the animal. Thirdly, it is expressly provided that this animal, whether it be a cow or a goat, must always be of the female sex. If therefore we translate, "the goat is thy share!," we place our hymn in direct contradiction with the tradition of the Sûtras. There is a still greater difficulty. If the poet really wished to say, "this goat is to be thy share," would he have left out the most important word, viz., thy? He does not say, "the goat is thy share," but only, "the goat share."

However, even if we retain the old translation, there is no lack of difficulties, though the whole meaning becomes more natural. The poet says, first, that the eye should go to the sun, the breath to the air, that the dead should return to heaven and earth, and his limbs rest among herbs. Everything therefore that was born, was to return to whence it came. How natural then that he should ask, what would become of the unborn, the eternal part of man. How natural that after such a question there should be a pause, and that then the poet should continue:

> Warm it with thy warmth!
> May thy glow warm it and thy flame!
> Assume thy kindest form, O Fire,
> Carry him away to the world of the Blessed!

Whom? Not surely the goat; not even the corpse, but the unborn, the eternal part of man.

It is possible, no doubt, and more than possible that from this passage by a very natural misunderstanding the idea arose that with the corpse a goat (*aga*) was to be burnt. We see in the Âtharvana, how eagerly the priests laid hold of that idea. We know it was owing to a similar misunderstanding that widows were burnt in India with their dead husbands, and that Yama, the old deity of the setting sun, was changed into a king of the dead, and lastly into the first of men who died. There are indeed vast distances beyond the hymns of the Veda, and many things even in the earliest hymns become intelligible only if we look upon them not as just arising, but as having passed already through many a metamorphosis.

This is only one instance of the numerous difficulties connected with a right understanding of a religion, even where that religion possesses a large literature. The fact, however, that scholars may thus differ, does not affect the really scientific character of their researches. They have to produce on either side the grounds for their opinions, and others may then form their own judgment. We are here on *terra firma*.

The mischief begins when philosophers, who are not scholars by profession, use the labours of Sanskrit, Zend, or classical scholars for their own purposes. Here, there is real danger. The same writers who, without any references, nay, it may be, without having inquired into the credibility of their witnesses, tell us exactly what Kaffers, Bushmen, and Hottentots believed on the soul, on death, on God and the world to come, seldom advance an opinion on the religion of Greeks, Romans, Persians, or Hindus, which a scholar would not at once challenge. Of this too I must give a few instances, not in a fault-finding spirit, but simply in order to point out a very real danger against which we ought all of us to guard most carefully in our researches into the history of religion.

There is no word more frequently used by the Brahmans than the word *Om*. It may stand for *avam*, and, like French *oui* for *hoc illud*, have meant originally Yes, but it soon assumed a solemn character, something like our *Amen*. It had to be used at the beginning, also at the end of every recitation, and there are few manuscripts that do not begin with it. It is even prescribed for certain salutations[20]; in fact, there were probably few words more frequently heard in ancient and modern India than *Om*. Yet we are told by Mr. H. Spencer that the Hindus avoid uttering the sacred name *Om*,[21] and this is to prove that semi-civilised races have been interdicted from pronouncing the names of their gods. It is quite possible that in a collective work, such as Dr. Muir's most excellent *Sanskrit Texts*, a passage may occur in support of such a statement. In the mystic philosophy of the Upanishads, *Om* became one of the principal names of the highest Brahman, and a knowledge of that Brahman was certainly forbidden to be divulged. But how different is that from stating that "by various semi-civilised races the calling of deities by their proper names has been interdicted or considered improper. It is so among the Hindus, who avoid uttering the sacred name *Om*; it was so with the Hebrews, whose pronunciation of the word Jehovah is not known for this reason; and Herodotus carefully avoids naming Osiris." The last statement again will surprise those who remember how it is Herodotus who tells us that, though Egyptians do not all worship the same gods, they all worship Isis and Osiris, whom they identify with Dionysus.[22]

Dr. Muir is no doubt perfectly right in saying that in some passages of the Veda "certain gods are looked upon as confessedly mere created beings,"[23] and that they,

like men, were made immortal by drinking soma. But this only shows how dangerous even such careful compilations as Dr. Muir's *Sanskrit Texts* are apt to become.
The gods in the Veda are called *agara* or *amartya*, immortal, in opposition to men,
who are *martya* or *mritu-bandhu*, mortal, and it is only in order to magnify the power
of soma, that this beverage, like the Greek ambrosia, is said to have conferred immortality on the gods. Nor did the Vedic poets think of their gods as what we mean by
"mere created beings," because they spoke of the dawn as the daughter of the sky, or
of Indra as springing from heaven and earth. At least we might say with much greater
truth that the Greeks looked upon Zeus as a mere created thing, because he was the
son of Kronos.

Again, what can be more misleading than, in order to prove that all gods were
originally mortals, to quote Buddha's saying, "Gods and men, the rich and poor,
alike must die?" In Buddha's time, nay, even before Buddha's time, the old Devas,
whom we choose to call gods, had been used up. Buddha believed in no Devas,
perhaps in no God. He allowed the old Devas to subsist as mere fabulous beings[24];
and as fabulous beings of much greater consequence than the Devas shared in the
fate of all that exists, viz., an endless migration from birth to death, and from death
to birth, the Devas could not be exempted from that common lot.

In forming an opinion of the mental capacities of people, an examination of their
language is no doubt extremely useful. But such an examination requires considerable care and circumspection. Mr. H. Spencer says, "When we read of an existing
South American tribe, that the proposition, 'I am an Abipone,' is expressible only in
the vague way—'I Abipone,' we cannot but infer that by such undeveloped grammatical structures only the simplest thoughts can be rightly conveyed."[25] Would not
some of the most perfect languages in the world fall under the same condemnation?

Study of the Religion of Savages

If such misunderstandings happen where they might easily be avoided, what shall we
think when we read broad statements as to the religious opinions of whole nations
and tribes who possess no literature, whose very language is frequently but imperfectly understood, and who have been visited, it may be, by one or two travellers only
for a few days, for a few weeks, or for a few years!

Let us take an instance. We are told that we may observe a very primitive state of
religion among the people of Fiji. They regard the shooting-stars as gods, and the
smaller ones as the departing souls of men. Before we can make any use of such a
statement, ought we not to know, first, what is the exact name and concept of god
among the Fijians; and secondly, of what objects besides shooting-stars that name is
predicated? Are we to suppose that the whole idea of the Divine which the Fijians
had formed to themselves is concentrated in shooting-stars? Or does the statement
mean only that the Fijians look upon shooting-stars as one manifestation out of
many of a Divine power familiar to them from other sources? If so, then all depends
clearly on what these other sources are, and how from them the name and concept
of something divine could have sprung.

When we are told that the poets of the Veda represent the sun as a god, we ask at
once what is their name for god, and we are told *deva*, which originally meant *bright*.

The biography of that single word *deva* would fill a volume, and not until we know its biography, from its birth and infancy to its very end, would the statement that the Hindus consider the sun as a *deva* convey to us any real meaning.

The same applies to the statement that the Fijians or any other races look upon shooting-stars as the departing souls of men. Are the shooting-stars the souls, or the souls the shooting-stars? Surely all depends here on the meaning conveyed by the word *soul*. How did they come by that word? What was its original intention? These are the questions which ethnological psychology has to ask and to answer, before it can turn with any advantage to the numerous anecdotes which we find collected in works on the study of man.

It is a well-known fact that many words for soul meant originally shadow. But what meaning shall we attach, for instance, to such a statement as that "Benin Negroes regard their shadows as their souls?" If soul is here used in the English sense of the word, then the Negroes could never believe their English souls to be no more than their African shadows. The question is, Do they simply say that *a* (shadow) is equal to *a* (shadow), or do they want to say that *a* (shadow) is equal to something else, viz., *b* (soul)? It is true that we also do not always see clearly what we mean by soul; but what we mean by it could never be the same as mere shadow only. Unless therefore we are told whether the Benin Negroes mean by their word for soul the *anima,* the breath, the token of life; or the *animus,* the mind, the token of thought; or the *soul,* as the seat of desires and passions; unless we know whether their so-called soul is material or immaterial, visible or invisible, mortal or immortal, the mere information that certain savage tribes look upon the shadow, or a bird, or a shooting-star as their soul seems to me to teach us nothing.

This was written before the following passage in a letter from the Rev. R. H. Codrington (dated July 3, 1877) attracted my attention, where that thoughtful missionary expresses himself in very much the same sense. "Suppose," he writes, "there are people who call the soul a shadow, I do not in the least believe they think the shadow a soul, or the soul a shadow; but they use the word shadow figuratively for that belonging to man, which is like his shadow, definitely individual, and inseparable from him, but unsubstantial. The Mota word we use for soul is in Maori a shadow, but no Mota man knows that it ever means that. In fact, my belief is, that in the original language this word did not definitely mean either soul or shadow, but had a meaning one can conceive but not express, which has come out in one language as meaning shadow, and in the other as meaning something like soul, i.e., second self."

What we must try to understand is exactly this transition of meaning, how from the observation of the shadow which stays with us by day and seems to leave us by night, the idea of a second self arose; how that idea was united with another, namely, that of breath, which stays with us during life, and seems to leave us at the moment of death; and how out of these two ideas the concept of a something, separate from the body and yet endowed with life, was slowly elaborated. Here we can watch a real transition from the visible to the invisible, from the material to the immaterial; but instead of saying that people, in that primitive stage of thought, believe their souls to be shadows, all we should be justified in saying would be that they believed that, after death, their breath, having left the body, would reside in something like the shadow that follows them during life.[26] The superstition that a dead body casts no shadow follows very naturally from this.

Nothing is more difficult than to resist the temptation to take an unexpected confirmation of any of our own theories, which we may meet with in the accounts of missionaries and travellers, as a proof of their truth. The word for God throughout Eastern Polynesia is *Atua* or *Akua*. Now *ata*, in the language of those Polynesian islanders, means shadow, and what would seem to be more natural than to see in this name of God, meaning originally shadow, a confirmation of a favourite theory, that the idea of God sprang everywhere from the idea of spirit, and the idea of spirit from that of shadow? It would seem mere captiousness to object to such a theory, and to advise caution where all seems so clear. Fortunately, the languages of Polynesia have in some instances been studied in a more scholarlike spirit, so that our theories must submit to being checked by facts. Thus Mr. Gill, who has lived twenty years at Mangaia, shows that *atua* cannot be derived from *ata*, shadow, but is connected with *fatu* in Tahitian and Samoan, and with *aitu*, and that it meant originally the core or pith of a tree. From meaning the core and kernel, *atu*, like the Sanskrit *sâra*, came to mean the best part, the strength of a thing, and was used at last in the sense of lord and master. The final *a* in *Atua* is intensive in signification, so that *Atua* expresses to a native the idea of the very core and life. This was the beginning of that conception of the deity which they express by *Atua*.[27]

When we have to deal with the evidence placed before us by a scholar like Mr. Gill, who has spent nearly all his life among one and the same tribe, a certain amount of confidence is excusable. Still even he cannot claim the same authority which belongs to Homer, when speaking of his own religion, or to St. Augustine, when giving us his interesting account of the beliefs of the ancient Romans. And yet, who does not know how much uncertainty is left in our minds after we have read all that such men have to say with regard to their own religion, or the religion of the community in the midst of which they grew up and passed the whole of their life!

The difficulties which beset travellers and missionaries in their description of the religious and intellectual life of savage tribes are far more serious than is commonly supposed, and some of them deserve to be considered before we proceed further.

Influence of Public Opinion on Travellers

First of all, few men are quite proof against the fluctuations of public opinion. There was a time when many travellers were infected with Rousseau's ideas, so that in their eyes all savages became very much what the Germans were to Tacitus. Then came a reaction. Partly owing to the influence of American ethnologists, who wanted an excuse for slavery, partly owing, at a later time, to a desire of finding the missing link between men and monkeys, descriptions of savages began to abound which made us doubt whether the Negro was not a lower creature than the gorilla, and whether he really deserved the name of man.

When it became a question much agitated, whether religion was an inherent characteristic of man or not, some travellers were always meeting with tribes who had no idea and name for gods[28]; others discovered exalted notions of religion everywhere. My friend Mr. Tylor has made a very useful collection of contradictory accounts given by different observers of the religious capacities of one and the same tribe. Perhaps the most ancient instance on record is the account given of the religion of the Germans by Caesar and Tacitus. Caesar states that the Germans count those only as gods whom they can perceive, and by whose gifts they are clearly benefited, such as the Sun, the Fire, and the Moon.[29] Tacitus declares "that they call by the names of gods that hidden thing which they do not perceive, except by reverence."[30]

It may, of course, be said that in the interval between Caesar and Tacitus the whole religion of Germany had changed, or that Tacitus came in contact with a more spiritual tribe of Germans than Caesar. But, granting that, do we always make allowance for such influences in utilising the accounts of early and later travellers?

Absence of Recognised Authorities among Savages

And even if we find a traveller without any scientific bias, free from any wish to please the leaders of any scientific or theological school, there remains, when he attempts to give a description of savage or half-savage tribes and their religion, the immense difficulty that not one of these religions has any recognised standards, that religion among savage tribes is almost entirely a personal matter, that it may change from one generation to another, and that even in the same generation the greatest variety of individual opinion may prevail with regard to the gravest questions of their faith. True, there are priests, there may be some sacred songs and customs, and there always is some teaching from mothers to their children. But there is no Bible, no prayer-book, no catechism. Religion floats in the air, and each man takes as much or as little of it as he likes.

We shall thus understand why accounts given by different missionaries and travellers of the religion of one and the same tribe should sometimes differ from each other like black and white. There may be in the same tribe an angel of light and a vulgar ruffian, yet both would be considered by European travellers as unimpeachable authorities with regard to their religion.

That there are differences in the religious convictions of the people is admitted by the Negroes themselves.[31] At Widah, Des Marchais was distinctly told that the nobility only knew of the supreme God as omnipotent, omnipresent, rewarding the evil and the good, and that they approached him with prayers, when all other appeals had failed. There is, however, among all nations, savage as well as civilised, another nobility—the divine nobility of goodness and genius—which often places one man many centuries in advance of the common crowd.

Think only what the result would be, if in England, the criminal drunkard and the sister of mercy who comes to visit him in his miserable den were both asked to give an account of their common Christianity, and you will be less surprised, I believe, at the discrepancies in the reports given by different witnesses of the creed of one and the same African tribe.

Authority of Priests

It might be said that the priests, when consulted on the religious opinions of their people, ought to be unimpeachable authorities. But is that so? Is it so even with us?

We have witnessed ourselves, not many years ago, how one of the most eminent theologians of this country declared that one whose bust now stands with those of Keble and Kingsley in the same chapel of Westminster Abbey, did not believe in the same God as himself! Need we wonder, then, if priests among the Ashantis differ as to the true meaning of their fetishes, and if travellers who have listened to different teachers of religion differ in the accounts which they give to us? In some parts of Africa, particularly where the influence of Mohammedanism is felt, fetishes and sellers of fetishes are despised. The people who believe in them are called *thiedos,* or infidels.[32] In other parts, fetish-worship rules supreme, and priests who manufacture fetishes and live by the sale of them shout very loudly, "Great is Diana of the Ephesians!"

Unwillingness of Savages to Talk of Religion

Lastly, let us consider that, in order to get at a real understanding of any religion, there must be a wish and a will on both sides. Many savages shrink from questions on religious topics, partly, it may be, from some superstitious fear—partly, it may be, from their helplessness in putting their own unfinished thoughts and sentiments into definite language. Some savage races are decidedly reticent. Speaking is an effort to them. After ten minutes' conversation, they complain of headache.[33] Others are extremely talkative, and have an answer to everything, little caring whether what they say is true or not.[34]

This difficulty is admirably stated by the Rev. R. H. Codrington, in a letter from Norfolk Island, July 3, 1877:

> But the confusion about such matters does not ordinarily lie in the native mind, but proceeds from the want of clear communication between the native and the European. A native who knows a little English, or one trying to communicate with an Englishman in his native tongue, finds it very much more easy to assent to what the white man suggests, or to use the words that he knows, without perhaps exactly knowing the meaning, than to struggle to convey exactly what he thinks is the true account. Hence visitors receive what they suppose trustworthy information from natives, and then print things which read very absurdly to those who know the truth. Much amusement was caused to-day when I told a Merlav boy that I had just read in a book (Capt. Moresby's on New Guinea) of the idols he had seen in his village, which it was hoped that boy would be able to teach the natives to reject. He had a hand in making them, and they are no more idols than the gurgoyles [*sic*] on your chapel; yet I have no doubt some native told the naval officers that they were idols, or devils, or something, when he was asked whether they were or not, and got much credit for his knowledge of English.

I mentioned in my first lecture the account of some excellent Benedictine missionaries,[35] who, after three years spent at their station in Australia, came to the conclusion that the natives did not adore any deity, whether true or false. Yet they found out afterwards that the natives believed in an omnipotent Being, who had created the world. Suppose they had left their station before having made this discovery, who would have dared to contradict their statements?

De Brosses, when he gave his first and fatal account of fetishism, saw none of these difficulties. Whatever he found in the voyages of sailors and traders was welcome to him. He had a theory to defend, and whatever seemed to support it, was sure to be true.

I have entered thus fully into the difficulties inherent in the study of the religions of savage tribes, in order to show how cautious we ought to be before we accept one-sided descriptions of these religions; still more, before we venture to build on such evidence, as is now accessible, far-reaching theories on the nature and origin of religion in general. It will be difficult indeed to eradicate the idea of a universal primeval fetishism from the text-books of history. That very theory has become a kind of scientific fetish, though, like most fetishes, it seems to owe its existence chiefly to ignorance and superstition.

Only let me not be misunderstood. I do not mean to dispute the fact that fetish-worship is widely prevalent among the Negroes of Western Africa and other savage races.

What I cannot bring myself to admit is that any writer on the subject, beginning with De Brosses, has proved, or even attempted to prove, that what they call fetishism is a primitive form of religion. It may be admitted to be a low form, but that, particularly in religion, is very different from a primitive form of religion.

Wide Extension of the Meaning of Fetish

One of the greatest difficulties we have to encounter in attempting to deal in a truly scientific spirit with the problem of fetishism, is the wide extension that has been given to the meaning of the word *fetish*.

De Brosses speaks of fetishes, not only in Africa, but among the Red Indians, the Polynesians, the northern tribes of Asia; and after his time hardly a single corner of the world has been visited without traces of fetish-worship being discovered. I am the last man to deny to this spirit which sees similarities everywhere, its scientific value and justification. It is the comparative spirit which is at work everywhere, and which has achieved the greatest triumphs in modern times. But we must not forget that comparison, in order to be fruitful, must be joined with distinction, otherwise we fall into that dangerous habit of seeing cromlechs wherever there are some upright stones and another laid across, or a dolmen wherever we meet with a stone with a hole in it.

We have heard a great deal lately in Germany, and in England also, of tree-worship and serpent-worship. Nothing can be more useful than a wide collection of analogous facts, but their true scientific interest begins only when we can render to ourselves an account of how, beneath their apparent similarity, there often exists the greatest diversity of origin.

It is the same in Comparative Philology. No doubt there is grammar everywhere, even in the languages of the lowest races; but if we attempt to force our grammatical terminology, our nominatives and accusatives, our actives and passives, our gerunds and supines upon every language, we lose the chief lesson which a comparative study of language is to teach us, and we fail to see how the same object can be realised, and was realised, in a hundred different ways, in a hundred different languages. Here, better than anywhere else, the old Latin saying applies, *Si duo dicunt idem, non est idem,* "If two languages say the same thing, it is not the same thing."

If there is fetish-worship everywhere, the fact is curious, no doubt; but it gains a really scientific value only if we can account for the fact. How a fetish came to be a fetish, that is the problem which has to be solved, and as soon as we attack fetishism in that spirit, we shall find that, though being apparently the same everywhere, its antecedents are seldom the same anywhere. There is no fetish without its antecedents, and it is in these antecedents alone that its true and scientific interest consists.

Antecedents of Fetishism

Let us consider only a few of the more common forms of what has been called fetishism; and we shall soon see from what different heights and depths its sources spring.

If the bones, or the ashes, or the hair of a departed friend are cherished as relics, if they are kept in safe or sacred places, if they are now and then looked at, or even spoken to, by true mourners in their loneliness, all this may be, and has been, called fetish-worship.

Again, if a sword once used by a valiant warrior, if a banner which had led their fathers to victory, if a stick, or let us call it a sceptre, if a calabash, or let us call it a drum, are greeted with respect or enthusiasm by soldiers when going to do battle themselves, all this may be called fetish-worship. If these banners and swords are blessed by priests, or if the spirits of those who had carried them in former years are invoked, as if they were still present, all this may be put down as fetishism. If the defeated soldier breaks his sword across his knees, or tears his colours, or throws his eagles away, he may be said to be punishing his fetish; nay, Napoleon himself may be called a fetish-worshipper when, pointing to the Pyramids, he said to his soldiers, "From the summit of these monuments forty centuries look down upon you, soldiers!"

This is a kind of comparison in which similarities are allowed to obscure all differences.

No, we cannot possibly distinguish too much, if we want not only to know, but to understand the ancient customs of savage nations. Sometimes a stock or a stone was worshipped, because it was a forsaken altar, or an ancient place of judgment[36]; sometimes because it marked the place of a great battle or a murder,[37] or the burial of a king; sometimes because it protected the sacred boundaries of clans or families. There are stones from which weapons can be made; there are stones on which weapons can be sharpened; there are stones, like the jade found in Swiss lakes, that must have been brought as heirlooms from great distances; there are meteoric stones fallen from the sky. Are all these simply to be labelled fetishes, because, for very good but very different reasons, they were all treated with some kind of reverence by ancient and even by modern people?

Sometimes the fact that a crude stone is worshipped as the image of a god may show a higher power of abstraction than the worship paid to the master-works of Phidias; sometimes the worship paid to a stone slightly resembling the human form may mark a very low stage of religious feeling. If we are satisfied with calling all this and much more simply fetishism, we shall soon be told that the stone on which all the kings of England have been crowned is an old fetish, and that in the coronation of Queen Victoria we ought to recognise a survival of Anglo-Saxon fetishism.

Matters have at last gone so far that people travelling in Africa actually cross-examined the natives whether they believed in *fetishes,* as if the poor Negro or the Hottentot, or the Papua could have any idea of what is meant by such a word! Native African words for fetish are *gri-gri, gru-gru,* or *ju-ju,* all of them possibly the same word.[38] I must quote at least one story, showing how far superior the examinee may sometimes be to the examiners. "A negro was worshipping a tree, supposed to be his fetish, with an offering of food, when some European asked whether he thought that the tree could eat. The negro replied: 'Oh, the tree is not the fetish, the fetish is a spirit and invisible, but he has descended into this tree. Certainly he cannot devour our bodily food, but he enjoys its spiritual part, and leaves behind the bodily part, which we see.'" The story is almost too good to be true, but it rests on the authority of Halleur,[39] and it may serve at least as a warning against our interpreting the sacrificial acts of so-called savage people by one and the same rule, and against our using technical terms so ill-chosen and so badly defined as fetishism.

Confusion becomes still worse confounded when travellers, who have accustomed themselves to the most modern acceptation of the word *fetish,* who use it, in fact, in the place of God, write their accounts of the savage races, among whom they have lived, in this modern jargon. Thus one traveller tells us that "the natives say that the great fetish of Bamba lives in the bush, where no man sees him or can see him. When he dies, the fetish-priests carefully collect his bones, in order to revive them and nourish them, till they again acquire flesh and blood." Now here "the great fetish" is used in the Comtian sense of the word; it means no longer *fetish,* but deity. A fetish that lives in the bush and cannot be seen is the very opposite of the *feitiço,* or the *gru-gru,* or whatever name we may choose to employ for those lifeless and visible subjects which are worshipped by men, not only in Africa, but in the whole world, during a certain phase of their religious consciousness.

Ubiquity of Fetishism

If we once go so far, we need not wonder that fetishes are found everywhere, among ancient and modern, among uncivilised and civilised people. The Palladium at Troy, which was supposed to have fallen from the sky, and was believed to make the town impregnable, may be called a fetish, and like a fetish it had to be stolen by Odysseus and Diomedes, before Troy could be taken. Pausanias states that in ancient times the images of the gods in Greece were rude stones, and he mentions such stones as still existing in his time, in the second century of our era.[40] At Pharae he tells us of thirty square stones (hermæ?), near the statue of Hermes, which the people worshipped, giving to each the name of a god. The Thespians, who worshipped Eros as the first among gods, had an image of him which was a mere stone.[41] The statue of Herakles at Hyettos was of the same character, according to the old fashion, as Pausanias himself remarks.[42] In Sicyon he mentions an image of Zeus Meilichios, and another of Artemis Patroa, both made without any art, the former a mere pyramid, the latter a column.[43] At Orchomenos again, he describes a temple of the Graces, in which they were worshipped as rude stones, which were believed to have fallen from the sky at the time of Eteokles. Statues of the Graces were placed in the temple during the lifetime of Pausanias.[44]

The same at Rome. Stones which were believed to have fallen from the sky were invoked to grant success in military enterprises.[45] Mars himself was represented by a spear. Augustus, after losing two naval battles, punished Neptune like a fetish, by excluding his image from the procession of the gods.[46] Nero was, according to Suetonius, a great despiser of all religion, though for a time he professed great faith in the Dea Syria. This, however, came to an end, and he then treated her image with the greatest indignity. The fact was that some unknown person had given him a small image of a girl, as a protection against plots, and as he discovered a plot against his life immediately afterwards, he began to worship that image as the highest deity, offering sacrifices to it three times every day, and declaring that it enabled him to foresee the future.[47]

If all this had happened at Timbuktu, instead of Rome, should we not call it fetishism?

Lastly, to turn to Christianity, is it not notorious what treatment the images of saints receive at the hands of the lower classes in Roman Catholic countries? Della Valle relates that Portuguese sailors fastened the image of St. Anthony to the bowsprit, and then addressed him kneeling, with the following words, "O St. Anthony, be pleased to stay there till thou hast given us a fair wind for our voyage."[48] Frezier writes of a Spanish captain who tied a small image of the Virgin Mary to the mast, declaring that it should hang there till it had granted him a favourable wind.[49] Kotzebue declares that the Neapolitans whip their saints if they do not grant their requests.[50] Russian peasants, we are told, cover the face of an image, when they are doing anything unseemly, nay, they even borrow their neighbours' saints, if they have proved themselves particularly successful.[51] All this, if seen by a stranger, would be set down as fetishism, and yet what a view is opened before our eye[s], if we ask ourselves, how such worship paid to an image of the Virgin Mary or of a saint became possible in Europe? Why should it be so entirely different among the Negroes of Africa? Why should all their fetishes be, as it were, of yesterday?

To sum up. If we see how all that can be called fetish in religions the history of which is known to us, is secondary, why should fetishes in Africa, where we do not know the earlier development of religion, be considered as primary? If everywhere else there are antecedents of a fetish, if everywhere else fetishism is accompanied by more or less developed religious idea, why should we insist on fetishism being the very beginning of all religion in Africa? Instead of trying to account for fetishism in all other religions by a reference to the fetishism which we find in Africa, would it not be better to try to account for the fetishism in Africa by analogous facts in religions the history of which is known to us?

No Religion Consists of Fetishism Only

But if it has never been proved, and perhaps, according to the nature of the case, can never be proved that fetishism in Africa, or elsewhere, was ever in any sense of the word a primary form of religion, neither has it been shown that fetishism constituted anywhere, whether in Africa or elsewhere, the whole of a people's religion. Though our knowledge of the religion of the Negroes is still very imperfect, yet I believe I may say that, wherever there has been an opportunity of ascertaining by long and patient intercourse the religious sentiments even of the lowest savage tribes, no tribe has ever been found without something beyond mere worship of so-called fetishes. A worship of visible material objects is widely spread among African tribes, far more widely than anywhere else. The intellectual and sentimental tendencies of the Negro may preëminently predispose him to that kind of degraded worship. All this I gladly admit. But I maintain that fetishism was a corruption of religion, in Africa as elsewhere, that the Negro is capable of higher religious ideas than the worship of stocks and stones, and that many tribes who believe in fetishes cherish at the same time very pure, very exalted, very true sentiments of the deity. Only we must have eyes to see, eyes that can see what is perfect without dwelling too much on what is imperfect. The more I study heathen religions, the more I feel convinced that, if we want to form a true judgment of their purpose, we must measure them, as we measure the Alps, by the highest point which they have reached. Religion is everywhere an aspiration rather than a fulfilment, and I claim no more for the religion of the Negro than for our own, when I say that it should be judged, not by what it appears to be, but by what it is—nay, not only by what it is, but by what it can be, and by what it has been in its most gifted votaries.

Higher Elements in African Religion: Waitz

Whatever can be done under present circumstances to gain an approximate idea of the real religion of the African Negroes, has been done by Waitz in his classical work on *Anthropology*. Waitz, the editor of Aristotle's *Organon*, approached his subject in a truly scholarlike spirit. He was not only impartial himself, but he carefully examined the impartiality of his authorities before he quoted their opinions. His work is well known in England, where many of his facts and opinions have found so charming an interpreter in Mr. Tylor. The conclusions at which Waitz arrived with regard to the true character of the religion of the Negroes may be stated in his own words:

> The religion of the negro is generally considered as a peculiar crude form of polytheism and marked with the special name of fetishism. A closer inspection of it, however, shows clearly that, apart from certain extravagant and fantastic features which spring from the character of the negro and influence all his doings, his religion, as compared with those of other uncivilised people, is neither very peculiar nor exceptionally crude. Such a view could only be taken, if we regarded the outward side only of the negro's religion or tried to explain it from gratuitous antecedents. A more profound investigation, such as has lately been successfully carried out by several eminent scholars, leads to the surprising result that several negro tribes, who cannot be shown to have experienced the influence of any more highly civilised nations, have progressed much further in the elaboration of their religious ideas than almost all other uncivilised races; so far indeed that, if we do not like to call them monotheists, we may at least say of them, that they have come very near to the boundaries of true monotheism, although their religion is mixed up with a large quantity of coarse superstitions, which with some other people seem almost to choke all pure religious ideas.[52]

Waitz himself considers Wilson's book on *West Africa, its History, Condition, and Prospects* (1856), as one of the best, but he has collected his materials likewise from many other sources, and particularly from the accounts of missionaries. Wilson was the first to point out that what we have chosen to call fetishism, is something very distinct from the real religion of the Negro. There is ample evidence to show that the same tribes, who are represented as fetish-worshippers, believe either in gods, or in a supreme good God, the creator of the world, and that they possess in their dialects particular names for him.

Sometimes it is said that no visible worship is paid to that Supreme Being, but to fetishes only. This, however, may arise from different causes. It may arise from an excess of reverence, quite as much as from negligence. Thus the Odjis or Ashantis call the Supreme Being by the same name as the sky, but they mean by it a personal God, who, as they say, created all things, and is the giver of all good things.[53] But though he is omnipresent and omniscient, knowing even the thoughts of men, and pitying them in their distress, the government of the world is, as they believe, deputed by him to inferior spirits, and among these again it is the malevolent spirits only who require worship and sacrifice from man.[54]

Cruickshank calls attention to the same feature in the character of the Negroes on the Gold Coast. He thinks that their belief in a supreme God, who has made the world and governs it, is very old, but he adds that they invoke him very rarely, calling him "their great friend," or "He who has made us." Only when in great distress they call out, "We are in the hands of God; he will do what seemeth right to him."[55] This view is confirmed by the Basle missionaries, who cannot certainly be suspected of partiality. They also affirm that their belief in a supreme God is by no means without influence on the Negroes. Often, when in deep distress, they say to themselves, "God is the old one, he is the greatest; he sees me, I am in his hand." The same missionary adds, "If, besides this faith, they also believe in thousands of fetishes, this, unfortunately, they share in common with many Christians."[56]

The Odjis or Ashantis, while retaining a clear conception of God as the high or the highest, the creator, the giver of sunshine, rain, and all good gifts, the omniscient, hold that he does not condescend to govern the world, but that he has placed created spirits as lords over hills and vales, forests and fields, rivers and the sea.[57] These are conceived as like unto men, and are occasionally seen, particularly by the priests. Most of them are good, but some are evil spirits, and it seems that in one respect at least these Negroes rival the Europeans, admitting the existence of a supreme evil spirit, the enemy of men, who dwells apart in a world beyond.[58]

Some of the African names given to the Supreme Being meant originally sun, sky, giver of rain; others mean Lord of Heaven, Lord and King of Heaven, the invisible creator. As such he is invoked by the Yebus, who, in praying to him, turn their faces to the ground. One of their prayers was: "God in Heaven, guard us from sickness and death; God, grant us happiness and wisdom."[59]

The Edîyahs of Fernando Po call the Supreme Being *Rupi,* but admit many lesser gods as mediators between him and man.[60] The Duallahs, on the Cameruns, have the same name for the Great Spirit and the sun.[61]

The Yorubas believe in a Lord of Heaven, whom they call *Olorun.*[62] They believe in other gods also, and they speak of a place called Ife in the district of Kakanda (5° E. L. Gr. 8° N. lat.) as the seat of the gods, a kind of Olympus, from whence sun and moon always return after having been buried in the earth, and from whence men also are believed to have sprung.[63]

Among the people of Akra, we are told by Römer that a kind of worship was paid to the rising sun.[64] Zimmerman denies that any kind of worship is paid there to casual objects (commonly called fetishes),[65] and we know from the reports of missionaries that their name for the highest god is *Jongmaa*, which signifies both rain and god. This Jongmaa is probably the same as *Nyongmo*, the name for God on the Gold Coast. There too it means the sky, which is everywhere, and has been from everlasting. A Negro, who was himself a fetish priest, said, "Do we not see daily how the grass, the corn, and the trees grow by the rain and the sunshine which he sends! How should he not be the creator?" The clouds are said to be his veil; the stars, the jewels on his face. His children are the *Wong*, the spirits which fill the air and execute his commands on earth.[66]

These Wongs, which have likewise been mistaken for fetishes, constitute a very important element in many ancient religions, and not in Africa only; they step in everywhere where the distance between the human and the divine has become too wide, and where something intermediate, or certain mediators, are wanted to fill the gap which man has created himself. A similar idea is expressed by Celsus when defending the worship of the genii. Addressing himself to the Christians, who declined to worship the old genii, he says: "God can suffer no wrong. God can lose nothing. The inferior spirits are not his rivals, that He can resent the respect which we pay to them. In them we worship only some attributes of Him from whom they hold authority, and in saying that One only is Lord, you disobey and rebel against Him."[67]

On the Gold Coast it is believed that these Wongs dwell between heaven and earth, that they have children, die, and rise again.[68] There is a Wong for the sea and all that is therein; there are other Wongs for rivers, lakes, and springs; there are others for pieces of land which have been inclosed [sic], others for the small heaps of earth thrown up to cover a sacrifice; others, again, for certain trees, for certain animals, such as crocodiles, apes, and serpents, while other animals are only considered as sacred to the Wongs. There are Wongs for the sacred images carved by the fetishman, lastly for anything made of hair, bones, and thread, and offered for sale as talismans.[69] Here we see clearly the difference between Wongs and fetishes, the fetish being the outward sign, the Wong the indwelling spirit, though, no doubt, here too the spiritual might soon have dwindled down into a real presence.[70]

In Akwapim the word which means both God and weather is *Jankkupong*. In Bonny, also, and in Eastern Africa among the Makuas, one and the same word is used to signify God, heaven, and cloud.[71] In Dahomey the sun is said to be supreme, but receives no kind of worship.[72] The Ibos believe in a maker of the world whom they call *Tshuku*. He has two eyes and two ears, one in the sky and one on the earth. He is invisible, and he never sleeps. He hears all that is said, but he can reach those only who draw near unto him.[73]

Can anything be more simple and more true? He can reach those only who draw near unto him! Could we say more?

Good people, it is believed, will see him after death, bad people go into fire. Do not some of us say the same?

That some of the Negroes are aware of the degrading character of fetish-worship is shown by the people of Akra declaring the monkeys only to be fetish-worshippers.[74]

I cannot vouch for the accuracy of every one of these statements for reasons which I have fully explained. I accept them on the authority of a scholar who was accustomed to the collation of various readings in ancient manuscripts: Professor Waitz. Taken together, they certainly give a very different impression of the Negroes from that which is commonly received. They show at all events that, so far from being a uniform fetishism, the religion of the Negro is many-sided in the extreme. There is fetish-worship in it, perhaps more than in other religions, but what becomes of the assertion that the religion of the Negro consists in fetishism and in fetishism only, and that the Negro never advanced beyond this, the lowest stage of religion? We have seen that there are in the religion of the Africans very clear traces of a worship of spirits residing in different parts of nature, and of a feeling after a supreme spirit, hidden and revealed by the sun or the sky. It is generally, if not always, the sun or the sky which forms the bridge from the visible to the invisible, from nature to nature's God. But besides the sun, the moon also was worshipped by the Negroes, as the ruler of months and seasons, and the ordainer of time and life.[75] Sacrifices were offered under trees, soon also to trees, particularly old trees which for generations had witnessed the joys and troubles of a family or a tribe.

Zoolatry

Besides all this which may be comprehended under the general name of physiolatry, there are clear indications also of zoolatry.[76] It is one of the most difficult problems to discover the motive which led the Negro to worship certain animals. The mistake which is made by most writers on early religions, is that they imagine there can be but one motive for each custom that has to be explained. Generally, however, there are many. Sometimes the souls of the departed are believed to dwell in certain animals. In some places animals, particularly wolves, are made to devour the dead bodies, and they may in consequence be considered sacred.[77] Monkeys are looked upon as men, slightly damaged at the creation, sometimes also as men thus punished for their sins. They are in some places believed to be able to speak, but to sham dumbness in order to escape labour. Hence, it may be, a reluctance arose to kill them, like other animals, and from this there would be but a small step to ascribing to them a certain sacro-sanctity. Elephants, we know, inspire similar feelings by the extraordinary development of their understanding. People do not like to kill them, or if they have to do it, they ask pardon from the animal which they have killed. In Dahomey, where the elephant is a natural fetish, many purificatory ceremonies have to be performed when an elephant has been slain.[78]

In some places it is considered lucky to be killed by certain animals, as for instance by leopards in Dahomey.

There are many reasons why snakes might be looked upon with a certain kind of awe, and even kept and worshipped. Poisonous snakes are dreaded, and may therefore be worshipped, particularly after they have been (perhaps secretly) deprived of their fangs. Other snakes are useful as domestic animals, as weather prophets, and may therefore have been fed, valued, and, after a time, worshipped, taking that word in that low sense which it often has and must have among uncivilised people. The idea that the ghosts of the departed dwell for a time in certain animals, is very widely prevalent; and considering the habits of certain snakes, hiding in deserted and even in inhabited houses, and suddenly appearing, peering at the inhabitants with their wondering eyes, we may well understand the superstitious awe with which they were treated. Again, we know that many tribes both in modern and ancient times have assumed the name of Snakes (Nâgas), whether in order to assert their autochthonic right to the country in which they lived, or because, as Diodorus supposes, the snake had been used as their banner, their rallying sign, or as we should say their totem or crest. As the same Diodorus points out, people may have chosen the snake for their banner, either because it was their deity, or it may have become their deity, because it was their banner. At all events nothing would be more natural than that people who, for some reason or other, called themselves Snakes, should in time adopt a snake for their ancestor, and finally for their god. In India the snakes assume, at an early time, a very prominent part in epic and popular traditions. They soon became what fairies or bogies are in our nursery tales, and they thus appear in company with Gandharvas, Apsaras, Kinnaras, &c., in some of the most ancient architectural ornamentations of India.

Totally different from these Indian snakes are the snake of the Zend-Avesta, and the snake of Genesis, and the dragons of Greek and Teutonic mythology. There is lastly the snake as a symbol of eternity, either on account of its shedding its skin, or because it rolls itself up into a complete circle. Every one of these creatures of fancy has a biography of his own, and to mix them all up together would be like writing *one* biography of all the people who were called Alexander.

Africa is full of animal fables, in the style of Æsop's fables, though they are not found among all tribes; and it is often related that, in former times, men and animals could converse together. In Bornu it is said that one man betrayed the secret of the language of animals to his wife, and that thenceforth the intercourse ceased.[79] Man alone is never, we are told, worshipped in Africa as a divine being; and if in some places powerful chiefs receive honours that make us shudder, we must not forget that during the most brilliant days of Rome divine honours were paid to Augustus and his successors. Men who are deformed, dwarfs, albinos and others, are frequently looked upon as something strange and uncanny, rather than what we should call sacred.

Psycholatry

Lastly, great reverence is paid to the spirits of the departed.[80] The bones of dead people also are frequently preserved and treated with religious respect. The Ashantis have a word *kla*,[81] which means the life of man. If used as a masculine, it stands for the voice that tempts man to evil; if used in the feminine, it is the voice that persuades us to keep aloof from evil. Lastly, *kla* [82] is the tutelary genius of a person who can be brought near by witchcraft, and expects sacrifices for the protection which he grants. When a man dies, his *kla* becomes *sisa,* and a *sisa* may be born again.

Many-Sidedness of African Religion

Now I ask, is so many-sided a religion to be classed simply as African fetish-worship? Do we not find almost every ingredient of other religions in the little which we know at present of the faith and worship of the Negro? Is there the slightest evidence to show that there ever was a time when these Negroes were fetish-worshippers, and nothing else? Does not all our evidence point rather in the opposite direction, viz., that fetishism was a parasitical development, intelligible with certain antecedents, but never as an original impulse of the human heart?

What is, from a psychological point of view, the really difficult problem, is how to reconcile the rational and even exalted religious opinions, traces of which we discovered among many of the Negro tribes, with the coarse forms of fetish-worship. We must remember, however, that every religion is a compromise between the wise and the foolish, the old and the young, and that the higher the human mind soars in its search after divine ideals, the more inevitable the symbolical representations, which are required for children and for the majority of people, incapable of realising sublime and subtle abstractions.

Much, no doubt, may be said in explanation, even in excuse of fetishism, under all its forms and disguises. It often assists our weakness, it often reminds us of our duties, it often may lead our thoughts from material objects to spiritual visions, it often comforts us when nothing else will give us peace. It is often said to be so harmless, that it is difficult to see why it should have been so fiercely reprobated by some of the wisest teachers of mankind. It may have seemed strange to many of us, that among the Ten Commandments which were to set forth, in the shortest possible form, the highest, the most essential duties of man, the second place should be assigned to a prohibition of any kind of images. "Thou shalt not make to thyself any graven image, nor the likeness of anything that is in heaven above, or in the earth beneath, or in the waters under the earth: thou shalt not bow down to them, nor worship them."

Let those who wish to understand the hidden wisdom of these words, study the history of ancient religions. Let them read the descriptions of religious festivals in Africa, in America, and Australia, let them witness also the pomp and display in some of our own Christian churches and cathedrals. No arguments can prove that there is anything very wrong in all these outward signs and symbols. To many people, we know, they are even a help and comfort. But history is sometimes a stronger and sterner teacher than argument, and one of the lessons which the history of religions certainly teaches is this, that the curse pronounced against those who would change the invisible into the visible, the spiritual into the material, the divine into the human, the infinite into the finite, has come true in every nation on earth. We may consider ourselves safe against the fetish-worship of the poor Negro; but there are few of us, if any, who have not their own fetishes, or their own idols, whether in their churches, or in their hearts.

The results at which we have arrived, after examining the numerous works on fetishism from the days of De Brosses to our own time, may be summed up under four heads:

1. The meaning of the word *fetish* (*feitiço*) has remained undefined from its first introduction, and has by most writers been so much extended, that it may include almost every symbolical or imitative representation of religious objects.

2. Among people who have a history, we find that everything which falls under the category of fetish points to historical and psychological antecedents. We are therefore not justified in supposing that it has been otherwise among people whose religious development happens to be unknown and inaccessible to us.
3. There is no religion which has kept itself entirely free from fetishism.
4. There is no religion which consists entirely of fetishism.

Supposed Psychological Necessity of Fetishism

Thus I thought I had sufficiently determined the position which I hold with regard to the theory of a universal primeval fetishism, or had at all events made it clear that the facts of fetish-worship, as hitherto known to us, can in no wise solve the question of the natural origin of religion.

The objection has, however, been raised by those who cling to fetishism, or at least to the Comtian theory of fetishism, that these are after all facts only, and that a complete and more formidable theory has to be encountered before it could be admitted that the first impulse to religion proceeded from an incipient perception of the infinite pressing upon us through the great phenomena of nature, and not from sentiments of surprise or fear called forth by such finite things as shells, stones, or bones—that is to say, by fetishes.

We are told that whatever the *facts* may be which, after all, by mere accident, are still within our reach, as bearing witness to the earliest phases of religious thought, there *must* have been a time, whether in historic or prehistoric periods, whether during the formation of quaternary or tertiary strata, when man worshipped stocks and stones, and nothing else.

I am far from saying that under certain circumstances mere argumentative reasoning may not be as powerful as historical evidence; still I thought I had done enough by showing how the very tribes who were represented to us as living instances of fetish-worship possessed religious ideas of a simplicity and, sometimes, of a sublimity such as we look for in vain even in Homer and Hesiod. Facts had been collected to support a theory, nay, had confessedly given the first impulse to a theory, and that theory is to remain, although the facts have vanished, or have at all events assumed a very different aspect. However, as it is dangerous to leave any fortress in our rear, it may be expedient to reply to this view of fetishism also, though in as few words as possible.

It may be taken for granted that those who hold the theory that religion must everywhere have taken its origin from fetishism, take fetish in the sense of casual objects which, for some reason or other, or it may be for no reason at all, were considered as endowed with exceptional powers, and gradually raised to the dignity of spirits or gods. They could not hold the other view, that a fetish was, from the beginning, an emblem or symbol only, an outward sign or token of some power previously known, which power, originally distinct from the fetish, was afterwards believed to reside in it, and in course of time came to be identified with it. For in that case the real problem for those who study the growth of the human mind would be the origin and growth of that power, previously known, and afterwards supposed to reside in a fetish. The real beginning of religious life would be there; the fetish would represent a secondary stage only. Nor is it enough to say (with Professor Zeller) that "fancy or imagination personifies things without life and without reason as gods."[83] The real question is: Whence that imagination?; and Whence, before all things, that unprovoked and unjustifiable predicate of God?

The theory therefore of fetishism with which alone we have still to deal is this, that a worship of casual objects is and must be the first inevitable step in the development of religious ideas. Religion not only does begin, but must begin, we are told, with a contemplation of stones, shells, bones, and such like things, and from that stage only can it rise to the conception of something else—of powers, spirits, gods, or whatever else we like to call it.

Whence the Supernatural Predicate of a Fetish?

Let us look this theory in the face. When travellers, ethnologists, and philosophers tell us that savage tribes look upon stones and bones and trees as their gods, what is it that startles us? Not surely the stones, bones, or trees; not the subjects, but that which is predicated of these subjects, viz., God. Stones, bones, and trees are ready at hand everywhere; but what the student of the growth of the human mind wishes to know is: Whence their higher predicates?; or, let us say at once, Whence their predicate God? Here lies the whole problem. If a little child were to bring us his cat and say it was a vertebrate animal, the first thing that would strike us would surely be: How did the child ever hear of such a name as a vertebrate animal? If the fetish-worshipper brings us a stone and says it is a god, our question is the same: Where did you ever hear of God?, and What do you mean by such a name? It is curious to observe how little that difficulty seems to have been felt by writers on ancient religion.

Let us apply this to the ordinary theory of fetishism, and we shall see that the problem is really this: Can spirits or gods spring from stones? Or, to put it more clearly: Can we understand how there should be a transition from the percept of a stone to the concept of a spirit or a god?

Accidental Origin of Fetishism

We are told that nothing is easier than this transition. But how? We are asked to imagine a state of mind when man, as yet without any ideas beyond those supplied to him by his five senses, suddenly sees a glittering stone or a bright shell, picks it up as strange, keeps it as dear to himself, and then persuades himself that this stone is not a stone like other stones, that this shell is not a shell like other shells, but that it is endowed with extraordinary powers, which no other stone or shell ever possessed before. We are asked to suppose that possibly the stone was picked up in the morning, that the man who picked it up was engaged in a serious fight during the day, that he came out of it victorious, and that he very naturally ascribed to the stone the secret of his success. He would afterwards, so we are told, have kept that stone for luck; it might very likely have proved lucky more than once; in fact, those stones only which proved lucky more than once would have had a chance of surviving as fetishes. It would then have been believed to possess some supernatural power, to be not a mere stone but something else, a powerful spirit, entitled to every honour and worship which the lucky possessor could bestow on it or on him.[84]

This whole process, we are assured, is perfectly rational in its very irrationality. Nor do I deny it; I only doubt whether it exhibits the irrationality of an uncultured mind. Is not the whole process of reasoning, as here described, far more in accordance with modern than with ancient and primitive thoughts? Nay, I ask, can we conceive it as possible except when men were already far advanced in their search after the infinite, and in full possession of those very concepts, the origin of which we want to have explained to us?

Are Savages like Children?

It was formerly supposed that the psychological problem involved in fetishism could be explained by a mere reference to children playing with their dolls, or hitting the chair against which they had hit themselves. This explanation, however, has long been surrendered, for, even supposing that fetishism consisted only in ascribing to material objects life, activity, or personality, call it figurism, animism, personification, anthropomorphism, or anthropopathism, the mere fact that children do the same as grown-up savages cannot possibly help us to solve the psychological problem. The fact, suppose it is a fact, would be as mysterious with children as with savages. Besides, though there is some truth in calling savages children, or children savages, we must here, too, learn to distinguish. Savages are children in some respects, but not in all. There is no savage who, on growing up, does not learn to distinguish between animate and inanimate objects, between a rope, for instance, and a serpent. To say that they remain childish on such a point is only to cheat ourselves with our own metaphors. On the other side, children, such as they now are, can help us but rarely to gain an idea of what primitive savages may have been. Our children, from the first awakening of their mental life, are surrounded by an atmosphere saturated with the thoughts of an advanced civilisation. A child, not taken in by a well-dressed doll, or so perfectly able to control himself as not to kick against a chair against which he had hit his head, would be a little philosopher rather than a savage, not yet emerging from fetishism. The circumstances or the surroundings are so totally different in the case of the savage and the child, that comparisons between the two must be carried out with the greatest care before they can claim the smallest scientific value.

I agree so far with the believers in primitive fetishism that if we are to explain religion as a universal property of mankind, we must explain it out of conditions which are universally present. Nor do I blame them if they decline to discuss the problem of the origin of religion with those who assume a primitive revelation, or a religious faculty which distinguishes man from the animal. Let us start, by all means, from common ground and from safe ground. Let us take man such as he is, possessing his five senses, and as yet without any knowledge except what is supplied to him by his five senses. No doubt that man can pick up a stone, or a bone or a shell. But then we must ask the upholders of the primitive fetish theory: How do these people, when they have picked up their stone or their shell, pick up at the same time the concepts of a supernatural power, of spirit, of god, and of worship paid to some unseen being?

The Four Steps

We are told that there are four steps—the famous four steps—by which all this is achieved, and the origin of fetishism rendered perfectly intelligible. First, there is a sense of surprise; secondly, an anthropopathic conception of the object which causes surprise; thirdly, the admission of a causal connection between that object and certain effects, such as victory, rain, health; fourthly, a recognition of the object as a power deserving of respect and worship. But is not this rather to hide the difficulties beneath a golden shower of words than to explain them?

Granted that a man may be surprised at a stone or a shell, though they would seem to be the very last things to be surprised at; but what is the meaning of taking an anthropopathic view of a stone or a shell? If we translate it into plain English, it means neither more nor less than that, instead of taking a stone to be a stone like all other stones, we suppose that a particular stone is not an ordinary stone, but endowed with the feelings of a man. Natural as this may sound, when clothed in technical language, when we use long names, such as anthropopathism, anthropomorphism, personification, figurism, nothing would really seem to do greater violence to common sense, or to our five senses, than to say that a stone is a stone, yet not quite a stone; and again, that the stone is a man, yet not quite a man. I am fully aware that, after a long series of intermediate steps, such contradictions arise in the human mind, but they cannot spring up suddenly; they are not there from the beginning, unless we admit disturbing influences much more extraordinary than a primeval revelation. It is the object of the science of religion to find out by what small and timid steps the human mind advanced from what is intelligible to what at first sight is almost beyond our comprehension. If we take for granted the very thing that has to be explained; if we once admit that it was perfectly natural for the primitive savage to look upon a stone as something human; if we are satisfied with such words as anthropopathism, or animism, or figurism, then all the rest no doubt is easy enough. The human stone has every right to be called superhuman, and that is not very far from divine; nor need we wonder that the worship paid to such an object should be more than what is paid to either a stone or to a man—that it too should be superhuman, which is not very far from divine.

Fetishism Not a Primary Form of Religion

My position then is simply this: It seems to me that those who believe in a primordial fetishism have taken that for granted which has to be proved. They have taken for granted that every human being was miraculously endowed with the concept of what forms the predicate of every fetish, call it power, spirit, or god. They have taken for granted that casual objects, such as stones, shells, the tail of a lion, a tangle of hair, or any such rubbish, possess in themselves a theogonic or god-producing character, while the fact that all people, when they have once risen to the suspicion of something supersensuous, infinite, or divine, have perceived its presence afterwards in merely casual and insignificant objects, has been entirely overlooked. They have taken for granted that there exists at present, or that there existed at any time, a religion entirely made up of fetishism; or that, on the other hand, there is any religion which has kept itself entirely free from fetishism. My last and most serious objection, however, is that those who believe in fetishism as a primitive and universal form of religion have often depended on evidence which no scholar, no historian, would feel justified to accept. We are justified therefore, I think, in surrendering the theory that fetishism either has been or must have been the beginning of all religion,[85] and we are bound to look elsewhere, if we wish to discover what were the sensuous impressions that first filled the human mind with a suspicion of the supersensuous, the infinite, and the divine.

Notes

1. C. Meiners, whose *Allgemeine Kritische Geschichte der Religionen,* 1806, was for many years the chief storehouse for all who wrote on the history of religion, says: "It cannot be denied that fetishism is not only the oldest, but also the most universal worship of gods."
2. "L'étranger qui arrive dans l'Inde, et moi-même je n'ai pas fait exception à cette règle, ne découvre d'abord que des pratiques religieuses aussi dégradantes que dégradées, un vrai polythéisme, presque du fétichisme."—*De la supériorité du Brahmanisme sur le Catholicisme,* Conférence donnée par M. Goblet d'Alviella. [Translation: "The foreigner who arrives in India—and I myself am no exception to this rule—discovers first of all that these religious practices are as degrading as they are debased, a genuine polytheism, bordering on fetishism"—Sondra Bacharach.]
3. *Synodus Pergam.,* ann. 1311, apud Muratorium, tom. 9. col. 561; incantationes, sacrilegia, auguria, vel maleficia, quæ facturæ sive præstigia vulgariter appellantur.
4. Similar customs mentioned by Waitz, *Anthropologie,* vol. 2, p. 177.
5. *Sociology,* p. 106. See also "On some Characteristics of Malayo-Polynesians," in *Journal of the Anthropological Institute,* February 1878.
6. "The savage are as old as the civilised races, and can as little be named primitive." A. M. Fairbairn, *Academy,* July 20, 1878.
7. A. B. Meyer, *On the Mafoor and other Papua Languages of New Guinea,* p. 11.
8. See Taplin, *The Narrinyeri, South Australian Aborigines,* p. 77.
9. Speaking of the Dahomans, Mr. Burton (*Memoirs of the Anthropological Society,* i, 314) says: "By perpetual cowrie-handling the people learn to be ready reckoners. Amongst the cognate Yorubas the saying, 'You cannot multiply nine by nine,' means 'you are a dunce.'"
10. Dobrizhofer, *Historia de Abiponibus,* 1784.
11. Benfey, *Das Indogermanische Thema von Zwei,* p. 14, sees in *ka* the *ka* of *e-ka,* one.
12. Winterbottom, *Account of the Native Africans in the Neighbourhood of Sierra Leone.* London, 1863, p. 230.
13. Many cases of forming the words eight and nine by ten, *minus* one or two, will be found in the Comparative Table of Numerals at the end of my "Essay on the Turanian Languages." See also Moseley, *On the Inhabitants of the Admiralty Islands,* p. 13, and Matthews, *Hidatsa Grammar,* p. 118; Marcel Devic, "Sur l'origine Etymologique de quelques noms de nombre," *Journal Asiatique,* 1879, p. 545.
14. "Things pass away very rapidly in a country where everything in the nature of a building soon decays, and where life is short, and there are no marked changes of seasons to make the people count by anything longer than months." R. H. Codrington, Norfolk Island, July 3, 1877.
15. This throws a curious light on the Buddhist literature, where we also find the same story told twice, once in metre (Gâthâ), and once in prose. See also Whitley Stokes, *Calendar of Oengus,* 1880, p. 24.
16. See Crevecœur, *Voyage dans la Haute-Pensylvanie,* Paris, 1801, i, p. 362; Schultze, *Fetischismus,* p. 48.
17. Beechey, *Voyage to the Pacific Ocean,* i, p. 337.

18. *Outlines,* p. 6.
19. *Chips from a German Workshop,* vol. 1, p. 140.
20. *Âpastamba-Sûtras,* i. 4, 13, 6; Prâtisâkhya, 832, 838.
21. *Sociology,* i, p. 298.
22. Her., ii. 42; 144; 156.
23. *Sanskrit Texts,* v, p. 12.
24. See M. M., "Buddhistischer Nihilismus."
25. *Sociology,* i, p. 149. Compare with this Hobbes, *Computation on Logic,* i, 3, 2. (*Works,* ed. Molesworth, vol. i, p. 31). "But there are, or certainly may be, some nations that have no word which answers to our verb *is,* who nevertheless form propositions by the position only of one name after another, as if instead of 'man is a living creature,' it should be said 'man a living creature;' for the very order of the names may sufficiently show their connection; and they are as apt and useful in philosophy, as if they were copulated by the verb *is.*"
26. Cf., Darmesteter, *Vendîdâd,* Introduction, p. xliii, note.
27. *Myths and Songs from the South Pacific,* p. 33.
28. M. M., *History of Ancient Sanskrit Literature,* p. 538.
29. *De Bello Gall.,* vi. 21.
30. Tac., *Germ.* 9.
31. Waitz, *Anthropologie,* vol. 2, 171.
32. Ibid., vol. 2, 200; "On Different Classes of Priests," vol. 2, 199.
33. Burchell, *Reisen in das Innere von Südafrika,* 1823, pp. 71, 281; Schultze, *Fetischismus,* p. 36; H. Spencer, *Sociology,* i, p. 94.
34. Mayer, *Papua-sprachen,* p. 19.
35. "A Benedictine Missionary's account of the natives of Australia and Oceania. From the Italian of Don Rudesindo Salvado (Rome, 1851)," by C. H. E. Carmichael, *Journal of the Anthropological Institute,* February 1878.
36. Paus., i. 28. 5.
37. Ibid., viii. 13. 3; x. 5. 5.
38. Waitz, ii, p. 175. F. Schultze states that the Negroes adopted that word from the Portuguese. Bastian gives *enquizi* as a name for fetish on the West Coast of Africa; also *mokisso* (Bastian, *St. Salvador,* pp. 254, 281).
39. *Das Leben der Neger West-Africa's,* p. 40; cf., Waitz, vol. 2, p. 188; Tylor, *Primitive Culture,* ii, 197.
40. Paus., vii. 22. 4.
41. Ibid., ix. 27. 1.
42. Ibid., ix. 24. 3.
43. Ibid., ii. 9. 6.
44. Ibid., ix. 38. 1.
45. Plin., *H. N.,* 37. 9.
46. Suet., *Aug.*
47. Suet., *Nero,* c. 56.
48. *Voyage,* vii. 409; Meiners, i, p. 181; F. Schultze, *Fetischismus,* p. 175.
49. *Relation du Voyage de la Mer du Sud,* p. 248; F. Schultze, loc. cit.
50. *Reise nach Rom,* i, p. 327.
51. Rig-veda IV. 24, 10.
52. *Anthropologie,* vol. 2, p. 167.
53. Waitz, vol. 2, p. 171.
54. Riis, *Baseler Missions-Magazin,* 1847, iv, 244, 248.
55. Cruickshank, p. 217, quoted by Waitz, vol. 2, p. 172.
56. *Baseler Missions-Magazin,* 1855, i, p. 88; 1853, ii, p. 86; Waitz, vol. 2, p. 173.
57. Waitz, vol. 2, p. 171.
58. Ibid., pp. 173, 174.
59. Ibid., p. 168; D'Avezac, p. 84, note 3.

60. Ibid., p. 168.
61. Allen and Thomson, *Narrative of the Expedition to the River Niger in 1841,* ii, pp. 199, 395, note.
62. Tucker, p. 192, note.
63. Tucker, *Abbeokuta, or an Outline of the Origin and Progress of the Yoruba Mission,* 1856, p. 248.
64. Römer, *Nachrichten von der Küste Guinea,* 1769, p. 84.
65. Zimmerman, *Grammatical Sketch of the Akra or Ga Language, Vocabulary,* p. 337.
66. *Baseler Missions-Magazin,* 1837, p. 559.
67. Froude, in *Fraser's Magazine,* 1878, p. 160.
68. Waitz, vol. 2, p. 183.
69. *Baseler Missions-Magazin,* 1856, ii, p. 131.
70. Waitz, vol. 2, pp. 174, 175.
71. Köler, *Einige Notizen über Bonny,* 1848, p. 61; Waitz, vol. 2, p. 169.
72. Salt, *Voyage to Abyssinia,* 1814, p. 41.
73. Schön and Crowther, *Journal of an Expedition up the Niger, in 1842,* pp. 51, 72; Waitz, vol. 2, p. 169.
74. Waitz, vol. 2, pp. 174–178.
75. Ibid., p. 175.
76. Ibid., p. 177.
77. Ibid., p. 177; Hostmann, *Zur Geschichte des Nordischen Systems der drei Culturperioden,* Braunschweig, 1875, p. 13, note.
78. Ibid., p. 178.
79. Kölle, *African Native Literature,* 1854, p. 145.
80. Waitz, vol. 2, p. 181.
81. *Baseler Missions-Magazin,* 1856, ii, pp. 134, 139; Waitz, vol. 2, p. 182.
82. Compare with this *kla* the *ka* of the ancient Egyptians; Renouf, *Hibbert Lectures,* p. 147.
83. *Vorträge und Abhandlungen,* Zweite Sammlung, 1877, p. 32.
84. Waitz, vol. 2, p. 187.
85. I am glad to find that both Dr. Happel, in his work, *Die Anlage des Menschen zur Religion,* 1878, and Professor Pfleiderer in his *Religionsphilosophie,* just published, take nearly the same view of the Fetish-theory.

CHAPTER 11
THE IDEAS OF INFINITY AND LAW

from *Lectures on the Origin and Growth of Religion* (1878)

Nihil in fide quod non ante fuerit in sensu.[1]

Every day, every week, every month, every quarter, the most widely read journals seem just now to vie with each other in telling us that the time for religion is past, that faith is a hallucination or an infantine disease, that the gods have at last been found out and exploded, that there is no possible knowledge except what comes to us through our senses, that we must be satisfied with facts and finite things, and strike out such words as infinite, supernatural, or divine from the dictionary of the future.

It is not my object in these lectures either to defend or to attack any form of religion: there is no lack of hands for either the one or the other task. My own work, as I have traced it out for myself, and as it seemed to be traced out for me by the spirit of the founder of these lectures, is totally different. It is historical and psychological. Let theologians, be they Brâhmanas or Sramanas, Mobeds or Mollahs, Rabbis or Doctors of Divinity, try to determine whether any given religion be perfect or imperfect, true or false; what we want to know is, how religion is possible; how human beings, such as we are, came to have any religion at all; what religion is, and how it came to be what it is.

When we are engaged in the science of language, our first object is not to find out whether one language is more perfect than another, whether one contains more anomalous nouns or miraculous verbs than another. We do not start with a conviction that in the beginning there was one language only, or that there is at present, or that there will be in the future, one only that deserves to be called a language. No: we simply collect facts, classify them, try to understand them, and hope thus to discover more and more the real antecedents of all language, the laws which govern the growth and decay of human speech, and the goal to which all language tends.

It is the same with the science of religion. Each of us may have his own feeling as to his own mother-tongue, or his own mother-religion; but as historians we must allow the same treatment to all. We have simply to collect all the evidence that can be found on the history of religion all over the world, to sift and classify it, and to try thus to discover the necessary antecedents of all faith, the laws which govern the growth and decay of human religion, and the goal to which all religion tends. Whether there ever can be one perfect universal religion, is a question as difficult to answer as whether there ever can be one perfect universal language. If we can only learn that even the most imperfect religion, like the most imperfect language, is something beyond all conception wonderful, we shall have learnt a lesson which is worth many a lesson in the various schools of theology.

It is a very old saying, that we never know a thing unless we know its beginnings. We may know a great deal about religion, we may have read many of the sacred books, the creeds, the catechisms, and liturgies of the world, and yet religion itself may be something entirely beyond our grasp, unless we are able to trace it back to the deepest sources from whence it springs.

In doing this, in trying to discover the living and natural springs of religion, we must take nothing for granted, except what is granted us by all philosophers, whether positive or negative. I explained in my first lecture how I was quite prepared to accept their terms, and I mean to keep to these terms to the very end of my course. We were told that all knowledge, in order to be knowledge, must pass through two gates and two gates only: the gate of the senses, and the gate of reason. Religious knowledge also, whether true or false, must have passed through these two gates. At these two gates therefore we take our stand. Whatever claims to have entered in by any other gate, whether that gate be called primeval revelation or religious instinct, must be rejected as contraband of thought; and whatever claims to have entered by the gate of reason, without having first passed through the gate of the senses, must equally be rejected, as without sufficient warrant, or ordered at least to go back to the first gate, in order to produce there its full credentials.

Having accepted these conditions, I made it the chief object of my lectures to lay hold of religious ideas on their passing for the first time through the gates of our senses; or, in other words, I tried to find out what were the sensuous and material beginnings of those ideas which constitute the principal elements of religious thought.

I endeavoured to show, first of all, that the idea of the infinite, which is at the root of all religious thought, is not simply evolved by reason out of nothing, but supplied to us, in its original form, by our senses. If the idea of the infinite had no sensuous percept to rely on, we should, according to the terms of our agreement, have to reject it. It would not be enough to say with Sir W. Hamilton, that the idea of the infinite is a logical necessity; that we are so made that wherever we place the boundary of space or time, we are conscious of space and time beyond. I do not deny that there is truth in all this, but I feel bound to admit that our opponents are not obliged to accept such reasoning.

I therefore tried to show that beyond, behind, beneath, and within the finite, the infinite is always present to our senses. It presses upon us, it grows upon us from every side. What we call finite in space and time, in form and word, is nothing but a veil or a net which we ourselves have thrown over the infinite. The finite by itself, without the infinite, is simply inconceivable; as inconceivable as the infinite without the finite. As reason deals with the finite materials, supplied to us by our senses, faith, or whatever else we like to call it, deals with the infinite that underlies the finite. What we call sense, reason, and faith are three functions of one and the same perceptive self; but without sense, both reason and faith are impossible, at least to human beings like ourselves.

The history of the ancient religion of India, so far as we have hitherto been able to trace it, is to us a history of the various attempts at naming the infinite that hides itself behind the veil of the finite. We saw how the ancient Aryans of India, the poets of the Veda, first faced the invisible, the unknown, or the infinite in trees, mountains and rivers, in the dawn and the sun, in the fire, the storm-wind, and the thunder; how they ascribed to all of them a self, a substance, a divine support, or whatever else we like to call it; and how, in doing so, they always felt the presence of something which they could not see behind what they could see, of something supernatural behind the natural, of something superfinite or infinite behind or within the finite. The names which they gave, the *nomina,* may have been wrong: but the search itself after the *numina* was legitimate. At all events, we saw how that search led the ancient Aryans as far as it has led most amongst ourselves, viz., to the recognition of a Father which is in heaven.

Nay, we shall see that it led them further still. The idea that God is *not* a father, then, *like* a father, and lastly a father, appears in the Veda at a very early time. In the very first hymn of the Rig-veda, which is addressed to Agni, we read: "Be kind to us, as a father to his son." The same idea occurs again and again in the Vedic hymns. Thus we read, Rig-veda, I. 104, 9: "Hear us, Indra, like a father!" In III. 49, 3 the poet says that Indra gives food, hears our call, and is kind to us, like a father. In VII. 54, 2, Indra is asked to be kind, as a father to his sons. Again, Rig-veda, VIII. 21, 14, we read: "When thou thunderest and gatherest the clouds, then thou art called like a father." Rig-Veda, X. 33, 3: "As mice eat their tails, sorrows eat me up, me thy worshipper, all-powerful god! For once, O mighty Indra, be gracious to us! Be to us like a father!" Rig-veda, X. 69, 10: "Thou borest him as a father bears his son in his lap." Rig-veda, III. 53, 2: "As a son lays hold of his father by his skirt, I lay hold of thee by this sweetest song." In fact, there are few nations who do not apply to their god or gods the name of Father.

But though it was a comfort to the early Aryans in the childhood of their faith, as it is to us in the faith of our childhood, to call God father, they soon perceived that this too was a human name, and that like all human names, it said but little, compared with what it was meant to say. We may envy our ancient forefathers, as we envy a child that lives and dies full of faith that he is going from one home to another home, from one father to another father. But as every child grows up to learn that his father is but a child, the son of another father, as many a child, on becoming a man, has to surrender one idea after another that seemed to form the very essence of father, so the ancients learnt, and we all of us have to learn it, that we must take out of that word father one predicate after another, all in fact that is conceivable in it, if we wish to apply it still to God. So far as the word is applicable to man, it is inapplicable to God; so far as it is applicable to God, it is inapplicable to man. "Call no man your father upon the earth: for one is your Father, which is in heaven" (Matt. 23.9). Comparison, as it began, so it often ends with negation. *Father* is, no doubt, a better name than fire, or the storm-wind, or the heaven, or the Lord, or any other name which man has tried to give to the infinite, that infinite of which he felt the presence everywhere. But father too is but a weak human name, the best, it may be, which the poets of the Veda could find, but yet as far from him whom they were feeling after, as the east is from the west.

Having watched the searchings of the ancient Aryans after the infinite in every part of nature, and having tried to understand the names which they gave to it, beginning with trees and rivers and mountains, and ending with their Heaven-father, we have now to consider the origin of some other ideas which, at first, might seem completely beyond the reach of our senses, but which nevertheless can be shown to have had their deepest roots and their true beginnings in that finite or natural world which, it is difficult to say why, we are so apt to despise, while it has been everywhere and is still the only royal road that leads us on from the finite to the infinite, from the natural to the supernatural, from nature to nature's God.

Theogony of the Veda

By imagining ourselves placed suddenly in the midst of this marvellous world, we tried to find out what would be the objects most likely to have startled, to have fascinated, to have awed our earliest forefathers—what would have roused and awakened them from mere staring and stolid wonderment, and have set them for the first time musing, pondering, and thinking on the visions floating past their eyes. And having done that, we tried to verify our anticipations by comparing notes with the poets of the Veda, in whose songs the most ancient records of religious thought are preserved to us, at least so far as that branch of humanity is concerned to which we ourselves belong. No doubt, between the first day-break of human thought and the first hymns of praise, composed in the most perfect metre and the most polished language, there may be, nay, there must be, a gap that can only be measured by generations, by hundreds, aye, by thousands of years. Yet such is the continuity of human thought, if once controlled by human language, that, on carefully examining the Vedic hymns, we found most of our anticipations realised, far beyond what we had any right to expect. The very objects which we had singled out as most likely to impress the mind with the sense that they were something more than what could be seen, or heard, or felt in them, had really served, if we might trust the Veda, as "windows through which the ancient Aryans first looked into infinitude."

The Infinite in Its Earliest Conception

When I say infinitude, do not let us take the infinite in its quantitative sense only, as the infinitely small or the infinitely great. Though this is perhaps the most general concept of the infinite, yet it is at the same time the poorest and emptiest. To the ancient Aryans the aspect of the infinite varied with the aspect of each finite object of which it, the infinite, was the ever-present background or complement. The more there was of the visible or audible or tangible or finite, the less there was of the invisible, the inaudible, the intangible, or the infinite in the consciousness of man. As the reach of the senses varied, so varied the suspicion of what might be beyond their reach.

The concept, for instance, of a river or a mountain would require far less of invisible background than the concept of the dawn or the storm-wind. The dawn approaches every morning, but what it is, and whence it comes, no one can tell. "The wind bloweth where it listeth, and thou hearest the sound thereof, but canst not tell whence it cometh and whither it goeth." It was easy to understand the ravages caused by the inundation of a river or by the fall of a mountain; it was more difficult to understand what causes the trees to bend before the approach of a hurricane, and who it is that, during a dark thunderstorm, breaks asunder the mountains and overthrows the stables and huts.

The so-called semi-deities, therefore, which always remained to a great extent within the reach of the senses, seldom assumed that dramatic character which distinguishes other deities; and among those deities again, those who were entirely invisible, and had nothing in nature to represent them, such as Indra, the rainer, Rudra, the howler, the Maruts, the pounders or storm-gods, even Varuna, the all-embracer, would soon assume a far more personal and mythological aspect than the bright sky, the dawn, or the sun. Again, what constitutes the infinite or supernatural character of all these beings, would at once be clothed in a simply human form. They would not be called infinite, but rather unconquerable, imperishable, undecaying, immortal, unborn, present everywhere, knowing everything, achieving everything, and at the very last only should we expect for them names of so abstract a nature as infinite.

I say, we should expect this, but I must say at the same time, that this expecting attitude is often very dangerous. In exploring new strata of thought, it is always best to expect nothing, but simply to collect facts, to accept what we find, and to try to digest it.

Aditi, the Infinite

You will be surprised, for instance, as I certainly was surprised when the fact first presented itself to me, that there really is a deity in the Veda who is simply called the boundless or the infinite, in Sanskrit A-diti.

Aditi is derived from diti, and the negative particle a. Diti again is regularly derived from a root DÂ (dyati), to bind, from which dita, the participle, meaning bound, and diti, a substantive, meaning binding and bond. Aditi therefore must originally have meant without bonds, not chained or enclosed, boundless, infinite, infinitude. The same root shows itself in Greek δέω, I bind, διάδημα, a diadem, that is bound round the head. The substantive diti would in Greek be represented by δέσις, a-diti by ά-δεσις.

It is easy to say that a deity, having such a name as Aditi, the infinite, must be of late origin. It is much wiser to try to learn what is, than to imagine what must be. Because the purely abstract concept of the infinite seemed modern, several of our most learned Vedic students have at once put down Aditi as a late abstraction, as being invented simply to account for the name of her sons, the well-known Âdityas or solar deities. From the fact that there are no hymns entirely addressed to her, they have concluded that Aditi, as a goddess, came in at the very last moments of Vedic poetry.

The same might be said of Dyaus, a name corresponding with the Greek Zεύς. He occurs even less frequently than Aditi amongst the deities to whom long hymns are addressed in the Veda. But so far from being a modern invention, we know now that he existed before a word of Sanskrit was spoken in India, or a word of Greek in Greece; that he is in fact one of the oldest Aryan deities, who at a later time was crowded out, if I may use that expression, by Indra, Rudra, Agni, and other purely Indian gods.

Aditi Not a Modern Deity

The same, I believe, is the case with Aditi. Her name occurs in invocations together with Dyaus, the sky, Prithivî, the earth, Sindhu, the rivers, and other really primitive

deities; and far from being a purely hypothetical mother of the Âdityas, she is represented as the mother of all the gods.

In order to understand this, we must try to find out what her own birthplace was,
what could have suggested the name of Aditi, the boundless, the infinite, and what
was the visible portion in nature to which that name was originally attached.

Natural Origin of Aditi

I believe that there can be little doubt that Aditi, the boundless, was one of the oldest
names of the dawn, or more correctly, of that portion of the sky from whence every
morning the light and life of the world flashed forth.

Look at the dawn, and forget for a moment your astronomy; and I ask you
whether, when the dark veil of the night is slowly lifted, and the air becomes transparent and alive, and light streams forth, you know not whence, you would not feel
that your eye, stretching as far as it can stretch, and yet stretching in vain, was looking into the very eye of the infinite? To the ancient seers the dawn seemed to open
the golden gates of another world, and while these gates were open for the sun to
pass in triumph, their eyes and their mind strove in their childish way to pierce
beyond the limits of this finite world. The dawn came and went, but there remained
always behind the dawn that heaving sea of light or fire, from which she springs. Was
not this the visible infinite? And what better name could be given than that which the
Vedic poets gave to it, Aditi, the boundless, the yonder, the beyond all and everything?

Thus, I believe, we can understand how a deity, which at first seemed to us so
abstract as to have no birthplace anywhere in nature, so modern that we could hardly
believe in its occurrence in the Veda, may have been one of the earliest intuitions and
creations of the Hindu mind.[2] In later times the boundless Aditi may have become
identified with the sky, also with the earth, but originally she was far beyond the sky
and the earth.

Thus we read in a hymn[3] addressed to Mitra and Varuna, representatives of day
and night, "O Mitra and Varuna, you mount your chariot which, at the dawning of
the dawn, is golden-coloured, and has iron poles at the setting of the sun:[4] from
thence you see Aditi and Diti"—that is, what is yonder and what is here, what is
infinite and what is finite, what is mortal and what is immortal.[5]

Another poet speaks of the dawn as the face of Aditi,[6] thus indicating that Aditi
is here not the dawn itself, but something beyond the dawn.

As the sun and all the solar deities rise from the east, we can well understand how
Aditi came to be called the mother of the bright gods, and more particularly of Mitra
and Varuna (Rig-veda, X. 36, 3), of Aryaman and Bhaga, and at last of the seven, or
even eight so-called Âdityas, that is, the solar deities, rising from the east. Sûrya, the
sun, is called not only Âditya (Rig-veda, VIII. 101, 11, *bat mahân asi sûrya, bat
âditya mahân asi,* "Truly, Sûrya, thou art great; truly, Âditya, thou art great"), but
also Âditeya (Rig-veda, X. 88, 11).

It was, no doubt, the frequent mention of these her sons that gave to Aditi almost
from the beginning a decidedly feminine character. She is the mother with powerful,
with terrible, with royal sons. But there are passages where Aditi seems to be
conceived as a male deity, or anyhow as a sexless being.

Though Aditi is more closely connected with the dawn, yet she is soon invoked, not only in the morning, but likewise at noon, and in the evening.[7] When we read in the Atharva-veda, X. 8, 16, "That whence the sun rises, and that where he sets, that I believe is the oldest, and no one goes beyond," we might almost translate "the oldest" by Aditi. Aditi soon receives her full share of veneration and worship, and she is implored not only to drive away darkness and the enemies that lurk in the dark, but likewise to deliver man from any sin which he may have committed.

Darkness and Sin

These two ideas—darkness and sin—which seem to us far apart, are closely connected with each other in the minds of the early Aryans. I shall read you some extracts to show how often one idea, the fear of enemies, evokes the other, the fear of sin, or what we should call our worst enemy. "O Âdityas, deliver us from the mouth of the wolves, like a bound thief, O Aditi!"[8] "May Aditi by day protect our cattle, may she, who never deceives, protect by night; may she, with steady increase, protect us from evil." (*Amhasah,* literally, from anxiety, from choking produced by the consciousness of sin.) "And may she, the wise Aditi, come with help to us by day! May she kindly bring happiness, and drive away all enemies!"[9]

Or again: "Aditi, Mitra, and also Varuna, forgive, if we have committed any sin against you! May I obtain the wide fearless light, O Indra! May not the long darkness come over us!"[10] "May Aditi grant us sinlessness!"[11]

One other idea seems very naturally to have sprung up from the concept of Aditi. Wherever we go, we find that one of the earliest imaginings of a future life arose from the contemplation of the daily coming and going of the sun and other heavenly bodies.[12] As we still say, "his sun has set," they said and believed that those who departed this life would go to the west, to the setting of the sun. The sun was supposed to be born in the morning and to die in the evening; or, if a longer life was given to him, it was the short life of one year. At the end of that the sun died, as we still say, the old year dies.

Immortality

But by the side of this conception, another would spring up. As light and life come from the east, the east, among many of the nations of antiquity, was looked upon as the abode of the bright gods, the eternal home of the immortals; and when the idea had once arisen that the departed or blessed among men joined the company of the gods, then they also might be transferred to the east.

In some such sense we see that Aditi is called "the birthplace of the immortals"; and in a similar sense one of the Vedic poets sings: "Who will give us back to the great Aditi; that I may see father and mother?"[13] Is not this a beautiful intimation of immortality, simple and perfectly natural; and if you look back to the steps which led to it, suggested by the ordinary events of everyday life, interpreted by the unassisted wisdom of the human heart?

Here is the great lesson which the Veda teaches us! All our thoughts, even the apparently most abstract, have their natural beginnings in what passes daily before our senses. *Nihil in fide nisi quod ante fuerit in sensu.* Man may for a time be unheedful of these voices of nature; but they come again and again, day after day, night after night, till at last they are heeded. And if once heeded, those voices disclose their purport more and more clearly, and what seemed at first a mere sunrise, becomes in the end a visible revelation of the infinite, while the setting of the sun is transfigured into the first vision of immortality.

Other Religious Ideas in the Veda

Let us examine one more of those ideas which to us seem too abstract and too artificial to be ascribed to a very early stratum of human thought, but which, if we may judge from the Veda, had risen in the human heart at the very first burst of its intellectual spring-tide. I do not mean to make the Veda more primitive than it is. I know full well the interminable vista of its antecedents. There is ring within ring in the old tree, till we can count no longer, and are lost in amazement at the long, slow growth of human thought. But by the side of much that sounds recent, there is much that sounds ancient and primitive. And here we ought, I think, to learn a lesson from archæology, and not try to lay down from the beginning a succession of sharply divided periods of thought. For a long time archæologists taught that there was first a period of stone, during which no weapons, no tools of bronze or iron, could possibly occur. That period was supposed to be followed by the bronze period, where the graves might yield both bronze and stone implements in abundance, but not a single trace of iron. Lastly, we were told, came the third period, clearly marked by the prevalence of iron instruments, which, when they had once been introduced, soon superseded both stone and bronze workmanship altogether.

This theory of the three periods, with their smaller subdivisions, contained no doubt some truth, but being accepted as a kind of archæological dogma, it impeded for a long time, like all dogma, the progress of independent observation; till at last it was discovered that much in the successive or contemporaneous use of the metals depended on local conditions, and that where mineral or palustric or meteoric iron existed in an easily accessible form, iron implements might be found and were found together with stone weapons, and previous to bronze workmanship.

This ought to be a warning to us against our preconceived theories as to the succession of intellectual periods. There are in the Veda thoughts as rude and crude as any paleolithic weapons, but by the side of them, we find thoughts with all the sharpness of iron and all the brilliancy of bronze. Are we to say that the bright and brilliant thoughts must be more modern than the rudely chipped flints that lie by their side? They may be, but let us remember who the workman is, and that there has been genius at all times, and that genius is not bound by years. To a man who has faith in himself and in the world around him, one glance is as good as a thousand observations; to a true philosopher, the phenomena of nature, the names given to them, the gods who represent them, all vanish by one thought like the mist of the morning, and he declares in the poetical language of the Veda, "There is but One, though the poets call it by many names," *Ekam sat viprâ bahudhâ vadanti.*

No doubt, we may say, the many names of the poets must have come first, before the philosophers could discard them. True, but the poets may have continued for ages invoking Indra, Mitra, Varuna, or Agni, while at the same time the philosophers of India protested, as Herakleitos protested and protested in vain, against the many names and the many temples and the many legends of the gods.

The Idea of Law

It has often been said that if there is an idea which we look for in vain among savage or primitive people, it is the idea of law. It would be difficult to find even in Greek and Latin a true rendering of "the reign of law" once chosen as the title of an important book by the Duke of Argyll. And yet that idea, in its first half-conscious form, is as old as almost anything in the Veda. Much has been written of late of unconscious cerebration, and most exaggerated accounts have been given of it. Yet there is a great deal of mental work going on, which we may call unconscious, viz., all mental work that has not yet found expression in language. The senses go on receiving thousands of impressions, most of which pass unheeded, and seem wiped out for ever from the tablets of our memory. But nothing is ever really wiped out, the very law of the conservation of force forbids it. Each impress leaves its mark, and by frequent repetition these marks accumulate until, from faint dots, they grow into sharp lines, and in the end determine the whole surface, the light, and shade, aye, the general character, of our mental landscape.

Thus we can understand that while the great, and at first overpowering, phenomena of nature were exciting awe, terror, admiration, and joy in the human mind, there grew up by the daily recurrence of the same sights, by the unerring return of day and night, by the weekly changes of the waning and increasing moon, by the succession of the seasons, and by the rhythmic dances of the stars, a feeling of relief, of rest, of security—a mere feeling at first, as difficult to express as it is still to express in French or Italian our "feeling at home," a kind of unconscious cerebration, if you like, but capable of being raised into a concept, as soon as the manifold perceptions which made up that feeling could be comprehended and, being comprehended, could be expressed in conscious language.

This feeling has found expression in various ways among the early philosophers of Greece and Rome. What did Herakleitos mean when he said, "The sun or Helios will not overstep the bounds"[14] (τὰ μέτρα), i.e., the path measured out for him; and what, if he said, the Erinys, the helpers of right, would find him out if he did? Nothing can show more clearly that he had recognised a law, pervading all the works of nature, a law which even Helios, be he the sun or a solar deity, must obey. This idea proved most fertile in Greek philosophy; as for religion, I believe we can trace in it the first germ of the Greek *moira* or fate.

Though we cannot expect to meet with any very ancient and original thoughts among the philosophers of Rome, yet I may quote here a well-known saying of Cicero's, containing a very true application of the thought indicated by Herakleitos: Cicero says that men were intended, not only to contemplate the order of the heavenly bodies, but to imitate it in the order and constancy of their lives;[15] exactly what, as we shall see, the poets of the Veda tried to express in their own simple language.

Let us ask now again, as we did when looking for the first germs of the concept of the infinite, what could have been the birthplace of the idea of order, measure, or law in nature? What was its first name, its first conscious expression?

I believe it was the Sanskrit *Rita,* a word which sounds like a deep key-note through all the chords of the religious poetry of India, though it has hardly ever been mentioned by writers on the ancient religion of the Brahmans.[16]

The Sanskrit *Rita*

Nearly all the gods have epithets applied to them, which are derived from this Rita, and which are meant to convey the two ideas, first, that the gods founded the order of nature, and that nature obeys their commands; secondly, that there is a moral law which man must obey, and for the transgression of which he is punished by the gods. Such epithets are far more important, as giving us an insight into the religion of ancient India, than the mere names of the gods, and their relation to certain phenomena in nature; but their accurate understanding is beset with many difficulties.

The primary, secondary, and tertiary meanings of such words as Rita occur sometimes in one and the same hymn; the poet himself may not always have distinguished very clearly between them; and few interpreters would venture to do for him what he has not done for himself. When *we* speak of law, do we always make it quite clear to ourselves what we mean by it? And can we expect that ancient poets should have been more accurate speakers and thinkers than modern philosophers?

No doubt, in most places where Rita occurs, a vague and general rendering of it such as law, order, sacred custom, sacrifice, may pass unchallenged; but if we look at any of the translations of the Vedic hymns, and ask ourselves what definite meaning we can connect with these high-sounding words, we shall often feel tempted to shut up the book in despair. If Agni, the god of fire, or some other solar deity is called "the firstborn of divine truth," what possible idea can such a translation convey? Fortunately, there is a sufficient number of passages left in which Rita occurs, and which enable us to watch the gradual growth of the word and its meanings.

Much, no doubt, in the reconstruction of such ancient buildings must of necessity be conjectural, and I offer my own ideas as to the original foundation of the word Rita and the superstructures of later periods, as no more than a guess and a first attempt.

The Original Meaning of Rita

Rita, I believe, was used originally to express the settled movement of the sun, and of all the heavenly bodies. It is a participle of the verb *Ri,* which may convey the sense either of joined, fitted, fixed; or of gone, the going, the path followed in going. I myself prefer the second derivation, and I recognise the same root in another word, *Nir-riti,* literally going away, then decay, destruction, death, also the place of destruction, the abyss, and in later times (like Anrita), the mother of Naraka, or hell.

The going, the procession, the great daily movement, or the path followed every day by the sun from his rising to his setting, followed also by the dawn, by day and night, and their various representatives, a path which the powers of night and darkness could never impede, would soon be regarded as the right movement, the good work, the straight path.[17]

It was not, however, so much the daily movement, or the path which it followed, as the original direction which determined it, the settled point from which it started and to which it returned, that became most prominent in the thoughts of the Vedic poets when speaking of Rita. Hence they speak of the path of Rita, which we can only translate by the right path; but which to them was the path determined by that unknown power which they had tried to grasp by the name of Rita.

If you remember how Aditi, the boundless, was at first meant for the east, which every morning seemed to reveal an endless distance beyond the sky from which the sun arose for his daily course, you will not be surprised to find that the Rita, the place or the power which determines the path of the sun, should occasionally in the Veda take the place of Aditi. As the dawn was called the face of Aditi, we find that the sun is called the bright face of Rita[18]; nay, we find invocations in which the great Rita[19] occupies a place next to Aditi, and heaven and earth. The abode of Rita is evidently the east,[20] where, according to a very ancient legend, the light-bringing gods are supposed every morning to break open the dark cave, the hiding-place of the robber, and to bring forth the cows,[21] that is to say, the days, each day being conceived as a cow, walking slowly from the dark stable across the bright pasture-ground of earth and the sky. When that imagery is changed, and the sun is supposed to yoke his horses in the morning and to run his daily course across the world, then Rita is called the place where they unharness his horses.[22] Sometimes it is said that the dawns dwell in the abyss of Rita,[23] and many stories are told, how either the dawns were recovered, or how the dawn herself assisted Indra and the other gods in recovering the stolen cattle, or the stolen treasure, hidden in the dark stable of the night.

Story of Saramâ

One of the best-known stories was that of Indra, who first sent Saramâ, the peep of day, to find out where the cows were hidden. When Saramâ had heard the lowing of the cows, she returned to tell Indra, who then gave battle to the robbers, and brought forth the bright cows. This Saramâ was afterwards represented as the dog of Indra, and the metronymic name given to her sons, Sârameya, having by Professor Kuhn been identified with Hermeias, or Hermes, was one of the first indications to point out to comparative mythologists the right path (the *panthâ ritasya*) into the dark chambers of ancient Aryan mythology. Well, this Saramâ, this old pointer of the dawn, is said to have found the cows, "by going on the path of Rita, the right path, or by going to the Rita, the right place."[24] One poet says: "When Saramâ found the cleft of the rock, she made the old great path to lead to one point. She, the quick-footed, led the way; knowing the noise of the imperishable (cows or days), she went first towards them" (Rig-veda, III. 31, 6).

In the preceding verse, the very path which was followed by the gods and their companions, the old poets, in their attempts to recover the cows, i.e., day-light, is called the path of the Rita; but in another place it is said that Indra and his friends tore Vala, the robber or his cave, to pieces, after finding out the Rita, the right place.[25]

That right, immoveable [sic], eternal place is likewise mentioned when a ποῦ στῶ[26] is looked for from which the gods could have firmly established both heaven and earth. Thus Varuna is introduced as saying, "I supported the sky in the seat of Rita";[27] and later on, Rita, like Satya, the true, is conceived as the eternal foundation of all that exists.

The path of Rita occurs again and again, as followed by the dawn, or the sun, or day and night, and the only way in which we can generally translate it, is the path of right, or the right path.

Thus we read of the dawn:

She follows the path of Rita, the right path; as if she knew them before, she never over-steps the regions.[28]

The dawn, who is born in the sky, dawned forth on the right path; she came near, revealing her greatness. She drove away the evil spirits, and the unkindly darkness.[29]

Of the sun it is said:

The god Savitri toils on the right way, the horn of the Rita is exalted far and wide; the Rita resists even those who fight well.[30]

When the sun rises, the path of Rita is said to be surrounded with rays,[31] and the same thought which was uttered by Herakleitos, "Helios will not overstep the bounds," finds expression in a verse of the Rig-veda, "Sûrya does not injure the appointed places."[32] This path, which is here called the path of Rita, is in other places called the broad walk, *gâtu*;[33] and like Rita, this *gâtu* also, the walk, finds sometimes a place among the ancient deities of the morning.[34] It is evidently the same path on which day and night are said to travel in turn,[35] and as that path varies from day to day, we also hear of many paths which are travelled on by the Asvinau, day and night, and similar deities.[36]

Another important feature is that this path, which is commonly called the path of Rita, is sometimes spoken of as the path which King Varuna, one of the oldest Vedic gods, made for the sun to follow (I. 24, 8); for we thus begin to understand why what in some places is called the law of Varuna, is in others called the law of Rita[37]; how, in fact, Varuna, the god of the all-embracing sky, could sometimes be supposed to have settled and determined what in other places is called the Rita, as an independent power.

When it had once been recognised that the gods overcame the powers of darkness by following the straight path or the path of right, it was but a small step for their worshippers to pray, that they also might be allowed to follow that right path. Thus we read: "O Indra, lead us on the path of *Rita*, on the right path over all evils."[38]

Or, "May we, O Mitra and Varuna, on your path of right, cross over all evils, as one crosses the waters in a ship."[39] The same gods, Mitra and Varuna, are said to proclaim the praises of the great Rita.[40] Another poet says: "I follow the path of Rita well."[41] Evil-doers, on the contrary, are said never to cross the path of Rita.[42]

Here is the first germ of thoughts and feelings which even now are not yet quite extinct:

Awake my soul, and with the sun
Thy daily course of duty run.

Rita, the Sacrifice

If we remember how many of the ancient sacrifices in India depended on the course of the sun, how there were daily sacrifices at the rising of the sun, at noon, and at the setting of the sun,[43] how there were offerings for the full moon and the new moon, while other sacrifices followed the three seasons, and the half-yearly or yearly progress of the sun, we may well understand how the sacrifice itself came in time to be called the path of Rita.[44]

At last Rita assumed the meaning of law in general. The rivers, which in some places are said to follow the path of Rita,[45] are spoken of in other hymns as following the Rita or law of Varuna. There are many more meanings or shades of meaning conveyed by Rita, which however are of less importance for our purpose. I have only to add, that as Rita came to express all that is right, good, and true, so Anrita was used to express whatever is false, evil, and untrue.

The Development of Rita

I do not know whether I have succeeded in giving you a clear idea of this Rita in the Veda, how it meant originally the firmly established movement of the world, of the sun, of morning and evening, of day and night; how the spring of that movement was localised in the Far East; how its manifestation was perceived in the path of the heavenly bodies, or, as we should say, in day and night; and how that right path on which the gods brought light out of darkness, became afterward the path to be followed by man, partly in his sacrifices, partly in his general moral conduct.[46] You must not expect in the development of these ancient conceptions too much accuracy and definiteness of thought. It was not there, it could not be there, and if we attempt to force those poetical imaginings into the various categories of rigorous thought, we shall only break their wings and crush out their soul: we shall have the dry bones, but no flesh, no blood, no life.

Difficulty of Translating

The great difficulty in all discussions of this kind arises from the fact that we have to transfuse thought from ancient into modern forms. In that process some violence is inevitable. We have no word so pliant as the Vedic *Rita*, so full of capability, so ready to reflect new shades of thought. All we can do is to find, if possible, the original focus of thought, and then to follow the various directions taken by the rays that proceeded from it. This is what I have endeavoured to do, and if in so doing I may seem to "have put a new garment upon an old," all I can say is that I see no other way, unless we all agree to speak not only Sanskrit, but Vedic Sanskrit.

A great English poet and philosopher has lately been much blamed for translating the old Hebrew belief in a personal Jehovah into a belief "in an eternal power, not ourselves, that makes for righteousness." It has been objected that it would be impossible to find in Hebrew an expression for so abstract, so modern, so purely English a thought as this. This may be true. But if the ancient poets of the Veda were to live to-day, and if they had to think modern thought and to speak modern speech, I should say that an eternal power, not ourselves, that makes for righteousness, would not be a very unlikely rendering they might feel themselves inclined to give of their ancient Rita.

Was *Rita* a Common Aryan Concept?

One more point, however, has to be settled. We have seen that in the Veda, *rita* belongs to one of the earliest strata of thought. The question now is, was *rita* a purely Vedic, or was it, like Dyaus, Zeus, Jupiter, a common Aryan concept?

It is difficult to speak confidently. There were, as we shall see, cognate ideas that found expression in Latin and German in words derived from the same root *ar,* but there is not sufficient evidence to show that, like the Rita of the Vedic poets, these words started from the conception of the daily, weekly, monthly, and annual movement of the heavenly bodies, and from nothing else.

In Sanskrit we have, besides *rita,* the common word for seasons, *ritu,* meaning originally the regular steps or movements of the year. In Zend, *ratu* is the same word, but it means not only order, but also he who orders.[47]

It has been frequently attempted to identify the Sanskrit *ritu,* season, and *rita,* settled, regular, particularly as applied to the course of the heavenly bodies and to the order of the ancient sacrifices, with the Latin *rîte,* according to religious usage, and *rîtus,* a rite, the form and manner of religious ceremonies. But *rî* in Latin never corresponds to Sanskrit *ri,* which is really a shortened form of *ar* or *ra,* and therefore represented in Latin by *or, er, ur,* and more rarely by *re.*

There seems, however, no difficulty in connecting the Latin *ordo* with our root *ar* or *ri;* and Benfey has shown that *ordo, ordinis,* would correspond to a Sanskrit form *ri-tvan. Ordior,* to weave, would seem to have meant originally a careful and orderly arrangement of anything, more particularly of threads.

The nearest approach to *rita* is to be found in the Latin *ratus,* particularly when we consider that *ratus* was originally referred in Latin also to the constant movement of the stars. Thus Cicero (Tusc. v. 24. 69) speaks of the *motus (stellarum) constantes et rati;* and again (N.D. ii. 20. 51) of the *astrorum rati immutabilesque cursus.* I incline myself to the opinion that this *ratus* in Latin is identical in origin and also in intention with Sanskrit *rita,* only that it never became developed and fixed in Latin as a religious concept, such as we saw in the Vedic *Rita.* But though I hold to this opinion, I do not wish to disguise its difficulties. Rita, if it was preserved in Latin, might have been *artus, ertus, ortus,* or *urtus,* but not *ratus,* not even *ritus,* as it appears in *irritus,* vain, i.e., unsettled. I fully admit, therefore, that phonetically Professor Kuhn's identification of Latin *ratus* with Sanskrit *râta* is far more regular. He derives it from *râ,* to give, and as from the root *dâ* we have in Latin *datum* and *redditum,* so from the root *râ* we should have quite regularly *ratum* and *irritum.* The difficulty in Professor Kuhn's etymology is the meaning. *Râta* means *given,* and though it assumes the meaning of granted, assigned, determined, and though in Zend too, *dâta,* law, comes from *dâ (dhâ),* both to give and to settle,[48] yet there is, as Corssen remarks, no trace of this having ever been the original meaning of Latin *ratum.*[49]

Nor are the phonetic difficulties in identifying Latin *ratus* with Sanskrit *rita* insurmountable. The Latin *ratis,* float, is generally connected with the Sanskrit root *ar,* to row, and Latin *gracilis* with Sanskrit *krisa.* If then Latin *ratus* is the same word as the Sanskrit *rita,* there is every reason to suppose that it too referred originally to the regular and settled movements of the heavenly bodies, and that like *considerare, contemplari,* and many such words, it became afterwards despecialised. In that case it would be interesting to observe that while in Sanskrit *rita,* from meaning the order of the heavenly movements, became in time the name for moral order and righteousness, *ratus,* though starting from the same source, lent itself in Latin and German to express logical order and reasonableness. For from the same root and closely connected with *ratus (pro ratâ)* we have the Latin *ratio,* settling, counting, adding and subtracting,—what we now call *reason,*—and Gothic *rathjo,* number, *rathjan,* to number; Old High German *radja,* speech, and *redjon,* to speak.[50]

Rita is Asha in Zend

But though we look in vain among the other Aryan languages for anything exactly corresponding to the Vedic *rita,* and cannot therefore claim for it, as in the case of Dyaus and Zeus, an antiquity exceeding the first separation of the Aryan races, we can show that both the word and the concept existed before the Iranians, whose religion is known to us in the Zend-Avesta, became finally separated from the Indians, whose sacred hymns are preserved to us in the Veda. It has long been known that these two branches of Aryan speech, which extended in a south-easterly direction, must have remained together for a long time after they had separated from all the other branches which took a north-westerly course. They share words and thoughts in common to which we find nothing analogous anywhere else. Particularly in their religion and ceremonial, there are terms which may be called technical, and which nevertheless are to be found both in Sanskrit and Zend. The word which in Zend corresponds to Sanskrit *rita* is *asha.* Phonetically *asha* may seem far removed from *rita,* but *rita* is properly *arta,* and the transition of Sanskrit *rt* into Zend *sh* is possible.[51] Hitherto *asha* in Zend has been translated by purity, and the modern Parsis always accept it in that sense. But this is a secondary development of the word, as has lately been shown by a very able French scholar, M. Darmesteter;[52] and by assigning to it the meaning which *rita* has in the Veda, many passages even in the Avesta receive for the first time their proper character. It cannot be denied that in the Avesta,[53] as in the Veda, *asha* may often be translated by purity, and that it is most frequently used in reference to the proper performance of the sacrifices. Here the Asha consists in what is called "good thoughts, good words, good deeds," good meaning ceremonially good or correct, without a false pronunciation, without a mistake in the sacrifice. But there are passages which show that Zoroaster also recognised the existence of a kosmos, governed by law or *rita.* He tells how the mornings go, and the noon, and the nights; and how they follow a law that has been traced for them. He admires the perfect friendship between the sun and the moon, and the harmonies of living nature, the miracles of every birth, and how at the right time there is food for the mother to give to her child. As in the Veda, so in the Avesta, the universe follows the Asha, the worlds are the creation of Asha. The faithful, while on earth, pray for the maintenance of Asha, while after death they will join Ormazd in the highest heaven, the abode of Asha. The pious worshipper protects the Asha, the world grows and prospers by Asha. The highest law of the world is Asha, and the highest ideal of the believer is to become an Ashavan, possessed of Asha, i.e., righteous.[54]

This will suffice to show that a belief in a cosmic order existed before the Indians and Iranians separated, that it formed part of their ancient, common religion, and was older therefore than the oldest Gâthâ of the Avesta and the oldest hymn of the Veda. It was not the result of later speculation, it did not come in, only after the belief in the different gods and their more or less despotic government of the world had been used up. No, it was an intuition which underlay and pervaded the most ancient religion of the Southern Aryans, and for a true appreciation of their religion it is far more important than all the stories of the dawn, of Agni, Indra, and Rudra.

Think only what it was to believe in a Rita, in an order of the world, though it be no more at first than a belief that the sun will never overstep his bounds. It was all the difference between a chaos and a kosmos, between the blind play of chance and an intelligible and therefore an intelligent providence. How many souls, even now, when everything else has failed them, when they have parted with the most cherished convictions of their childhood, when their faith in man has been poisoned, and when the apparent triumph of all that is selfish, ignoble, and hideous, has made them throw up the cause of truth, of righteousness, and innocence as no longer worth fighting for, at least in this world; how many, I say, have found their last peace and comfort in a contemplation of the Rita, of the order of the world, whether manifested in the unvarying movement of the stars, or revealed in the unvarying number of the petals, and stamens, and pistils of the smallest forget-me-not! How many have felt that to belong to this kosmos, to this beautiful order of nature, is something at least to rest on, something to trust, something to believe, when everything else has failed! To us this perception of the Rita, of law and order in the world, may seem very little; but to the ancient dwellers on earth, who had little else to support them, it was everything—better than their bright beings, their Devas, better than Agni and Indra; because, if once perceived, if once understood, it could never be taken from them.

What we have learnt then from the Veda is this, that the ancestors of our race in India did not only believe in divine powers more or less manifest to their senses, in rivers and mountains, in the sky and the sun, in the thunder and rain, but that their senses likewise suggested to them two of the most essential elements of all religion, the concept of the infinite, and the concept of order and law, as revealed before them, the one, in the golden sea behind the dawn, the other in the daily path of the sun. These two concepts which sooner or later must be taken in and minded by every human being, were at first no more than an impulse, but their impulsive force would not rest till it had beaten into the minds of the fathers of our race the deep and indelible impression that "all is right," and filled them with a hope, and more than a hope, that "all will be right."

Notes

1. "There is nothing in faith that has not first come through the senses."—The Editor.
2. I have treated fully of Aditi in the Rig-veda, in my translation of the Rig-veda Sanhitâ, vol. 1, pp. 230–251. There is an excellent essay by Dr. Alfred Hillebrandt, "Über die Göttin Aditi," 1876. He derives the word from *dâ,* "to bind" (p. 11), but prefers to explain Aditi by imperishableness, and guards against the idea that Aditi could mean omnipresent.
3. Rig-veda, V. 62, 8.
4. The contrast between the light of the morning and the evening seems expressed by the colour of the two metals, gold and iron.
5. Rig-veda, I. 35, 2.

6. Ibid., I. 113, 19, *áditer ánîkam.*
7. Ibid., V. 69, 3.
8. Ibid., VIII. 67, 14.
9. Ibid., VIII. 18, 6, 7.
10. Ibid., II. 27, 14.
11. Ibid., I. 162, 22.
12. H. Spencer, *Sociology,* i, p. 221.
13. Rig-veda, I. 24, 1.
14. *Heracliti Reliquiæ,* xxix.
15. *De Senectute,* xxi.
16. Ludwig, *Anschauungen des Veda,* p. 15, has given the best account of Rita.
17. Rig-veda, VII. 40, 4.
18. Ibid., VI. 51, 1.
19. Ibid., X. 66, 4.
20. Ibid., X. 68, 4.
21. Sometimes these cows seem to be meant also for the clouds carried off from the visible sky to the dark abyss beyond the horizon.
22. Rig-veda, V. 62, 1.
23. Ibid., III. 61, 7.
24. Ibid., V. 45, 7, *ritám yatî sarámâ gâh avindat;* V. 45, 8.
25. Ibid., X. 138, 1.
26. Or, *locus standi,* a place to stand, after Archimedes' famous saying: "δός μοι ποῦ στῶ καὶ κινῶ τὴν γῆν," "Give me but a place to stand and I will move the earth"—The Editor.
27. Rig-veda, IV. 42, 4.
28. Ibid., I. 124, 3; cf., V. 80, 4.
29. Ibid., VII. 75, 1.
30. Ibid., VIII. 86, 5; X. 92, 4; VII. 44, 5.
31. Ibid., I. 136, 2; I. 46, 11.
32. Ibid., III. 30, 12; cf., I. 123, 9 and 124, 3.
33. Ibid., I. 136, 2.
34. Ibid., III. 31, 15; Indra produced together the sun, the dawn, the walk, and Agni.
35. Ibid., I. 113, 3.
36. Ibid., VIII. 22, 7.
37. Ibid., I. 123, 8, 9, *varunasya, dhâma* and *ritasya dhâma.*
38. Ibid., X. 133, 6.
39. Ibid., VII. 65, 3.
40. Ibid., VIII. 25, 4; cf., I. 151, 4–6.
41. Ibid., X. 66, 13.
42. Ibid., IX. 73, 6.
43. Manu, IV. 25, 26.
44. Rig-veda, I. 128, 2; X. 31, 2; 70, 2; 110, 2; etc.
45. Ibid., II. 28, 4; I. 105, 12; VIII. 12, 3.
46. There is a similar development to be observed in the Hebrew *yâshâr,* straight, from *âshar,* to go forward, a root which has supplied some mythical germs in Hebrew also. See Goldziher, *Mythology among the Hebrews,* p. 123. Still more striking are the coincidences between the Vedic Rita and the Egyptian Maât, so eloquently described in Mr. Le Page Renouf's *Hibbert Lectures,* p. 119 seq. The *Tao* of Lao-tse also seems to be of the same kith and kin.
47. Darmesteter, *Ormazd et Ahriman,* p. 12. It seems to have the same meaning in Kaush. Up. I. 2, where *ritavah,* as a gen. of *ritu,* means the moon, the ordainer of the seasons.
48. Darmesteter, loc. cit., p. 253.
49. Kuhn ingeniously compares the superlative *râtatamâ brahmâni* with the *beneficia ratissima et gratissima,* in *Festus,* ed. Lindemann, p. 236.

50. For further derivatives see Corssen, *Aussprache des Lateinischen,* i, p. 477.
51. The identity of *arta* (*rita*) and *asha* was first pointed out by de Lagarde (*Gesammelte Abhandlungen,* p. 152), and by Opper (*Inscriptions des Achémenides,* p. 105). It was accepted by Haug (*Das 18 Capitel des Vendidad, Sitzungsberichte der Kgl. Bayer. Akad. der Wissenschaften,* 1868, p. 526), and supported by Hübschmann (*Ein Zoroastrisches Lied,* p. 76). Thus Skt. *martya* = Zend *mashya*; Skt. *pritanâ* = Zend *peshanâ*; Skt. *bhartar* = Zend *bâshar*; Skt. *mrita* = Zend *mesha*; Zend [*sic*] *peretu* = Zend *peshu.* Spiegel (*Arische Studien,* p. 33) challenges some of these identifications, and explains them differently. Still he too admits the possible interchange of Skt. *rt* and Zend *sh.* See Pischel, *Gött. gel. Anzeigen,* 1877, p. 1554. In Parsi, Asha Vahista, the excellent Asha, is Ardibehest. Darmesteter, *Vendidâd,* Intr., lxx.
52. *Ormazd et Ahriman, leurs origines et leur histoire,* par James Darmesteter, Paris, 1877.
53. Darmesteter, loc. cit., p. 14.
54. This view of the origin of *asha* has been criticised by M. de Harlez, in the *Journal Asiatique,* 1878, pp. 157–176, and vindicated by M. Darmesteter, *Journal Asiatique,* 1881, p. 492.

CHAPTER 12
FORGOTTEN BIBLES (1884)

The first series of translations of the *Sacred Books of the East,* consisting of twenty-four volumes, is nearly finished, and a second series, which is to comprise as many volumes again, is fairly started. Even when that second series is finished, there will be enough material left for a third and fourth series, and though I shall then long have ceased from my labours as editor, I rejoice to think that the reins when they drop out of my hands will be taken up and held by younger, stronger, and abler conductors.

I ought indeed to be deeply grateful to all who have helped me in this arduous, and, as it seemed at first, almost hopeless undertaking. Where will you get the Oriental scholars, I was asked, willing to give up their time to what is considered the most tedious and the most ungrateful task, translating difficult texts that have never been translated before, and not being allowed to display one scrap of recondite learning in long notes and essays, or to skip one single passage, however corrupt or unintelligible?

And if you should succeed in assembling such a noble army of martyrs, where in these days will you find the publisher to publish twenty-four or forty-eight portly volumes, volumes which are meant to be studied, not to be skimmed, which will never be ordered by Mudie or Smith, and which conscientious reviewers may find it easier to cut up than to cut open?

It was no easy matter, as I well knew, to find either enthusiastic scholars or enthusiastic publishers, but I did not despair, because I felt convinced that sooner or later such a collection of translations of the Fathers of the Universal Church would become an absolute necessity. My hope was at first that some very rich men who are tired of investing their money would come forward to help in this undertaking, but though they seem willing to help in digging up mummies in Egypt or oyster-shells in Denmark, they evidently do not think that much good could come from digging up the forgotten Bibles of Buddhists or Fire-worshippers. I applied to learned societies and academies, but, of course, they had no disposable funds. At last the Imperial Academy of Vienna—all honour be to it—was found willing to lend a helping hand. But in 1875, just when I had struck my tent at Oxford to settle in Austria, the then Secretary of State for India, Lord Salisbury, and the Dean of Christ Church, Dr. Liddell, brought their combined influence and power of persuasion to bear on the Indian Council and the University Press at Oxford. The sinews of war were found for at least twenty-four volumes. In October 1876, the undertaking was started, and, if all goes well, in October 1884, the first series of twenty-four volumes will stand on the shelves of every great library in Europe, America, and India. And more than that. Such has been the interest taken in this undertaking by the students of ancient language, religion, and philosophy, that even the unexpected withdrawal of the patronage of the India Office under Lord Salisbury's successor could not endanger the successful continuation of this enterprise, at least during the few years that I may still be able to conduct it.

But while personally I rejoice that all obstacles which were placed in our way, sometimes from a quarter where we least expected it, have been removed, and that with the generous assistance of some of the best Oriental scholars of our age, some at least of the most important works illustrating the ancient religions of the East have been permanently rescued from oblivion and rendered accessible to every man who understands English, some of my friends, men whose judgement I value far higher than my own, wonder what ground there is for rejoicing. Some, more honest than the rest, told me that they had been great admirers of ancient Oriental wisdom till they came to read the translations of the *Sacred Books of the East.* They had evidently expected to hear the tongues of angels, and not the babbling of babes. But others took higher ground. What, they asked, could the philosophers of the nineteenth century expect to learn from the thoughts and utterances of men who had lived one, two, three, or four thousand years ago? When I humbly suggested that these books had a purely historical interest, and that the history of religion could be studied from no other documents, I was told that since Comte's time it was perfectly known how religion arose, and through how many stages it had to pass in its development from fetishism to positivism, and that whatever facts might be found in the Sacred Books of the East, they must all vanish before theories which, like all Comtian theories, are infallible and incontrovertible. If anything more was to be discovered about the origin and nature of religion, it was not from dusty historical documents, but from psycho-physiological experiments, or possibly from the creeds of living savages.

I was not surprised at these remarks. I had heard similar remarks many years ago, and they only convinced me that the old antagonism between the historical and theoretical schools of thought was as strong to-day as ever. This antagonism applies not only to the study of religion, but likewise to the study of language, mythology, and philosophy, in fact of all the subjects to which my own labours have more specially been directed for many years, and I therefore gladly seize this opportunity of clearly defining once for all the position which I have deliberately chosen from the day that I was a young recruit to the time when I have become a veteran in the noble army of research.

There have been, and there probably always will be, two schools of thought, the *Historical* and the *Theoretical.* Whether by accident or by conviction I have been through life a follower of the Historical School, a school which in the study of every branch of human knowledge has but one and the same principle, namely, *Learn to understand what is by learning to understand what has been.*

That school was in the ascendant when I began life. It was then represented in Germany by such names as Niebuhr for history, Savigny for law, Bopp for language, Grimm for mythology; or, to mention more familiar names, in France by Cuvier for natural history; in England by a whole school of students of history and nature, who took pride in calling themselves the only legitimate representatives of the Baconian school of thought.

What a wonderful change has come over us during the last thirty or forty years! The Historical School which, in the beginning of our century, was in the possession of nearly all professorial chairs, and wielded the sceptre of all the great academies, has almost dwindled away, and its place has been taken by the Theoretical School, best known in England by its eloquent advocacy of the principles of evolution. This Theoretical School is sometimes called the *synthetic,* in opposition to the Historical School, which is *analytic.* It is also characterised as *constructive,* or as reasoning *à priori.* In order to appreciate fully the fundamental difference between the two schools, let us see how their principles have been applied to such subjects as the science of language, religion, or antiquities.

The Historical School, in trying to solve the problem of the origin and growth of language, takes language as it finds it. It takes the living language in its various dialects, and traces each word back from century to century, until from the English now spoken in the streets, we arrive at the Saxon of Alfred, the Old Saxon of the Continent, and the Gothic of Ulfilas, as spoken on the Danube in the fifth century. Even here we do not stop. For finding that Gothic is but a dialect of the great Teutonic stem of language, that Teutonic again is but a dialect of the great Aryan family of speech, we trace Teutonic and its collateral branches, Greek, Latin, Celtic, Slavonic, Persian, and Sanskrit, back to that proto-Aryan form of speech which contained the seeds of all we now see before us, as germs, plants, flowers, fruits in the languages of the Aryan race.

After having settled this historical outline of the growth of our family of speech, the Aryan, we take any word, or a hundred, or a thousand words, and analyse them, or take them to pieces. That words can be taken to pieces, every grammar teaches us, though the process of taking them to pieces scientifically and correctly, dissecting limb from limb, is often as difficult and laborious as any anatomical preparation. Well, let us take quite a modern word: the American *cute,* sharp. We all know that *cute* is only a shortening of *acute,* and that *acute* is the Latin *acutus,* sharp. In *acutus,* again, we easily recognise the frequent derivative *tus,* as in *cornutus,* horned, from *cornu,* horn. This leaves us *acu,* as in *acu-s,* a needle. In this word, the *u* can again be separated, for we know it is a very common derivative, in such words as *pec-u,* cattle, Sanskrit pasú, from PAS, to tether; or *tanú,* thin, Greek τανύ, Lat. *tenu-i-s,* from TAN, to stretch. Thus we arrive in the end at AK, and here our analysis must stop, for if we were to divide AK into A and K, we should get, as even Plato knew (*Theaetetus,* 205), mere letters, and no longer significant sounds or syllables. Now what is this AK? We call it a root, which is, of course, a metaphor only. What we mean by calling it a root is that it is the residuum of our analysis, and a residuum which itself resists all further analysis. But what is important is that it is not a mere theoretic postulate, but a fact, an historical fact, and at the same time an ultimate fact.

With these ultimate facts, that is, with a limited number of predicative syllables, to which every word in any of the Aryan languages can be traced back, or, as we may also express it, from which every word in these languages can be derived, the historical school of comparative philology is satisfied, at least to a certain extent; for it has also to account for certain pronouns and adverbs and prepositions, which are not derived from predicative, but from demonstrative roots, and which have supplied, at the same time, many of those derivative elements, like *tus* in *acu-tus,* which we generally call suffixes or terminations.

After this analysis is finished, the historical student has done his work. AK, he says, conveys the concept of sharp, sharpness, being sharp or pointed. How it came to do that we cannot tell, or, at least, we cannot find out by historical analysis. But that it did so, we can prove by a number of words derived from AK in Sanskrit, Persian, Greek, Latin, Celtic, Slavonic, and Teutonic speech. For instance: Sanskrit *âsu,* quick (originally sharp), Greek ὠκύς, Lat. *oc-ior,* Lat. *ac-er,* eager, *acus, acuo, acies, acumen*; Greek ἀκμή, the highest point, our *edge,* A.-S. *ecg;* also to *egg* on; ἄκων, a javelin, *acidus,* sharp, bitter, *ague,* a sharp fever, *ear* of corn, Old High German *ahir,* Gothic *ahs,* Lat. *acus, aceris,* husk of grain, and many more.

Let us now look at the Theoretical School and its treatment of language. How could language arise?, it says; and it answers: Why, we see it every day. We have only to watch a child, and we shall see that a child utters certain sounds of pain and joy, and very soon after imitates the sounds which it hears. It says *Ah!,* when it is surprised or pleased; it soon says *Baa!,* when it sees a lamb, and *Bow-wow!,* when it sees a dog. Language, we are told, could not arise in any other way; so that interjections and imitations must be considered as the ultimate, or rather the primary facts of language, while their transition into real words is, we are assured, a mere question of time.

This theory seems to be easily confirmed by a number of words in all languages, which still exhibit most clearly the signs of such an origin; and still further, by the fact that these supposed rudiments of human speech exist, even it an earlier stage, in the development of animal life, namely, in the sounds uttered by many animals; though, curiously enough, far more fully and frequently by our most distant ancestors, the birds, than by our nearest relation, the ape.

It is not surprising, therefore, that all who believe in a possible transition from an ape to a man should gladly have embraced this theory of language. The only misfortune is that such a theory, though it easily explains words which really require no explanation, such as crashing, cracking, creaking, crunching, scrunching, leaves us entirely in the lurch when we come to deal with real words—I mean words expressive of general concepts, such as man, tree, name, law—in fact, nine-tenths of our dictionary.

I certainly do not wish to throw unmerited contempt on this Theoretical School. Far from it. We want the theorist quite as much as the historian. The one must check the other, nay, even help the other, just as every government wants an opposition to keep it in order, or, I ought perhaps to say, to give it from time to time new life and vigour. I only wished to show by an example or two, what is the real difference between these two schools, and what I meant when I said that, whether by temperament, or by education, or by conviction, I myself had always belonged to the Historical School.

Take now the science of religion, and we shall find again the same difference of treatment between the historian and the theorist.

The *theorist* begins by assuring us that all men were originally savages, or, to use a milder term, children. Therefore, if we wish to study the origin of religion, we must study children and savages.

Now at the present moment some savages in Africa, Australia, and elsewhere are supposed to be fetish-worshippers. Therefore we are assured that five thousand or ten thousand years ago religion must have begun with a worship of fetishes—that is, of stones, and shells, and sticks, and other inanimate objects.

Again, children are very apt not only to beat their dolls, but even to punish a chair or a table if they have hurt themselves against it. This shows that they ascribe life and personality—nay, something like human nature—to inanimate objects and hence we are told that savages would naturally do the same. A savage, in fact, is made to do everything that an anthropologist wishes him to do; but, even then, the question of all questions, why he does what he is supposed to do, is never asked. We are told that he worships a stone as his God, but how he came to possess the idea of God, and to predicate it of the stone, is called a metaphysical question of no interest to the student of anthropology—that is, of man. If, however, we press for an answer to this all-important question, we are informed that *animism, personification,* and *anthropomorphism* are the three well-known agencies which fully account for the fact that the ancient inhabitants of India, Greece, and Italy believed that there was life in the rivers, the mountains, and the sky; that the sun, and the moon, and the dawn were cognizant of the deeds of men, and, finally, that Jupiter and Juno, Mars and Venus, had the form and the beauty, the feelings and passions of men. We might as well be told that all animals are hungry because they have an appetite.

We read in many of the most popular works of the day how, from the stage of fetishism, there was a natural and necessary progress to polytheism, monotheism, and atheism, and after these stages have been erected one above the other, all that remains is to fill each stage with illustrations taken from every race that ever had a religion, whether these races were ancient or modern, savage or civilised, genealogically related to each other, or perfect strangers.

Again, I must guard most decidedly against being supposed to wish to throw contempt or ridicule on this school. Far from it. I differ from it; I have no taste for it; I also think it is often very misleading. But to compare the thoughts and imaginations of savages and civilised races, of the ancient Egyptians, for instance, and the modern Hottentots, has its value, and the boldest combinations of the Theoretic[al] School have sometimes been confirmed in the most unexpected manner by historical research.

Let us see now how the Historical School goes to work in treating of the origin and growth of religion. It begins by collecting all the evidence that is accessible, and classifies it. First of all, religions are divided into those that have sacred books, and those that have not. Secondly, the religions which can be studied in books of recognised or canonical authority, are arranged genealogically. The New Testament is traced back to the Old, the Koran to both the New and Old Testaments. This gives us one class of religions, the *Semitic.*

Then, again, the sacred books of Buddhism, of Zoroastrianism, and of Brâhmanism are classed together as Aryan, because they all draw their vital elements from one and the same proto-Aryan source. This gives us a second class of religions, the *Aryan.*

Outside the pale of the Semitic and Aryan religions, we have the two book-religions of China, the old national traditions collected by Confucius, and the moral and metaphysical system of Lao-tse. This gives us a class of *Turanian* religions. The study of those religions which have sacred books is in some respects easy, because we have in these books authoritative evidence on which our further reasonings and conclusions can be safely based. But, in other respects, the very existence of these books creates new difficulties, because, after all, religions do not live in books only, but in human hearts, and where we have to deal with Vedas, and Avestas, and Tripitakas, Old and New Testaments, and Korans, we are often tempted into taking the book for the religion.

Still, the study of book-religions, if we once have mastered their language, admits, at all events, of more definite and scientific treatment than that of native religions which have no books, no articles, no tests, no councils, no pope. Any one who attempts to describe the religion of the ancient Greeks and Romans—I mean their real faith, not their mythology, their ceremonial, or their philosophy—knows the immense difficulty of such a task. And yet we have here a large literature, spread over many centuries, we know their language, we can even examine the ruins of their temples.

Think after that, how infinitely greater must be the difficulty of forming a right conception, say, of the religion of the Red Indians, the Africans, the Australians. Their religions are probably as old as their languages, that is, as old as our own language; but we know nothing of their antecedents, nothing but the mere surface of to-day, and that immense surface explored in a few isolated spots only, and often by men utterly incapable of understanding the language and the thoughts of the people. And yet we are asked to believe by the followers of the Theoretic[al] School that this mere surface detritus is in reality the granite that underlies all the religions of the ancient world, more primitive than the Old Testament, more intelligible than the Veda, more instructive than the mythological language of Greece and Rome. It may be so. The religious map of the world may show as violent convulsions as the geological map of the earth. All I say to the enthusiastic believers in this contorted evolution of religious thought is, let us wait till we know a little more of Hottentots and Papúans; let us wait till we know at least their language, for otherwise we may go hopelessly wrong.

The Historical School, in the meantime, is carrying on its more modest work by publishing and translating the ancient records of the great religions of the world, undisturbed by the sneers of those who do not find in the Sacred Books of the East what they, in their ignorance, expected—men, who, if they were geologists would no doubt turn up their noses at a kitchen-midden, because it did not contain their favourite lollypops. Where there are no sacred texts to edit and to translate, the true disciples of the Historical School—men such as, for instance, Bishop Caldwell or Dr. Hahn in South Africa, Dr. Brinton or Horatio Hale in North America—do not shrink from the drudgery of learning the dialects spoken by savage tribes, gaining their confidence, and gathering at last from their lips some records of their popular traditions, their ceremonial customs, some prayers, it may be, and some confession of their ancient faith. But even with all these materials at his disposal, the historical student does not rush at once to the conclusion that either in the legends of the Eskimos or in the hymns of the Vedic Âryas, we find the solution of all the riddles in the science of religion. He only says that we are not likely to find any evidence much more trustworthy, and that therefore we are justified in deriving certain lessons from these materials. And what is the chief lesson to be learnt from them? It is this, that they contain certain words and concepts and imaginations which are as yet inexplicable, which seem simply irrational, and require for their full explanation antecedents which are lost to us; but that they contain also many words and concepts and imaginations which are perfectly intelligible, which presuppose no antecedents, and which, whatever their date may be, may be called primary and rational. However strange it may seem to us, there can be no doubt that the perception of the Unknown or the Infinite was with many races as ancient as the perception of the Known or the Finite, that the two were, in fact, inseparable. To men who lived on an island, the ocean was the Unknown, the Infinite, and became in the end their God. To men who lived in valleys, the rivers that fed them and whose sources were unapproachable, the mountains that protected them, and whose crests were inaccessible, the sky that overshadowed them, and whose power and beauty were unintelligible, these were their unknown beings, their infinite beings, their bright and kind beings, what they called their Devas, their "Brights," the same word which, after passing through many changes, still breathes in our *Divinity*.

This unconscious process of theogony is historically attested, is intelligible, requires no antecedents, and is, so far, a primary process. How old it is, who would venture to ask or to tell? All that the Historical School ventures to assert is that it explains one side of the origin of religion, namely, the gradual process of naming or conceiving the Infinite. While the Theoretic[al] School takes the predicate of God, when applied to a fetish, as granted, the Historical School sees in it the result of a long-continued evolution of thought, beginning with the vague consciousness of something invisible, unknown, and unlimited, which gradually assumes a more and more definite shape through similes, names, myths, and legends, till at last it is divested again of all names, and lives within us as the invisible, inconceivable, unnameable [*sic*]—the infinite God.

I need hardly say that though in the science of religion as in the science of language, all my sympathies are with the Historical School, I do not mean to deny that the Theoretical School has likewise done some good work. Let both schools work on, carefully and honestly, and who knows but that their ways, which seem so divergent at present, may meet in the end.

Nowhere, perhaps, can we see the different spirit in which these two schools, the Historical and the Theoretical, set to work, more clearly than in what is called by preference the Science of Man, *Anthropology;* or the Science of People, *Ethnology;* or more generally the science of old things, of the works of ancient men, *Archaeology*. The Theoretic[al] School begins, as usual, with an ideal conception of what man must have been in the beginning. According to some, he was the image of his Maker, a perfect being, but soon destined to fall to the level of ordinary humanity. According to others, he began as a savage, whatever that may mean, not much above the level of the beasts of the field, and then had to work his way up through successive stages which are supposed to follow each other by a kind of inherent necessity. First comes the stage of the hunter and fisherman, then that of the breeder of cattle, the tiller of the soil, and lastly that of the founder of cities.

As man is defined as an animal which uses tools, we are told that according to the various materials of which these tools were made, man must again by necessity have passed through what are called the three stages or ages of stone, bronze, and iron, raising himself by means of these more and more perfect tools to what we might call the age of steel and steam and electricity, in which for the present civilisation seems to culminate. Whatever discoveries are made by excavating the ruins of ancient cities, by opening tombs, by ransacking kitchen-middens, by exploring once more the flint-mines of prehistoric races, all must submit to the fundamental theory, and each specimen of bone or stone or bronze or iron must take the place drawn out for it within the lines and limits of an infallible system.

The Historical School takes again the very opposite line. It begins with no theoretical expectations, with no logical necessities, but takes its spade and shovel to see what there is left of old things; it describes them, arranges them, classifies them, and thus hopes in the end to understand and explain them. When a Schliemann begins his work at Hissarlik he digs away, notes the depth at which each relic has been found, places similar relics side by side, unconcerned whether iron comes before bronze, or bronze before flint. Let me quote the words of a young and very careful archaeologist, Mr. Arthur Evans, in describing this kind of work, and the results which we obtain from it. "In the topmost stratum of Hissarlik," he writes:

> (which some people like to call Troy), extending six feet down, we find remains of the Roman and Macedonian Ilios, and the Aeolic colony; and the fragments of archaic Greek pottery discovered (hardly distinguishable from that of Sparta and Mykenai) take us back already to the end of the first millennium before our era.
>
> Below this, one superposed above the other, lie the remains of no less than six successive prehistoric settlements, reaching down to over fifty feet below the surface of the hill. The formation of this vast superincumbent mass by artificial and natural causes must have taken a long series of centuries; and yet, when we come to examine the lowest deposits, the remains of the first and second cities, we are struck at once with the relatively high state of civilisation at which the inhabitants of this spot had already arrived.
>
> The food-remains show a people acquainted with agriculture and cattle-rearing, as well as with hunting and fishing. The use of bronze was known, though stone implements continued to be used for certain purposes, and the bronze implements do not show any of the refined forms—notably the *fibulae*—characteristic of the later Bronze Age.
>
> Trade and commerce evidently were not wanting. *Articles de luxe* of gold, enamel, and ivory were already being imported from lands more directly under Babylonian and Egyptian influence, and jade axeheads came by prehistoric trade-routes from the Kuen-Lun, in China. The local potters were already acquainted with the use of the wheel, and the city walls and temples of the second city evince considerable progress in the art of building.[1]

Such is the result of the working of the Historical School. It runs its shaft down from above; the Theoretical School runs its shaft up from below. It may be that they are both doing good work, but such is the strength of temperament and taste, even among scientific men, that you will rarely see the same person working in both mines; nay, that not seldom you hear the same disparaging remarks made by one party and the other, which you may be accustomed to hear from the promoters of rival gold-mines in India or in the south of Africa.

I might show the same conflict between Historical and Theoretical research in almost every branch of human knowledge. But, of course, we are all most familiar

with it through that important controversy, which has occupied the present generation more than anything else, and in which almost every one of us has taken part and
taken sides—I mean the controversy about Evolution.

It seems almost as if I myself had lived in prehistoric times, when I have to confess
that, as a young student, I witnessed the downfall of the theory of Evolution which,
for a time, had ruled supreme in the Universities of Germany, particularly in the
domain of Natural History and Biology. In the school of Oken, in the first philosophy of Schelling, in the eloquent treatises of Goethe, all was Evolution,
Development, or as it was called in German, *Das Werden,* the Becoming. The same
spirit pervaded the philosophy of Hegel. According to him, the whole world was an
evolution, a development by logical necessity, to which all facts must bow. If they
would not, *tant pis pour les faits.*

I do not remember the heyday of that school, but I still remember its last despairing struggles. I still remember at school and at the university rumours of Carbon,
half solid, half liquid, the famous *Urschleim,* now called Protoplasm, the Absolute
Substance out of which everything was evolved. I remember the more or less amus-
ing discussions about the loss of the tail, about races supposed to be still in possession of that ancestral relic. I well remember my own particular teacher, the great
Greek scholar Gottfried Hermann,[2] giving great offence to his theological colleagues
by publishing an essay in 1840 in which he tried to prove the descent of man from
an ape. Allow me to quote a few extracts from this rare and little noticed essay. As
the female is always less perfect than the male, Hermann argued that the law of
development required that Eve must have existed before Adam, not Adam before
Eve. Quoting the words of Ennius: "Simia quam similis, turpissima bestia, nobis";
he goes on in his own peculiar Latin:

> Ex hac nobili gente quid dubitemus unam aliquando simiam exortam putare, quae
> paullo minus belluina facie et indole esset? Ea, sive illam Evam sive Pandoram appel-
> lare placet, quum ex alio simio gravida facta esset, peperit, ut saepenumero fieri constat,
> filium matri quam patri similiorem, qui primus homo fuit.
> Haec ergo est hominis generisque humani origo, non illa quidem valde honesta, sed
> paullo tamen honestior multoque probabilior, quam si ex luto aqua permixto, cui
> anima fuerit inspirata, genus duceremus.

Surely Gottfried Hermann was a bolder man than even Darwin, and to me who
had attended his lectures at Leipzig in 1841, Darwin's *Descent of Man,* published in
1871, was naturally far less novel and startling by its theory than by the facts by
which that theory was once more supported. Kant's philosophy also had familiarised
students of Anthropology with the same ideas. For he, too, towards the end of his
Anthropologie, had spoken of a third period in the development of nature, when an
Oran-Utang or Chimpanzee may develop his organs of locomotion, touch, and
speech to the perfection of human organs, raise his brain to an organ of thought, and
slowly elevate himself by social culture.

But this was not all. Oken (1779–1851) and his disciples taught that the transition from inorganic to organic nature was likewise a mere matter of development.
The first step, according to him, was the formation of rising bubbles, which he called
infusoria, and the manifold repetition of which led, as he taught, to the formation of

plants and animals. The plant was represented by him as an imperfect animal, the animal as an imperfect man. To doubt that the various races of men were descended from one pair was considered at that time, and even to the days of Prichard, not only a theological, but a biological heresy. All variety was traced back to unity—and in the beginning there was nothing but Being; which Being, coming in conflict with Not-being, entered upon the process of Becoming, of development, of evolution. While this philosophy was still being preached in some German universities, a sharp reaction took place in others, followed by the quick ascendency of that Historical School of which I spoke before. It was heralded in Germany by such men as Niebuhr, Savigny, Bopp, Grimm, Otfried Müller, Johannes Müller, the two Humboldts, and many others whose names are less known in England, but who did excellent work, each in his own special line.

I have tried to describe the general character of that school, and I have to confess that during the whole of my life I have remained a humble disciple of it. I am not blind to its weak points. It fixes its eye far too much on the individual; it sees differences everywhere, and is almost blind to similarities. Hence the bewildering mass of species which it admitted in Botany and Zoology. Hence its strong protest against the common origin of mankind; hence its still stronger protest against the transition from inorganic to organic life, from the plant to the beast, from the beast to the man. Hence, in the science of language, its reluctance to admit even the possibility of a common origin of human speech, and, in the science of religion, its protest against deriving the religion of civilised races from a supposed anterior stage of fetishism. Hence in Geology its rejection of plutonic and volcanic theories, and its careful observation of the changes that have taken place, or are still taking place, on the surface of the earth, within, or almost within, the historical recollection of man.

In the careful anatomy of the eye by Johannes Müller, and his philosophical analysis of the conditions of the process of seeing, we have a specimen of what I should call the best work of the Historical School, even in physical science. In Mr. Herbert Spencer's account of the origin of the eye, we have a specimen of what I call the best work of the Theoretical School. Mr. Spencer tells us that what we now call the eye consisted originally of a few pigmentary grains under the outermost dermal layer, and that rudimentary vision is constituted by the wave of disturbance which a sudden change in the state of these pigmentary grains propagates through the body; or, to put it into plain English, that the eye began with some sore place in the skin, sensitive to light, which smarted or tickled, and thus developed in time into what is now the most wonderful mechanism, as described by Johannes Müller, Helmholtz, and others.

Now I have little doubt that many of my readers who have patiently followed my argument up to this point, will say to themselves: "What then about Darwinism?" Is that *historical* or *theoretic[al]*? Is it a mere phase in the evolution of thought, or is it something permanent, and beyond the reach of further development? Such a question is not easy to answer. Nothing is so misleading as names—I mean, even such names as materialism, idealism, realism, and all the rest—which, after all, admit of some kind of definition. But when we use a proper name—the name of a philosopher—and then speak of all he has been and thought and taught, as his *ism,* such as Puseyism or Darwinism, the confusion becomes quite chaotic. And with no one is this more the case than with Darwin. The difference between Darwin and many who call themselves Darwinians is as great at least as that between the horse and the mule. But Darwin himself is by no means a man who can be easily defined and classified. The very greatness and power of Darwin seem to me to consist in his combining the best qualities of what I have called the Historical and Theoretical Schools. So long as he observes and watches the slow transition of individual peculiarities into more or less permanent varieties; so long as he exhibits the changes that take place before our very eyes by means of artificial breeding, as in the case of pigeons; so long as he shows that many of the numberless so-called species among plants or animals share all that is essential in common, and differ by accidental peculiarities only; so long as he traces living species back to extinct species, the remains of which have been preserved to us in the geological archives of our globe; so long, in fact, as he goes backward, step by step, and opens to us page after page in the forgotten book of life, he is one of the greatest and most successful representatives of the Historical School. But when his love of systematic uniformity leads him to postulate four beginnings for the whole realm of organic life, though not yet *one*, like his followers; when he begins to sketch a possible genealogical tree of all generations of living things, though not yet with the heraldic minuteness of his pupil, Professor Haeckel; when he argues that because natural selection can account for certain very palpable changes, as between the wolf and the spaniel, it may also account for less palpable differences, as between the ape and the man, though no real man of science would venture to argue in that way; when, in fact, he allows his hopes to get the better of his fears, he becomes a follower and a very powerful supporter of the Theoretic[al] School.

It may be the very combination of these two characters which explains the enormous influence which Darwin's theories have exercised on the present generation; but, if so, we shall see in that combination the germs of a new schism also, and the conditions of further growth. Great as was Darwin's conscientiousness, we cannot deny that occasionally his enthusiasm, or his logical convictions, led him to judge of things of which he knew nothing, or very little. He had convinced himself that man was genealogically descended from an animal. That was yet merely a theoretical conviction, as all honest zoologists—I shall only mention Professor Virchow—now fully admit. As language had been pointed out as a Rubicon which no beast had ever crossed, Darwin lent a willing ear to those who think that they can derive language, that is, real *logos,* from interjections and mimicry, by a process of spontaneous evolution, and produced himself some most persuasive arguments. We know how able, how persuasive a pleader Darwin could be. When he wished to show how man could have descended from an animal which was born hairy and remained so during life,[3] he could not well maintain that an animal without hair was fitter to survive than an animal with hair. He therefore wished us to believe that our female semi-human progenitors lost their hair by some accident, were, as Hermann said, "minus belluina facie et indole," and that in the process of sexual selection this partial or complete baldness was considered an attraction, and was thus perpetuated from mother to son. It was difficult, no doubt, to give up Milton's Eve for a semi-human progenitor, suffering, it may be, from leprosy or leucoderma, yet Darwin, like Gottfried Hermann, nearly persuaded us to do so. However, in defending so hopeless, or, at all events, so unfortified a position as the transition of the cries of animals into the language of man, even so great a general, as Darwin undoubtedly was, will occasionally encounter defeat, and, I believe I may say without presumption, that, to speak of no other barrier between man and beast, the barrier of language remains as unshaken as ever, and renders every attempt at deriving man genealogically from any known or unknown ape, for the present at least, impossible, or, at all events, unscientific.

After having described, however briefly and imperfectly, the salient features of the two great schools of thought, the *Historical* and the *Theoretical,* I wish in a few words to set forth the immense advantage which the followers of the Historical School enjoy over the mere theorist, not only in dealing with scientific problems, but likewise in handling the great problems of our age, the burning questions of religion, philosophy, morality, and politics.

History, as I said before, teaches us to understand what is by teaching us to understand what has been. All our present difficulties are difficulties of our own making. All the tangles at which we are pulling were made either by ourselves, or by those who came before us. Who else should have made them? The Historical School, knowing how hopeless it is to pull and tear at a tangled reel by main force, quietly takes us behind the scenes, and shows us how first one thread and then another and a third, and in the end hundreds and thousands of threads went wrong, but how in the beginning they lay before man's eyes as even and as regular as on a weaver's loom.

Men who possess the historical instinct, and who, whenever they have to deal with any of the grave problems of our age, always ask how certain difficulties and apparent contradictions first arose, are what we should call practical men, and, as a rule, they are far more successful in unravelling knotty questions than the man who has a theory and a remedy ready for everything, and who actually prides himself on his ignorance of the past. I think I can best make my meaning clear by taking an instance. Whether Dean Stanley was what is now called a scientific historian, a very laborious student of ancient chronicles and charters, is not for me to say; but if I were asked to define his mind, and his attitude towards all the burning questions of the day, whether in politics, or morality, or religion, I should say it was historical. He was a true disciple of the Historical School. I could show it by examining the position he took in dealing with some of the highest questions of theology. But I prefer, as an easier illustration, to consider his treatment of one of the less exciting questions, the question of vestments.

Incredible as it may seem, it is a fact nevertheless that not many years ago a controversy about surplices, and albs, and dalmatics, and stoles raged all over England. The question by whom, at what time, and in what place, the surplice should be worn, divided brother from brother, and father from child, as if that piece of white linen possessed some mysterious power, or could exercise some miraculous influence on the spirit of the wearer. Any one who knew Stanley would know how little he cared for vestments or garments, and how difficult he would have found it to take sides, either right or left, in a controversy about millinery or ritual. But what did he do? "Let us look at the surplice *historically,*" he said. What is a surplice?; and, first of all, What is the historical origin or the etymology of the word? *Surplice* is [from] the Latin *super-pellicium. Super-pellicium* means what is worn over a fur or fur-jacket. Now this fur-jacket was not worn by the primitive Christians in Rome, or Constantinople, or Jerusalsm, nor is there any mention of such a vestment at the time of the Apostles. What, then, is the history of that fur-jacket? So far as we know, it was a warm jacket worn by German peasants in the colder climate of their country, and it was worn by laity and clergy alike, as in fact all garments were which we now consider exclusively ecclesiastical. As this fur-jacket was apt to get dirty and unsightly, a kind of smock-frock, that could be washed from time to time, was worn over it—and this was called the *super-pellicium,* the surplice.

Stanley thought it sufficient gently to remind the wearer of the surplice that what he was so proud of was only the lineal descendant of a German peasant's smock-frock; and I believe he was right, and his historical explanation certainly produced a better effect on all who had a sense of history and of humour than the most elaborate argument on the mystical meaning of that robe of purity and innocence.

He did the same with other vestments. Under the wand of the historian, the *alb* turned out to be the old Roman tunic or shirt, and the deacon officiating in his alb was recognised as a servant working in his shirt-sleeves. The *dalmatic,* again, was traced back to the shirt with long sleeves worn by the Dalmatian peasants, which became recognised as the dress of the deacon about the time of Constantine. The *cassock* and *chasuble* turned out to be great coats, worn originally by laity and clergy alike, while the *cope,* descended from the *copa* or *capa,* also called *pluviale,* was translated by Stanley as a "waterproof." The *mitre* was identified with the caps and turbans worn in the East by princes and nobles, and to this day by the peasant women. The division into two points was shown to be the mark of the crease which is the consequence of its having been folded and carried under the arm, like an opera-hat. The *stole,* lastly, in the sense of a scarf, had a still humbler origin. It was the substitute for the *orarium* or handkerchief, used for blowing the nose. No doubt, the possession and use of a handkerchief was in early times restricted to the "higher circles." It is so to the present day in Borneo, for instance, where only the king is allowed to carry a handkerchief and to blow his nose. In like manner then as in Borneo the handkerchief became the insignia of royalty, it rose in the Roman Church to become the distinctive garment of the deacon.

I know that some of these explanations have been contested, and rightly contested, but the general drift of the argument remains unaffected by such reservations. I only quote them in order to explain what I meant by Stanley's historical attitude, an attitude which all who belong to the Historical School, and are guided by an historical spirit, like to assume when brought face to face with the problems of the day.

But what applies to small questions applies likewise to great. Instead of discussing the question whether the mystic marriage between Church and State can ever be dissolved, the historian looks to the register and to the settlements, in order to find out how that marriage was brought about. Instead of discussing the various theories of inspiration, the historian asks: Who was the first to coin the word? In what sense did he use it? Did he claim inspiration for himself or for others? Did he claim it for one book only, or for all truth? How much light can be thrown on this subject by a simple historical treatment may be seen in some excellent lectures, delivered lately before a Secularist audience by Mr. Wilson, the Head Master of Clifton College, in the presence of the Bishop of Exeter, and published under the title, *The Theory of Inspiration, or, Why men do* not *Believe the Bible.*

And this historical treatment seems to me the best, not only for religious and philosophical, but also for social problems. Who has not read the eloquent pages of Mr. Henry George on *Progress and Poverty*? Who has not pondered on his social panacea, the nationalisation of the land? It is of little use to grow angry about these questions, to deal in blustering rhetoric, or hysterical invective. So long as Mr. Henry George treats the question of the tenure of land historically, his writings are extremely interesting, and, I believe, extremely useful, as reminding people that a great portion of the land in England was not simply bought for investment, but was granted by the sovereign on certain conditions, such as military service, for instance. Those who held the land had to defend the land, and it may well be asked why that duty, or why the taxes for army and navy, should now fall equally on the whole country. It might be said that all this happened a long time ago. But the reign of Charles the Second does not yet belong so entirely to the realm of fable that the nation might not trace its privileges back to that time quite as much as certain families whose wealth dates from the same period. Again, if Mr. Henry George shows that in more recent times common land was enclosed in defiance of historical right, he is doing useful work, if only by reminding lords of the manor that they should not court too close an inspection of their title-deeds. If there are historical rights, there are historical rights on both sides, on the side of those who have no land quite as much as on the side of those who have, and surely we are all of us most thankful that at the time of Charles the Second, and earlier still, at the time of Henry the Eighth, some large tracts of land were nationalised—were confiscated, in fact—that is, transferred from the hands of former proprietors to the fiscus, the national treasury. What would our national universities be without nationalised land? They would have to depend, as in Germany, on taxation, and be administered, as in Germany, by a government board. If, at the same time, some more land had been nationalised in support of schools, hospitals, almshouses, aye, even in support of army and navy, instead of being granted to private individuals, should we not all be most grateful? But though we may regret the past, we cannot ignore it, and, to quote Mr. Henry George's own words, "instead of weakening and confusing the idea of property, we should surround it with stronger sanctions."

So far all historical minds would probably go with Mr. Henry George. But when he joins the Theoretical School, and tells us that every human being born into this world has a divine right to a portion of God's earth, it is difficult to argue with him, for how does he know it? Again, how does he know how much it should be, and, what is more important still, in what part of the world it should be? An acre of land in the city of London is very different from an acre of land in Australia. Besides, what is the use of land unless it has been cleared? An old Indian lawgiver says very truly, "The deer belongs to him who sticks his arrow into him, and the land to him who digs the stumps out of it."[4] If a man by his spade has made a piece of waste land worth having, surely it belongs to him as much as a sheet of paper belongs to the man who has made it worth having by his pen.

But, though I do not see how, with any regard for the rights of property, which Mr. Henry George regards as sacred, the nationalisation of the land could ever be carried out in an ancient country, such as England, without fearful conflicts, or without a religious revival, nor how it could effect, by itself alone, the cure of the crying evils of the present state of our society, I admire Mr. Henry George for the truths, the bitter truths, which he tells us, and it seems to me sheer intellectual cowardice to say that his ideas are dangerous, and should not be listened to. The facts which he places before us are dangerous, but there is far less danger in his theories, even if we all accepted them. We all hold theories which might be called dangerous, if we ever thought of carrying them out. We all hold the theory that we ought to love our neighbour exactly as ourselves; but no one seems afraid that we should ever do so.

One more question still waits for an answer. Although the historical treatment may be the best, and the only efficacious treatment of all problems affecting religion, philosophy, morality, and politics, should we not follow up our tangles in a straight line, from knot to knot, from antecedent to antecedent? And if so, what can be the use of the Sacred Books of the East for the religious problems of the West? What light can the Rig-veda or the Vedânta philosophy of India throw on Kant's *Critique of Pure Reason*? How can the Koran help us in facing modern problems of morality? How can the Laws of Manu, applicable to the village system of ancient India, help us in answering the social problems of Mr. Henry George?

Perhaps the readiest answer I can give is: Look at the sciences of Language, of Mythology, of Religion. What would they be without the East? They would not even exist. We have learnt that history does not necessarily proceed from the present to the past in one straight line only. The stream of history runs in many parallel branches, and each generation has not only fathers and grandfathers, but also uncles and great-uncles. In fact, the distinguishing character of all scientific research in our century is comparison. We have not only comparative philology, but also comparative jurisprudence, comparative anatomy, comparative physiology. Many points in English law become intelligible only by a comparison with German law. Many difficulties in German law are removed by a reference to Roman or Greek law. Many even of the most minute rules of German, Roman, and Greek law become intelligible only by a reference to the ancient customs and traditions preserved in the law-books of India.

This being so, it follows that a real historical study of the ancient language, the ancient philosophy, and the ancient religion of the East and, more particularly of India, may have its very important bearing on the questions nearest to our own hearts. The mere lesson that we are not the only people who have a Bible, that our theologians are not the only theologians who claim for their Bible a divine inspiration, that our Church is not the only Church which has declared that those who do not hold certain doctrines cannot be saved, may have its advantages, if rightly understood.

These indirect lessons are often far more impressive than any more direct teaching. We see them ourselves, or we must draw them for ourselves, and that is always a better discipline than when we have simply to accept what we are told. It may seem a roundabout way, and yet it often leads to the end far more rapidly than a more direct route, nay, in some cases it is the only practicable route.

Let us take comparative anatomy as an illustration:

We all of us want to know what our bodily organism is like, how we see or hear, how we breathe, how we digest—in fact, how we live. But for a long time people shrank from dissecting a human body. They then took a mollusk, or a fish, or a bird, or a dog, or even so man-like an animal as an ape, and they soon grew accustomed to the idea that the muscles, bones, nerves, or even brains in the anatomical preparations correspond to their own muscles, their own bones, their own nerves, even their own brains. They gladly listened to an explanation how all these organs work together in the bodies of animals, and produce results very similar to those which they know from their own experience. Their mind thus grew stronger, larger, and more comprehensive—it may be, more tolerant.

If after a time you go a step further, and bring a dead human body before them to dissect it before their eyes, there will be at first a little shudder creeping over them, something like the feeling which a young curate might have when recognising for the first time the smock-frock of a German peasant as the prototype of his own beloved surplice. However, even that shudder might possibly be overcome, and in the end some useful lesson might be learned from seeing ourselves as we are in the flesh.

But now suppose some bold vivisectionist were to venture beyond, and to dissect before our eyes a living man, in order to show us how we really breathe, and digest, and live, or in order to make us see what is right and wrong in his system. We should all say it was horrible, intolerable. We should turn away, and stop the proceedings.

If we apply all this, *mutatis mutandis,* to a study of religion, we shall readily understand the great advantages not only of an historical study of our own religion, but also of a comparative study of Eastern religions as they can be studied now in the translations of the Sacred Books of the East. Those who are willing to learn may learn from a comparative study of Eastern religions all that can be known about religions—how they grow, how they decay, and how they spring up again. They may see all that is good and all that is bad in various forms and phases of ancient faith, and they must be blinder than blind if they cannot see how the comparative anatomy of those foreign religions throws light on the questions of the day, on the problems nearest to our own hearts, on our own philosophy, and on our own faith.

Notes

1. *Academy,* December 29, 1883.
2. "Evam ante Adamum creatam fuisse, sive de quodam communi apud Mosen et Hesiodum errore circa creationem generis humani," in Ilgen's *Zeitschrift für die histor. Theologie,* 1840, B. X., pp. 61–70.
3. *Descent of Man,* ii, p. 377, where more details may be found as to the exact process of baldness or denudation in animals.
4. In Australia, if two or more spears are found in the same animal when killed, it is the property of him who threw the first. Nicolay, *Account of the Natives of Western Australia,* Perth, 1879, p. 11.

CHAPTER 13

PHYSICAL RELIGION

from *Gifford Lectures on Physical Religion* (1890)

Definition of Physical Religion

Physical Religion is generally defined as a worship of the powers of nature. We hear it said of ancient as well as of modern nations, that their gods were the sun or the moon, the sky with its thunder and lightning, the rivers and the sea, the earth, and even the powers under the earth. As Aaron said to the Israelites, the poets and prophets of the heathens are supposed to have said to their people: "These be thy gods."

There are some well-known philosophers who go even further, and who, repeating again and again the old mistake of De Brosses and Comte, maintain that the earliest phase of all religion is represented by people believing in stones and bones and fetishes of all kinds as their gods.

God, as a Predicate

As their gods! Does it never strike these theorisers that the whole secret of the origin of religion lies in that predicate, *their gods*. Where did the human mind find that concept and that name? That is the problem which has to be solved; everything else is mere child's play.

We ourselves, the heirs of so many centuries of toil and thought, possess, of course, the name and concept of God, and we can hardly imagine a human mind without that name and concept. But, as a matter of fact, the child's mind is without that name and concept, and such is the difference of meaning assigned by different religions, nay, even by members of the same religion, to the name of God, that a general definition of it has almost become an impossibility. Nevertheless, however our ideas of God may differ, for us to say that the sun or the moon, or a pebble, or the tail of a tiger was God, would be absurd and self-contradictory.

The Greeks also, at least the more enlightened among them, who had arrived at the name and concept of God—men, I mean, like Socrates and Plato—could never have brought themselves to say that any one of their mythological deities, such as Hermes or Apollo, was God, ὁ θεός. The Greeks, however, had likewise the name and concept of gods in the plural, but even that name, which has a meaning totally different from that of God in the singular, could never have been applied by them to what are called fetishes, bones, feathers, or rags. Most of the Negro tribes, who are so glibly classed as fetish-worshippers, possess a name of God, quite apart from their fetishes; nay, their concept of God is often very pure and simple and true. But they would never apply that name to what we, not they, have called their fetish-gods. All they really do is to preserve with a kind of superstitious awe some casual objects, just as we nail a horse-shoe on our stable-doors, or keep a farthing for luck in our purse. These objects they call *gri-gri*, or *ju-ju*.[1] This may mean anything, but certainly it does not mean fetish in the sense given to this word by De Brosses and others, neither does it mean God.

It has led to the greatest confusion of thought that our modern languages had to take the singular of the Greek plural, θεοί, the gods, and use it for θεός, God. It is quite true historically that the idea of θεός, God, was evolved from the idea of θεοί, gods; but in passing through that process of intellectual evolution, the meaning of the word became changed as completely as the most insignificant seed is changed when it has blossomed into a full-blown rose. Θεός, God, admits of no plural, θεοί always implies plurality.

The problem of Physical Religion has now assumed a totally different aspect, as treated by the Historical School. Instead of endeavouring to explain how human beings could ever worship the sky as a god, we ask, how did any human being come into possession of the predicate god? We then try to discover what that predicate meant when applied to the sky, or the sun, or the dawn, or the fire. With us the concept of God excludes fire, the dawn, the sun, and the sky; at all events, the two concepts no longer cover each other. What we want to study therefore is that ever-varying circumference of the predicate god, which becomes wider or narrower from century to century, according to the objects which it was made to include, and after a time to exclude again.

This problem, and a most difficult problem it is, can be studied nowhere so well as in the Veda, that is, in the ancient hymns of the Rig-veda. I doubt whether we should ever have understood the real nature of the problem with which we have to deal, unless we had become acquainted with the Rig-veda.

Deification

It is quite clear that other nations also passed through the same phases of thought as the Aryan conquerors of India. We see the results of that process everywhere. In Africa, in America, in the Polynesian islands, everywhere we catch glimpses of the process of deification. But the whole of that process is nowhere laid open before our eyes in such fulness and with such perspicuity as in the Veda. Deification, as we can watch it in the Veda, does not mean the application of the name and concept of god to certain phenomena of nature. No, it means the slow and inevitable development of the concept and name of God out of these very phenomena of nature—it means the primitive theogony that takes place in the human mind as living in human language.

It has always been perfectly well known that *Zeus*, for instance, had something to do with the sky, *Poseidon* with the sea, *Hades* with the lower regions. It might have been guessed that *Apollo*, like *Phœbos* and *Helios*, had a solar, *Artemis*, like *Mene*, a lunar character. But all this remained vague, the divine epithet applied to them all remained unintelligible, till the Veda opened to us a stratum of thought and language in which the growth of that predicate could be watched, and its application to various phenomena of nature be clearly understood.

It will be the chief object of this course of lectures to elucidate this process of religious evolution, to place clearly before you, chiefly from the facts supplied by the hymns of the Veda, the gradual and perfectly intelligible development of the predicate god from out of the simplest perceptions and conceptions which the human mind gained from that objective nature by which man found himself surrounded.

The Natural and the Supernatural

We have now classified the whole of our experience which we derive from nature under two heads, as either *natural* or *supernatural, natural* comprising all that seems to us regular, conformable to rule, and intelligible, *supernatural* all that we consider as yet or altogether as beyond the reach of rule and reason. This, however, as you will see, is but the last result of a long succession of intellectual labour. At first sight, nothing seemed less natural than nature. Nature was the greatest surprise, a terror, a marvel, a standing miracle, and it was only on account of their permanence, constancy, and regular recurrence that certain features of that standing miracle were called natural, in the sense of foreseen, common, intelligible. Every advance of natural science meant the wresting of a province from the supernatural, if we may use that word in the sense of what remains as yet a surprise, a terror, a marvel, or a miracle in man's experience of objective nature.

It was that vast domain of surprise, of terror, of marvel, and miracle, the unknown, as distinguished from the known, or, as I like to express it, the infinite, as distinct from the finite, which supplied from the earliest times the impulse to religious thought and language, though in the beginning these thoughts and names had little of what we now call religions about them. You remember that the very name of *deva* in Sanskrit, of *deus* in Latin, which afterwards became the name of God, meant originally bright, and no more. It came to mean God after a long process of evolution, which took place even before the Aryan separation, and of which we can only just catch the last glimpses in the phraseology of the Vedic poets.

Agni, Fire, as One of the Devas

How this came about we shall, I think, best learn to understand if we analyse the growth of one of the many Devas or gods who form the Pantheon of the Veda. Many of these Vedic Devas appear likewise under more or less puzzling disguises in the mythology and religion of the other Aryan nations. Some, however, exist in the Veda only as real Devas, while we find no trace of them, as mythological or divine beings, in other countries of the Aryan world. I shall begin my analysis of Physical Religion with a Deva, belonging to this latter class, with the god of fire, called Agni in the Veda, but unknown under that name in any of the other Aryan mythologies, though the word *agni*, in the sense of fire, occurs in Latin as *ignis,* in Lituanian as *ugnì,* in old Slavonic as *ogni.*

When I say the god of fire, I use an expression which has become familiar to us from classical mythology. We speak of a god of the sky, or of the wind, or of the rain.

But you will see that in the Veda we can watch this god of fire long before he is a god at all; and, on the other hand, we shall be able to trace his further growth till he is no longer a god of fire merely, but a supreme god, a god above all other gods, a creator and ruler of the world.

In fact we shall learn to understand by this one instance the authentic history of that long psychological process which, beginning with the simplest and purely material perceptions, has led the human mind to that highest concept of deity which we have inherited together with our language, as members of the great Aryan, and not of the Semitic family.

Early Conceptions of Fire

If you can for a moment transfer yourselves to that early stage of life to which we must refer not only the origin, but likewise the early phases of Physical Religion, you can easily understand what an impression the first appearance of Fire must have made on the human mind. Fire was not given as something permanent or eternal, like the sky, or the earth, or the water. In whatever way it first appeared, whether through lightning or through the friction of the branches of trees, or through the sparks of flints, it came and went, it had to be guarded, it brought destruction, but at the same time it made life possible in winter, it served as a protection during the night, it became a weapon of defence and offence, and last, not least, it changed man from a devourer of raw flesh into an eater of cooked meat. At a later time it became the means of working metal, of making tools and weapons, it became an indispensable factor in all mechanical and artistic progress, and has remained so ever since. What should we be without fire even now?

The Etymological Meaning of Agni

What then did the early Âryas think of it, or, what is the same, how did they name it? Its oldest name in Sanskrit is Agni, and this has been preserved in Latin as *ignis,* in Lituanian as *ugnì,* in old Slavonic as *ogni.* It was therefore a very old name. So far as we can venture to interpret such ancient names, Agni seems to have expressed the idea of quickly moving, from a root AG or AG, to drive. The nearest approach would be the Latin *ag-ilis.* Another Sanskrit name for fire is *vah-ni,* and this, too, coming from the same root which we have in *veho* and *vehemens,* would have meant originally what moves about quickly. In the Veda Agni is called *raghupatvan,* quickly flying (X. 6, 4).

Names of Fire

It will be useful to examine some more of the old names of fire, because every one of them, if we can still interpret it etymologically, will enable us to see in how many different ways fire was conceived by the Âryas, how it struck them, what they thought of it.

Dahana means simply the burner.

Anala, from *an,* to breathe, would seem to mean the breathing, or blowing fire, just as *anila* is a name for wind. The root AN, to breathe, is the same which we have in *animus, anima,* and in Greek ἄνεμος. In the Veda the fire is often said to be breathing (*abhi-svasan,* I. 140, 5).

Pâvaka, a frequent name of Agni, conveys the meaning of cleaning, clearing, illuminating. Some scholars have derived πῦρ and *fire* from the same root.

Tanûnapât is a Vedic name of Agni. It is explained as meaning "offspring of himself." It is possible, no doubt, to conceive Agni as self-born. He is called *sva-yoni* in the Mahâbhârata (19, 13931). But the usual idea in the Veda is that he has a father and mother, namely, the two fire-sticks.

Gâtavedas, another name for Agni, means all-seeing, all-knowing, like *visvavedas.*

Vaisvânara seems to convey the meaning of kept by all men, or useful and kind to all, universal.

Another epithet applied to Agni is *Bhuranyu.* Bhuranyu means quick, and is formed on the same lines as Agni and Vahni. Derived as it is from a root BHAR, to bear, to carry, it seems to have meant originally, carried along headlong, borne away, or possibly, bearing away, like the Greek φερόμενος. This Sanskrit word *bhuranyu* is almost the same word as the Greek Φορωνεύς, who is supposed to have brought to men the gift of fire, and to have become the founder of cities (Paus. ii. 15, 5).[2]

Fire, Named as Active

We ourselves occupy, of course, a totally different position from those who had first to conceive and to name fire. We learn the name mechanically from our parents, and the sound fire is a mere outward sign for what burns and hurts, or warms and cheers us. In after life we may learn to call fire with the ancient Greek philosophers one of the four elements; and, later on, a study of natural philosophy may teach us that fire consists of luminous and calorific rays, that it is a natural force, or, it may be, a motion of something unknown which we call ether. But in all this we deal with predicates only, and the underlying substance remains as unknown as the underlying agent whom the, as yet, undivided Âryas called simply Agni, the mover.

At all events we may well understand that the early inhabitants of the earth were puzzled by the fire. There was nothing like it in the whole world—now visible, now invisible, tangible, yet dangerous to touch, destroying whole forests and the habitations of men, and yet most welcome on the hearth, most cheerful in winter.

We can well understand how, after the senses had once taken note of this luminous apparition in its ever-varying aspects, a desire arose in the human mind, and in the human mind only, to know it; to know it, not simply in the sense of seeing or feeling it, but to know it in the sense of conceiving and naming it, which is a very different thing.

How could that be done? I cannot explain here once more the whole of the process of conceiving and naming, or naming and conceiving. You will find that subject treated in my first course of Gifford Lectures, and more fully in my work on *The Science of Thought,* published in 1887.

I can here only state it as a fact that the only instruments by which man could achieve this process of naming were what we call *roots,* and that all these roots, owing to the manner in which they first came into existence, expressed actions, the ordinary actions performed by men in an early state of society. There were roots expressive of striking, pushing, carrying, binding, lifting, squeezing, rubbing, and all the rest, and with these roots all that we now call naming and conceiving, the whole of our language, the whole of our thought, has been elaborated.

This is a fact, simply a fact, and not a mere theory. To doubt it, as has been done of late again, is to doubt the laws of thought. We may differ as to the exact form in which those roots existed from the first. Such doubts are allowable with regard to roots, as elements of speech, they are allowable with regard to letters, as the elements of sound, nay even with regard to the chemical elements, as constituting the whole material world. But to doubt the existence of any of these three classes of elements is either ignorance or unreason.

No one denies that we name and conceive by means of signs. These signs might have been anything, but, as a matter of fact, they were sounds; and again, as a matter of fact, these sounds were what in the Science of Language we call roots. When we examine these roots, as the actual elements of speech, we find that they signify acts, and we conclude that their sound was originally the involuntary *clamor concomitans* of the simplest acts of man. This last conclusion may no doubt be called an hypothesis only, and I have never represented it as anything else; but, till a better hypothesis has been suggested, I retain it as the best working hypothesis.

If then the Âryas possessed a root, such as AG, by which they expressed their own acts of marching, running, jumping, and, at last, moving in general, all they did in naming and conceiving the marching, running, jumping, or quickly moving luminous appearances of fire, was to say to each of them: "Moving here," "Moving there," or in Sanskrit *Ag-ni-s*.[3]

Agni therefore meant originally the mover, and no more. Many more qualities of the mover might be recalled by the name of Agni, but they were not definitely expressed by that one name. We must remember, however, that by calling him Agni, or the quick mover, the ancient people knew no more who or what that mover was than we do when speaking of fire as an element, or as a force of nature, or, as we do now, as a form of motion. It sounds very learned when we say that "a mass of matter becomes a source of light and heat in consequence of an extremely rapid vibratory movement of its smallest particles, which is propagated as a series of undulations into the surrounding ether, and is felt by our tactile nerves as heat, and by our optic nerves, if the undulations are sufficiently rapid, as light."

I confess, from a philosophical point of view, I see little difference between this Ether, and Agni, the god of fire. Both are mythological. Professor Tyndal asks quite rightly: "Is it in the human mind to imagine motion, without at the same time imagining something moved? Certainly not. The very conception of motion includes that of a moving body. What then is the thing moved in the case of sunlight? The undulatory theory replies that it is a substance of determinate mechanical properties, a body which may or may not be a form of ordinary matter, but to which, whether it is or not, we give the name of Ether."

May not the ancient Âryas say with the same right: "Is it in the human mind to imagine motion without at the same time imagining some one that moves?" Certainly not. The very conception of motion includes that of a mover, and, in the end, of a prime mover. Who then is that mover? The ancient Âryas reply that it is a subject of determinate properties, a person who may or may not be like ordinary persons, but to whom, whether he is or not, we give the name of Agni.

Agni as a Human or Animal Agent

When that step had once been made, when the word *Agni,* Fire, had once been coined, the temptation was great, nay, almost irresistible, as Agni was conceived as an agent, to conceive him also as something like the only other agents known to man, as either an animal or human agent.

We often read in the Veda of the tongue or the tongues of Agni, which are meant for what we call his lambent flames. We read of his bright teeth (*sukidan,* VII. 4, 2), of his jaws, his burning forehead (*tapuh-mûrdhan,* VII. 3, 1), nay, even of his flaming and golden hair *(sokih-kesa,* V. 8, 2; *hiranyakesa,* I. 79, 1), and of his golden beard (*hirismasru,* V. 7, 7). His face (*anîkam*) is mentioned, but that means no more than his appearance, and when he is called winged (I. 58, 5; VIII. 32, 4), or even the hawk of the sky (*divah syenah,* VII. 15, 4), that is simply intended to express, what his very name expresses, his swift movement.

This may help to explain how some nations, particularly the Egyptians, were led on to conceive some of their gods in the shape of animals. It arose from a necessity of language. This was not the case, however, in India. Agni and the other gods of the Veda, if they are imagined at all in their bodily shape, are always imagined as human, though never as so intensely human as the gods and goddesses of the Greeks. Beauty, human, superhuman, ideal beauty, is not an Indian conception. When in later times the Indians also invented plastic representations of their gods, they did not shrink from unnatural and monstrous combinations, so long as they helped to convey the character of each god.

All this is perfectly intelligible, and a careful study of language supplies us with the key to almost all the riddles of ancient mythology.

New Explanation of Animism, Personification, and Anthropomorphism

Formerly the attribution of movement, of life, of personality, and of other human or animal qualities to the great phenomena of nature, was explained by names such as *Animism, Personification, Anthropomorphism.* It seemed as if people imagined that to name a process was to explain it.

Mr. Herbert Spencer, against Animism

Here we owe a debt of gratitude to Mr. Herbert Spencer for having stood up for once as the champion of primitive man. I have often pointed out the bad treatment which these poor primitive creatures receive at the hands of anthropologists. Whatever the anthropologists wish these primitives to do or not to do, to believe or not to believe, they must obey, like silent Karyatides supporting the airy structures of ethnological psychology (*Völkerpsychologie*). If *Animism* is to be supported, they must say, "Of course, the storm has a soul." If *Personification* is doubted, they are called in as witnesses that their fetish is very personal indeed. If *Anthropomorphism* has to be proved as a universal feature of early religion, primitive man is dragged in again, and has to confess that the uncouth stone which he worships is certainly a man, and a great deal more than a man.

Whenever I protested against this system of establishing Animism, Personification, and Anthropomorphism as the primeval springs of all religion, I was told that I knew nothing of primitive man, nor of his direct descendants, the modern savages. I have always pleaded guilty of a complete want of acquaintance with primitive man, and have never ventured to speak about savages, whether ancient or modern, unless I knew something, however little, of the nature of their language. Mr. H. Spencer, however, cannot be disposed of so easily. If any one knows the savages, surely he does. But even he has had to protest at last against the theory that the primitive man is a kind of maid-of-all-work, at the beck and call of every anthropologist. "The assumption," he writes, "tacit or avowed, that the primitive man tends to ascribe life to things which are not living, is clearly an untenable assumption" (*Sociology,* p. 143). He defends even the child, which has likewise had to do service again and again for what I called Nursery-psychology, against the charge of animism. When a child says, "Naughty chair to hurt baby—beat it," Mr. Herbert Spencer shows that this burst of anger admits of very different explanations, and that no one would be more frightened than the child if the chair, on being beaten, began to kick, to bite, or to cry.

But though Mr. Herbert Spencer does not believe that any human being ever mistook an inanimate for an animate object, for even animals have learnt to make that distinction, he still considers them capable of very wonderful follies. He thinks that they do not distinguish between what they see in dreams and what they see while awake (p. 147), nay, he considers them capable of mistaking their actual shadows for their souls. On this point we shall have to touch at a later time.

At present it suffices to state that all these processes have now been traced back to their *vera causa,* namely, to language, and more particularly to what are called the roots of language. As every one of these roots expressed, owing to their very origin, one of the many acts with which men in an early state of society were most familiar, the objects thus named could not be named and conceived except as agents of such acts or as subjects.

If the Aryan nations wished to speak of fire, they could only speak of it as doing something. If they called it Agni, they meant the agent of fire. Instead of this understood agent, implied in the name of Agni, we hear other nations speak of the heart, the soul, the spirit, the lord, or the god of fire.[4] But all these expressions belong to a later phase of thought, for they presuppose the former elaboration of such concepts as soul, spirit, god, or they are based on metaphor, as in the case of heart.

Professor Tiele's Theory of the Gods as *Facteurs*

Professor Tiele in his *Le Mythe de Kronos,* 1886, came nearest to my own view on the development of the concept of God. "The ancient gods," he says, "are what, according to our abstract manner of speaking, we should call '*des facteurs, des forces, des sources de vie*'" (p. 9). He does not indeed lay stress on the fact that there was in our very language and thought an irresistible necessity of our speaking of the sky, the sun, the fire, if we speak of them at all, as agents. He only warns us against supposing that "the gods are ever the phenomena of nature themselves, considered as acting persons, but always what he calls souls or spirits, represented as analogous to the soul of man, that impart movement to the celestial bodies and produce all the effects for good or evil which appear in nature." This is most true, but does it not explain one difficulty by another? Was the soul of man a matter of more easy discovery than the soul of the sky? When we have once arrived at the concept of a spirit, as something substantial, yet different from the material body, the task of the religious and mythological poet is easy enough. In another place, Professor Tiele most rightly defines the physical deities, not as *"des objets naturels que l'on a personnifiés,"* but as *"des êtres positifs, des esprits, que l'on a vus à l'œuvre dans la nature, où ils se manifestent par leur action"* (p. 30). All this is perfectly true in our modern languages, which supply us with such terms as *esprits* and *êtres positifs,* ready made, but if we have to account for the more ancient formations and the earliest strata of religious thought, the science of language alone will solve the riddle why the great phenomena of nature were named as agents, as *facteurs,* nay, it will show that what at first seemed a mere freak of fancy was in reality a necessity of language. While I accept Professor Tiele's *facteurs,* I cannot, for the early periods of human thought, accept his *forces* or *sources de vie.* While I gladly accept Mr. H. Spencer's *agents,* I cannot accept his *agencies.*[5]

The Agents in Nature

Facts are stronger than theories, and unless the facts as collected in my *Science of Thought* can be shown to be no facts, the fact remains and will remain for ever, that all objects which were named and conceived at all, were named and conceived at first as agents. The sky was he who covers, the sun he who warms, the moon he who measures night and day, the cloud he who rains, the fire he who moves, the horse he who runs, the bird he who flies, the tree he who grows or shades, even the stone he who cuts. We need not wonder at this, for we ourselves still speak of a cutter, a tender, a sucker, a slipper, of clinkers and splinters, without thinking of the activities ascribed to all these objects by the primitive framers of words.

Though the agents of the different acts of nature remained unknown, yet as the agents of the light of the sun or of the rain of the clouds, they were conceived as very real agents. All this was the work, the almost inevitable work of language, provided always that we take language in the sense of the Greek *logos,* comprehending both speech and thought as one.

The Categories of the Understanding

If we once have accustomed ourselves to speak of thought as something different from language, then, of course, instead of appealing to the necessities of language as a whole, we should, with Kant, have to appeal to the categories of the understanding. We should then have to recognise the category of substance as embodied in the active character of roots. We should thus gain, perhaps, a clearer insight into the abstract process of thought, but we should lose all that is most important to us, namely, the historical growth of the human mind.

I have neither forgotten Kant, nor surrendered my belief in his categories. But the study of language, as the embodiment of thought, has made it clear to me that Kant's categories are abstractions only. They have no existence by themselves. They are not pigeon-holes made of a pine and covered with cloth—they are simply the inside of language.

The Categories of Language

Justice has at last been done to language. At first Aristotle learnt from language what he very properly called the categories, that is, the predicaments, or what we can predicate of our experience. Afterwards these categories, though originally abstracted from language, claimed complete independence and became extremely masterful in their relation to language and grammar. At last, however, language has now resumed her proper position as the only possible embodiment of deliberate thought, and the categories, so far from being the moulds in which language was cast, are recognised once more as the inherent forms of thought-language.

We shall thus understand why fire, if it was to be named at all, could at first be named in one way only, namely, as an agent.

Fire, as a Deva

We may now advance a step further, and ask how it was that Agni in the Veda is not conceived as an agent only, but as a god, or, if not, as yet, as a god in the Greek sense of the word, at least as a *Deva*. How shall we account for that?

Here we touch at once on the most vital point in our analysis. Certainly in the Veda Agni was called *deva*, perhaps more frequently than any other god. But fortunately in the Veda we can still discover the original meaning of the word *deva*. It did not mean divine, for how should such a concept have been suddenly called into being? *Deva* is derived from the root DIV, and meant originally bright. From the same root we have in Sanskrit *diva*, sky, *divasa*, day, in Latin *dies*, and many more, all originally expressive of light and brightness. In many passages where Agni, or the Dawn, or the Sky, or the Sun are called *deva*, it is far better to translate *deva* by bright than by divine, the former conveying a natural meaning in harmony with the whole tenour of the Vedic hymns, the latter conveying hardly any meaning at all.

But it is true nevertheless that this epithet *deva*, meaning originally bright, became in time, in the Vedic, nay even in the Aryan period already, the recognised name of those natural agents whom we have been accustomed to call gods. We can watch the evolutionary process before our very eyes. When the different phenomena of nature representing light, such as the morning, the dawn, the sun, the moon, the sky, had been invoked each by its own name, they could all be spoken of by the one epithet which they shared in common, namely *deva*, bright. In this general concept of those Bright ones, all that was special and peculiar to each was dropt, and there remained only the one epithet, *deva*, to embrace them all. Here then there arose, as if by necessity, a new concept, in which the distinctive features of the various bright beings had all been merged in that of brightness, and in which even the original meaning of brightness, being shared by so many very different beings, had been considerably dimmed or generalised, so that there remained little more than the concept of agent which, as modified by brightness, had been from the beginning contained in the root DIV.

You will now perceive the difference between our saying that the ancient Âryas applied the name of gods to the fire, the sun, or the sky, or our watching the process by which these Âryas were brought to extract or abstract from the concepts of fire, sun, moon, and sky, all being bright beings, the general concept of *Deva-hood*. But, though we cannot help translating *deva* by *god*, you will easily understand what a distance there is from *Deva-hood* to *God-hood*. A Deva is as yet no more than a bright agent, then a kind agent, then a powerful agent, a more than human agent, nay, if you like, a superhuman agent; and then only, by another step, by what may be called a step in the dark, a divine agent.

Greek and Roman Gods

In Greece the process was slightly different. The Greeks very soon endowed these powerful agents with human qualities, to such an extent that immortality seems almost the only quality which they do not share in common with human beings. In Italy the old gods had less of that anthropomorphic character which they had in Greece. It is, in fact, a distinguishing feature of ancient Roman mythology that there are few family ties that hold the gods together, while the Greek gods are all related with one another most intimately, if not always, most correctly.

The early Christians invented still another concept for these Greek and Roman gods. They did not deny their substantial existence, but they accepted them as living beings, as spirits, as they called them, but as evil spirits. This idea has remained till almost to our own time, when the study of ancient religion and ancient language has enabled us to see what the Devas of the Âryas really were—not evil spirits, not human or superhuman beings, but names given to the most prominent phenomena of nature, which naturally and necessarily implied the idea of agents. With the progress of language and thought *we* are now able to speak instead of agents, of agencies, of forces, forces of nature, as we call them; but what is behind those agencies, what is behind warmth or light or ether, we know as little as the Vedic Rishis knew what was behind their Agni or their other Devas.

Ruskin on the Ancient Gods

How powerful the influence of words may be, how long they may continue to charm and to mislead even the wisest, we may see from an eloquent passage in Mr. Ruskin's *Præterita*, vol. 3, p. 172. He tries to explain to himself and to others what he means when he speaks, as he often does, half poet, half philosopher as he is, of gods. "By gods in the plural," he writes,

> I mean the totality of spiritual powers, delegated by the Lord of the universe to do, in their several heights, or offices, parts of His will respecting man, or the world that man is imprisoned in; not as myself knowing, or in security believing, that there are such, but in meekness accepting the testimony and belief of all ages, to the presence, in heaven and earth, of angels, principalities, powers, thrones, and the like—with genii, fairies, or spirits ministering and guarding, or destroying and tempting, or aiding good work and inspiring the mightiest. For all these I take the general word 'gods,' as the best understood in all languages, and the truest and widest in meaning, including the minor ones of seraph, cherub, ghost, wraith, and the like; and myself knowing for indisputable fact, that no true happiness exists, nor is any good work ever done by human creatures, but in the sense or imagination of such presences.

Does not this confirm the words of Rosmini when he said: "The deeper we penetrate into this matter, the more do we find that all our intellectual errors, all the pernicious theories, the deceptive sophistries by which individuals and nations have been deluded, can be traced back to the vague and improper use of words."[6]

Evolution of the Word *Deva*

It is very important that you should clearly apprehend this process by which the word *deva,* originally meaning bright, assumed in time the meaning of god, in that sense at least in which the Hindus, like the Greeks and Romans, would speak of *Agni,* the fire, *Ushas,* the dawn, *Dyaus,* the sky, as their *Devas,* or their gods. It is one of the most interesting cases of intellectual evolution, for it shows us how a word, having originally the purely material meaning of brightness, came in the end by the most natural process to mean divine. There was nothing intentional in that process. It was impossible that there should have been a conscious intention to express the divine, for, if there had been such a conscious intention, there would have been already in the human mind a pre-existent name and concept of the divine. The process was one of the most natural evolution. You may say that nothing could be evolved that was not involved in the word *deva,* and in one sense this is perfectly true. In the idea of agency, which was involved in every root, there lay the germ which, as one outside envelope after the other was removed, came out in the end in all its simplicity and purity. But it came out nevertheless *after* it had been coloured or determined by these former envelopments. It had passed through an historical process, and had thus grown into an historical concept.

Nor must we suppose that the evolution of the word *deva* was the only evolution which gave us in the end the idea of divine. That idea was evolved in many different ways, but nowhere can we watch every stage in the evolution so well as in the history of the word *deva.* Our own word *God* must have passed through a similar evolution, provided it be an old word. But unfortunately nearly all its antecedents are lost, and its etymology is quite unknown.

We have as yet traced the history, or, if you like, the evolution of the word *deva* to that stage only when it signifies a number of bright, kind, powerful agents, such as Mr. Ruskin declared he could still accept on the testimony and belief of all ages. But its history, as we shall see, does not end there. It gradually rises to the highest concept of deity, to a belief in a God above all gods, a creator, a ruler of the world, a judge, and yet a compassionate father, so that what seems at first a mere matter of linguistic archaeology, will stand before us in the end as the solution of one of the most vital questions of religious philosophy. How many times has the question been asked, Whence comes the idea of God?, and how many different answers has it elicited! Some people maintain it is inherent in the human mind, it is an innate idea, or a precept, as it has lately been called. Others assert that it could have come to man by a special revelation only. Others again, like Professor Gruppe, maintain that it is a mere hallucination that took possession of one man, and was then disseminated through well-known channels over the whole world. We do not want any of these guesses. We have a guide that does not leave us in the dark when we are searching for the first germs of the idea of God. Guided by language, we can see as clearly as possible how, in the case of *deva,* the idea of God grew out of the idea of light, of active light, of an awakening, shining, illuminating, and warming light. We are apt to despise the decayed seed when the majestic oak stands before our eyes, and it may cause a certain dismay in the hearts of some philosophers that the voice of God should first have spoken to man from out the fire. Still as there is no break between *deva,* bright, as applied to Agni, the fire, and many other powers of nature, and the *Deus Optimus Maximus* of the Romans, nay, as the God whom the Greeks ignorantly worshipped was the same God whom St. Paul declared unto them, we must learn the lesson, and a most valuable lesson it will turn out to be—that the idea of God is the result of an unbroken historical evolution, call it a development, an unveiling, or a purification, but not of a sudden revelation.

Natural Revelation of God

It seems almost incredible that in our days such a lesson, confirmed as it is by the irrefragable evidence of historical documents, should be objected to as dangerous to the interests of religion, nay, should form the object of virulent attacks.

For some reason or other, our opponents claim for their own theories the character of orthodoxy, while they try to prejudge the whole question by stigmatising our own argument as heterodox. Now I should like to ask our opponents, first of all, by what authority such metaphysical theories as that of innate ideas can possibly claim the name of orthodox, or where they can point to chapter and verse in support of what they call either a special or a universal primeval revelation, imparting to human beings the first concept and name of God? To a student of the religions of the world, in their immense variety and their constant divisions, the names of orthodox and heterodox, so freely used at all times and on all sides, have lost much both of their charm and their terror. What right have we to find fault with the manner in which the Divine revealed itself, first to the eyes, and then to the mind of man? Is the revelation in nature really so contemptible a thing that we can afford to despise it, or at the utmost treat it as good enough for the heathen world? Our eyes must have grown very dim, our mind very dull, if we can no longer perceive how the heavens declare the glory of God. We have now named and classified the whole of nature, and nothing seems able any longer to surprise, to terrify, to overwhelm us. But if the mind of man had to be roused for the first time, and to be lifted up to the conception of something beyond itself, what language could have been more powerful than that which spoke in mountains and torrents, in clouds and thunderstorms, in skies and dawns, in sun and moon, in day and night, in life and death? Is there no voice, no meaning, is there no revelation in all this? Was it possible to contemplate the movements of the heavenly bodies, the regular return of day and night, of spring and winter, of birth and death, without the deepest emotion?

Of course, people may say now: We know all this, we can account for it all, and philosophy has taught us *Nil admirari.* If that is so, then it may be true indeed that the sluggish mind of man had to be stirred once more by a more than natural revelation. But in the early days of the world, the world was too full of wonders to require any other miracles. The whole world was a miracle and a revelation, there was no need for any special disclosure. At that time the heavens, the waters, the sun and moon, the stars of heaven, the showers and dew, the winds of God, fire and heat, winter and summer, ice and snow, nights and days, lightnings and clouds, the earth, the mountains and hills, the green things upon the earth, the wells, and seas, and floods—all blessed the Lord, praised Him, and magnified Him for ever.

Can we imagine a more powerful revelation? Is it for us to say that for the children of men to join in praising and magnifying Him who revealed Himself in His own way in all the magnificence, the wisdom, and order of nature, is mere paganism, polytheism, pantheism, and abominable idolatry? I have heard many blasphemies, I have heard none greater than this.

It may be said, however, that the road from nature leads only to nature's *gods*, to a belief in many, not in *one* supreme God. It certainly leads through that gate, but it does not stop there. If we return to the Veda, the oldest record of a polytheistic faith, and if we take up once more the thread where we left it, we shall be able to see how *Agni*, the god of fire, being at first but one by the side of many other gods, develops into something much higher. He does not remain one out of many gods. He becomes in the end a supreme god, *the* Supreme God, till his very name is thrown away, or is recognised as but one out of many names by which ancient seers in their helpless language called that which is, the One and All. You may remember the passage from the Veda which I quoted before: "That which is one, the seers call in many ways, they call it Agni, Yama, Mâtarisvan."

The Biography of Agni

This process, which I call the theogonic process, is so important that we must study it carefully, and step by step, in the case of at least one of the ancient gods. If I select for that purpose the god of fire, Agni, and not Dyaus, Zeus, Jupiter, the supreme god of the Aryan Pantheon, it is because the biography of Dyaus, having been fully worked out by me on former occasions, need not be gone through again in full detail.[7] It is my chief object at present to show how many roads, starting from different beginnings, all converged and met in the end in the same central point, the belief in one Supreme Agent, manifested in all that is and moves and lives, and how the perception of the Infinite was revealed everywhere in what we call the perceptions of the Finite.

Notes

1. *Hibbert Lectures*, p. 103. The names *fitiso, fetish*, and *fitisero*, priest, are traced back to Portuguese sailors in Africa by W. J. Müller, *Die Afrikanische Landschaft Fetu*, 1675.
2. Kuhn, *Mythologische Studien*, i, p. 211.
3. From the same root we have in Greek ἄγω, to drive, ἄγρα, the chase; in Latin *ago, agmen*. The Sk. *agra*, Gr. ἀγρός, Lat. *ager*, Goth. *akr-s*, mean meadow and field, possibly from the cattle being driven over it. The German *Trift* comes likewise from *treiben*. The words for goat also may be referred to this root, if they meant originally quickly moving or agile; Sk. *aga*, Greek αἴξ, Lit. *ožys*. Consider the *drift* of an argument, and what are you *driving* at.
4. Brinton, *Myths of the New World*, p. 48 seq.
5. *Sociology*, p. 237.
6. *The Ruling Principle of Method, applied to Education by Rosmini*. Translated by Mrs. W. Grey, 1887, p. 262.
7. *Science of Language*, vol. 2, chap. 11.

Chapter 14

Religion, Myth, and Custom

from *Gifford Lectures on Physical Religion* (1890)

Difference between Religion and Mythology

I was anxious to explain to you in my last lecture how the same source which supplied the ancient world with religious concepts, produced also a number of ideas which cannot claim to be called religious in any sense, least of all in that which *we* ourselves connect with the name of religion.

We saw how in the Veda the concept of Fire had been raised higher and higher, till at last it became synonymous with the Supreme Deity of the Vedic poets. But in the amorous vagaries of Agni, as related in the later poetry of India, or in Greece in the monstrous birth of Hephaestos, likewise a representative, or, as we sometimes say, likewise a god of fire, in his disgraceful ejection from the sky, in his marriage with Aphrodite, to say nothing of the painful *dénouement* of that ill-judged union, there is very little of religion, very little of "the perception of the infinite under such manifestations as are able to influence the moral conduct of man."

These mythological stories are, no doubt, chips and splinters from the same block out of which many a divine image has been chiselled by the human mind, but their character, their origin and purpose are totally different. This distinction, however, has not only been neglected, it seems often to have been wilfully [*sic*] neglected. Whenever it was necessary to criticise any of the non-Christian religions in a hostile spirit, these stories, the stories of Venus and Vulcan and Mars, have constantly been quoted as showing the degraded character of ancient gods and heroes, and of pagan religion in general.

This is most unfair. Neither does this mythological detritus, not to say rubbish, represent the essential elements of the religion of Greeks and Romans, nor did the ancients themselves believe that it did. We must remember that the ancient nations had really no word or concept as yet for religion in the comprehensive sense which we attach to it. It would hardly be possible to ask the question in any of the ancient languages, or even in classical Greek, whether a belief in Hephaestos and Aphrodite constituted an article of religious faith.

It is true that the ancients, as we call them rather promiscuously, had but one name for their gods, whether they meant Jupiter, the *Deus Optimus Maximus,* or Jupiter, the faithless husband of Juno. But when we speak of the ancients in general, we must not forget that we are speaking, not only of Homer and Hesiod, but likewise of men like Herakleitos, Aeschylos, and Plato. These ancient thinkers knew as well as we do that nothing unworthy of the gods could ever have been true of them, still less of the supreme God; and if they tolerated mythology and legends, those who thought at all about these matters looked upon them as belonging to quite a different sphere of human interests.

If we once understand how mythology and legends arose, how they represent an inevitable stage in the growth of ancient language and thought, we shall understand not only their outward connection with religious ideas, but likewise their very essential difference.

Secular Ideas become Religious

While on the one hand it is perfectly true that the sources of religion and mythology are conterminous, nay, that certain concepts which in their origin might be called religious wither away into mere mythology and romance, we shall see that it likewise happens, and by no means unfrequently [*sic*], that ideas, at first entirely unconnected with religion, are attracted into the sphere of religion, and assume a religious character in the course of time. This is an important subject, but beset with many difficulties.

Of course, the deification of an animal, such as an Egyptian Apis, or the apotheosis of a human being, such as Romulus or the Emperor Augustus, presupposes the previous existence in the human mind of the concept of divinity, a concept which, as we saw, required many generations for its elaboration. Again, the attribution of a divine sanction bestowed either on customs or laws, presupposes a belief in something superhuman or divine. But, after a time, all this is forgotten, and these later corruptions of religious thought are mixed up with the more primitive elements of religion in a hopeless confusion.

Let us consider to-day a few instances of secular customs being afterwards invested with a religious authority.

Lighting and Keeping of Fire

When we remember how difficult it must have been in early times to light a fire at a moment's notice, and what fearful consequences might follow if a whole community was left during the winter without a fire burning on the hearth, we require no far-fetched explanations for a number of time-hallowed customs connected with the lighting, and still more with the guarding of the fire. It was not necessary that every tribe which kept a sacred fire should have a belief in fire as a god, as was the case with the Vedic poets. Quite apart from any deeper religious convictions, mere common sense would have led men in a primitive state of society to value any new discovery for striking fire and to adopt measures for preserving it, whether for private or for public use. If the Romans appointed vestal virgins to keep a fire always burning, the Damaras in Africa did exactly the same.[1]

It is the custom, or it was till very lately, among German peasants, for a man when he married and left his father's house, to take a burning piece of wood from the paternal home and to light with it the fire on his own hearth. Exactly the same is told us of many uncivilised races. Among the Damaras, for instance, when a tribe migrated from one place to another, they took some burning logs from the old to the new home.

Nowhere, however, do we find this custom more fully described than in India. In the Vedic hymns fire is the *griha-pati,* the lord of the house. A house was a fire, and "so many fires" are mentioned even now in the census of half-nomadic tribes in Russia, nay, even in Italy (*fuochi*), as meaning "so many families, or houses." In ancient India, as described to us in the Grihya-sûtras, the most important act when a man married and founded his own household, was the kindling of the fire in his own house, with fire brought from the home of his bride, or with fire newly rubbed. In the fourth night after the wedding the husband has to establish the fire within his house. He assigns his seat to the south of it to the Brahman, places a pot of water to the north, cooks a mess of sacrificial food, sacrifices the two âgya portions, and then makes five oblations to Agni, Vâyu, Sûrya, Kandra, and Gandharva. Here Agni, Fire, holds the first place among the domestic gods. After him follow Wind, Sun, and Moon, and lastly the Gandharva, whoever he may be.[2]

This domestic fire, when once lit, remained the friend and protector of the family in every sense of the word, and we see the most touching superstitions arising from this in India, and in every part of the world. Many years ago, in my article on "Funeral Ceremonies" (1855),[3] I translated a passage from Âsvalâyana's Grihya-sûtras (IV. 1), in which it is said that if a disease befall one who has set up sacred fires, he should leave his village (with his fires) and go in an eastern, northern, or north-eastern direction. And why? Because there is a saying, "Fires love the village." It is understood therefore that the fires, longing to return to the village, will bless him and make him whole.[4] Here we see how a mere proverb, "The fires love the village," may lead without any effort to a metamorphosis of the fire into a friend, a friend with all the feelings of other friends, willing even to render a service and to restore a man to health, if thereby they may themselves be enabled to return to their beloved hearth.

Besides the fire in each house, the custom of keeping a public fire also is alluded to at an early date.

According to the Dharma-sûtras of Âpastamba (II. 10, 25), a king has to build a palace, a hall, and a house of assembly, and in every one of them a fire is to be kept, a kind of *ignis foci publici sempiternus,* and daily oblations to be offered in it, just the same as in every private house.

There are many sayings among civilised and uncivilised nations, implying a respect for fire and a recognition of its value for domestic purposes. The Ojibways, for instance, have a saying that one ought not to take liberties with fire, but we are never told that the Ojibways worshipped the fire as a god.[5]

There is a very wide-spread feeling against spitting, or throwing anything unclean into the fire or into the water. We saw it mentioned by Herodotus and by Manu. It is a godless thing, they say in Bohemia, to spit into the fire. The Mongolians, as Schmidt tells us, consider it sinful to extinguish fire by water, to spit into the fire, or to defile it in any other way.[6]

Such rules, though evidently intended at first for a very definite and practical object, were soon invested with a kind of sacred authority. If the Bohemian says it is a godless thing to spit into the fire, he soon adds a reason: Because it is God's fire. This is, of course, a very modern idea; it may be called a Christian idea, based on a belief that all good and perfect gifts come from God—but it is nevertheless a very natural after-thought.

Religious Sanction for Customs

What therefore we must try to find out in all these observances is, whether at first there was not an intelligible object in them, whether they did not serve some useful purpose, and whether the religious sanction did not come much later in the day. When there once existed a belief in divine beings, any custom or law, and particularly those which it was difficult to enforce by mere human authority, were naturally placed by the ancient lawgivers under the protection of the gods. Professor von Ihering, one of the highest German authorities on the history of law, has traced many of these sacred commandments back to their true origin, namely their *Zweck,* their practical object.

It is quite clear, for instance, that in early times it was necessary to guard the purity of rivers by some kind of religious protection. No sanitary police could have protected them in their long meandering courses. Pausanias (iii. 25, 4) tells a story of a spring on the promontory of Tainaron in Laconia (Cape Matapan) which possessed some miraculous qualities, but lost them because a woman had dared to wash dirty linen in it.

In a primitive household, where the central fire was, as it were, the property of all, a similar restriction against defilement was equally necessary. And when with the change of domestic arrangements the original object of such restrictions ceased to be understood, they became what we find them to be in many countries, mere unmeaning customs, and, for that very reason, often invested with a sacred authority. When the real purpose (*Zweck*) was forgotten, a new purpose had to be invented.

Baptism by Water and Fire

For instance, people wonder why the inhabitants of Mexico, as well as of Peru,[7] should have been acquainted with baptism by water and fire. Originally, however, these seem to have been very simple and useful acts of purification, which in later time only grew into sacramental acts. The nurse had to bathe the child immediately after birth, and to invoke the so-called goddess of water to cleanse the child from everything unclean, and to protect it against all evil. That is to say, every new-born child had to be washed. Afterwards there followed a more brilliant baptism. Friends and relations were invited to a feast, the child was carried about in the house, as if to present it to the domestic deities, and while the nurse placed it in water she recited the following words: "My child, the gods, the lords of heaven, have sent thee into this miserable world; take this water which will give you life." Then she sprinkled water on the mouth, the head, and the chest of the child, bathed the whole body, rubbed every limb, and said: "Where art thou, ill luck? In which limb dost thou hide? Move away from this child!" Prayers were then offered to the gods of the water, the earth, and the sky. The child had to be dressed, to be put in a cradle, and to be placed under the protection of the god of cradles and the god of sleep. At the same time a name was given to the infant.

All this is full of elements which remind us of similar practices among the Romans, the *Amphidromias* of the Greeks, and the name-giving ceremonies described in the Vedic Grihya-sûtras.

Next followed the baptism of fire. This also was originally nothing but an act of purification. Like water, fire also was conceived by many nations as purifying. "Fire," as Plutarch says in his *Quæstiones Romanæ* (cap. i), "purifies, water hallows."[8] Its very name in Sanskrit, *pâvaka*, means purifier. In India we were met by two trains of thought. Either fire was conceived as purifying everything, or it was represented as shrinking from contact with all that is impure. In Mexico the former idea prevailed. It had probably been observed that fire consumed deleterious substances, and that the fumes of fire served as a preservative against miasma and illness. Hence in the baptism of fire in Mexico the child was carried four times through a fire, and was then supposed to have been purified.

Purification by Fire

Whether there is some truth in this belief in the purifying powers of fire, we must leave to medical men to determine. Anyhow it is a belief or a superstition which has lasted for many centuries. When cholera rages in India, we still receive our letters well smoked. Menander tells us that Zemarchus, the ambassador of Justinian, was led by the Turks round a fire, so that he might be purified.[9] According to Plano Carpini, a foreign ambassador was actually led through two fires by the Mongolians. Castrén traces all these customs back to a religious reverence for the fire. It seems, however, much more plausible that the custom had a purely utilitarian foundation, that it was in fact the forerunner of our modern quarantine, which many medical authorities now look upon as equally superstitious.

Nor was the purificatory or disinfecting power of fire restricted to human beings. Cattle were often submitted to the same process of lustration. The object was originally purely practical, though superstitious ideas began soon to cluster around it.

Lustration of Animal

The Romans had their annual lustrations. On the twenty-first of April, after a sacrifice had been offered, hay and straw were piled up in rows, and when they had been lighted, the flocks were driven through the burning fire. The shepherds often jumped through the flames, following their flocks.[10]

This purely disinfecting character is still more clearly visible in the so-called *Need-fire* of the Teutonic nations.

Need-fire

Joh. Reiskius, in a book published in 1696, tells us that whenever pestilence broke out among small or large cattle the peasants determined to have a *Not-feuer*. All other fires in the village had then to be put out, and by the usual method of rubbing pieces of wood, covered with pitch, a new fire was lighted. When it had grown large enough, horses and cattle were twice or thrice driven through it.[11] Afterwards the fire was extinguished, but each householder carried home a burning log to light his own fire, or dipped it afterwards in the wash-tub, and then let it lie in the manger.

This ceremony of the *Need-fire* might in fact have been witnessed in Scotland as late as the last century.[12] A Miss Austin relates that in the year 1767, in the isle of Mull, in consequence of a disease among the black cattle, the people agreed to perform an incantation, though they esteemed it a wicked thing. They carried to the top of Carnmoor a wheel and nine spindles of oak-wood. They extinguished every fire in every house within sight of the hill. The wheel was then turned from east to west over the nine spindles, long enough to produce fire by friction. If the fire were not produced by noon, the incantation lost its effect. If they failed for several days running, they attributed this failure to the obstinacy of one householder, who would not let his fires be put out for what he considered so wrong a purpose. However, by bribing his servants, they contrived to have them extinguished, and on that morning raised the fire. They then sacrificed a heifer, cutting in pieces and burning, while yet alive, the diseased part. They then lighted their own hearths from the pile, and ended by feasting on the remains. Words of incantation were repeated by an old man from Morven, who came over as master of the ceremonies, and who continued speaking all the time the fire was being raised. This man was living a beggar at Bellochrog. Asked to repeat the spell, he said the sin of repeating it once had brought him to beggary, and that he dared not say those words again. The whole country believed him accursed.

Tinegin in Ireland

In Ireland also, according to Martin,[13] the same heathenish custom might have been witnessed within the memory of men. The inhabitants made use of a fire called *tinegin,* i.e., a forced fire, or fire of necessity. This word is formed from the Irish *teine,* fire, and *eigin,* violence. It is either a simple translation of the English need-fire, or it expresses the same idea which is conveyed by the Vedic name for fire, *sahasah sûnuh,* son of strength or effort. This Tinegin was used as an antidote against the plague or murrain in cattle, and it was performed thus: "All the fires in the parish were extinguished, and then eighty-one (9×9) married men, being thought the necessary number for effecting this design, took two great planks of wood, and nine of them were employed by turns, who by their repeated efforts rubbed one of the planks against the other until the heat thereof produced fire; and from this forced fire each family is supplied with new fire, which is no sooner kindled than a pot full of water is quickly set on it, and afterwards sprinkled upon the people infected with the plague, or upon the cattle that have the murrain. And this they all say they find successful by experience. It was practised on the mainland, opposite to the south of Skie, within these thirty years."

Now suppose some Portuguese priests had visited Scotland and Ireland, as they visited the West Coast of Africa, and had described the religion of the natives from what they saw with their own eyes, as they described the fetish worship of the Negroes. They might have described them, first of all, as fire-worshippers; secondly, as fetish-worshippers, for the fire, we are told, is a fetish when it is invoked for help; thirdly, as performing animal sacrifices, for they sacrificed a heifer and feasted on it; fourthly, as sorcerers, for they repeated unintelligible incantations; and lastly, as animists, for they believed that there was some kind of spirit in the fire. That these people were Christians, and that their religion was something quite different from these popular amusements, they could never have guessed. Yet it is on the strength of some stray observations made on the West Coast of Africa that we are asked to believe that the religion of the Negroes is pure Fetishism, nay, that Fetishism was the primitive religion of all mankind.

It might be said that such heathenish customs existed in Scotland and Ireland only, and if, as careful travellers ought to do, our Portuguese missionaries had explored England also, they would have found there no traces of fetishism and sorcery. But no, in England also they might have witnessed similar heathenish ceremonies, for we are told by fair authorities that not long ago two ladies in Northamptonshire saw a fire in the field and a crowd round it. They said, "What is the matter?"—"Killing a calf." "What for?"—"To stop the murrain." The ladies went away as quickly as possible. On speaking to the clergyman he made inquiries. The people did not like to talk of the affair, but it appeared that when there is a disease among the cows, or the calves are born sickly, they sacrifice, that is, they kill and burn one for good luck.

We have still later testimony of the permanence of similar superstitious customs. They seem to have survived to the present day. At a meeting of the Society of Antiquarians of Scotland held at the Royal Institution, Edinburgh, and reported in the *Scotsman,* Tuesday, March 11, 1890, the Rev. Alexander Stewart, Nether Lochaber, gave an account of some examples which had recently come to his knowledge of the survival of certain superstitions relating to fire in the Highlands of Scotland and in Wigtownshire. The first case happened in March last, and was accidentally witnessed by Dr. Stewart's informant. Having gone to a small hamlet in a remote glen to leave a message for the shepherd, he was surprised to find there was no one in the houses, but seeing a slight smoke in a hollow at some distance, he concluded that he would find the women there washing. On reaching the bank above the hollow he was astonished to see five women engaged in the ceremony of passing a sick child through the fire. Two of the women standing opposite each other held a blazing hoop vertically between them, and two others standing on either side of the hoop were engaged in passing the child backwards and forwards through the opening of the hoop. The fifth woman, who was the mother of the child, stood at a little distance, earnestly looking on. After the child had been several times passed and repassed through the fiery circle, it was returned to its mother, and the burning hoop was thrown into a pool of water close by. The child, which was about eighteen months old, was a weakling, and was supposed to have come under the baleful influence of an evil eye. When taken home a bunch of bog-myrtle was suspended over its bed. The somewhat analogous superstition of putting a patient in the centre of a cart-wheel when the red-hot fire was put on it at the door of the smithy was practised in Wigtownshire half a century ago.

Purpose of Customs often Forgotten

Now, I ask, is all this to be called religion? If Christians can perform these vagaries, why should not the Negroes of Africa indulge in superstitious practices without therefore deserving to have their religion represented as nothing but fetish-worship? The Negroes of Africa, and, in fact, most uncivilised races, are most unwilling to speak about what we mean by religion; they often have not even a name for it. They are proud, on the contrary, of their popular amusements, feastings, dances, and more or less solemn gatherings, and welcome strangers who come to see them. Some of these gatherings may in time have assumed a religious character. But the wide prevalence of many of the customs which we described, such as the ceremonial observed in the lighting and keeping of the fire, the purification of children, and the lustration of cattle, shows that in many of them there was originally a definite and practical object. Sometimes we can still discover it, but in other cases the real object has completely disappeared. We cannot tell, for instance, why, when the new fire was lighted, it should have been thought necessary to extinguish the fires in every house. Yet we find exactly the same custom which we met with in Germany in the island of Lemnos also, the very island on which Hephaestos was believed to have been precipitated by Zeus. Here all fires had to be extinguished during nine days, till a ship arrived from Delos, bringing the new sacred fire from the hearth of Apollo. This fire was afterwards distributed among all the families, and a new life was supposed to begin.[14]

When, after the battle of Plataeae, the Greeks sent to Delphi to ask what sacrifices they ought to offer, they were told by the Pythian god to erect an altar to Zeus Eleutherios, but not to sacrifice till all the fires in the country had been extinguished, because they had been contaminated by the barbarians, and till new fire had been fetched from the common hearth at Delphi.[15]

During the Middle Ages a similar custom prevailed in Germany. At Marburg and in Lower Saxony the fire was lighted once every year by rubbing two pieces of wood. This was the new fire which was to take the place of the old fires. These were supposed to have been contaminated by contact with impurities during the year.[16]

Nor is it necessary that there should always have been a very deep motive for these customs. We can hardly imagine, for instance, a very stringent reason why the guardianship of the public fire should have been committed to vestal virgins. It was so not only in Rome. In Ireland also the fire of St. Brigida at Kildar was not allowed to be approached by men. The Damaras, an African tribe, entrusted their fire, as we saw, to young maidens. In Mexico, in Peru, in Yucatan, the sacred fire was likewise guarded by a company of virgins.[17] All we can say in this and similar cases is that in a primitive state of society the watching over the fire on the hearth would naturally fall to the unmarried daughters of a family who stayed at home, while other duties called their brothers into the field. The mere continuation of such an arrangement would in time impart to it something of a time-honoured and venerable character, and the less the original purpose of such ancient customs was understood, the more likely it was that a kind of religious sanction should be claimed for them.

Essential Difference between Religion, Mythology, and Ceremonial

What I am anxious to place in the clearest light is that a great deal of what we class as religious, whether among ancient or modern peoples, had really in the beginning very little or nothing to do with what we ourselves mean by religion. Mythology affects ancient religion—in one sense it may be said to affect all religion. But mythology by itself is never religion, as little as rust is iron. Ceremonial again affects religion; it may be, that in the world we live in, ceremonial has become inseparable from religion. But ceremonial by itself is never religion, as little as shade is light.

I wanted to show you how out of the same materials both religious and non-religious concepts may be formed. It was for that purpose that I chose Fire and tried to exhibit its threefold development, either as truly theogonic, or as mythological, or as ceremonial and sacrificial.

Theogonic Development of Agni

In India we are able to prove by documentary evidence that the concept of Fire, embodying the concepts of warmth, light, and life, was raised gradually to that of a divine and supreme being, the maker and ruler of the world. And if in the Veda we have the facts of that development clearly before us, it seems to me that we have a right to say that in other religions also where Fire occupies the same supreme position, it may have passed through the same stages through which Agni passed in the Veda.

Mythological Development of Agni

By the side of this theogonic process, however, we can likewise watch in the Veda the beginning at least of a mythological development which becomes wider and richer in the epic and paurânic literature of India. This side is most prominent in Greece and Rome, where the legends told of Hephaestos retain but few grains of Agni as the creator and ruler of the world.

Ceremonial Development of Agni

Lastly, the ceremonial development of Fire is exhibited to us in what has sometimes been called fire-worship, but is in most cases merely a recognition of the usefulness of fire for domestic, sacrificial, and even medicinal purposes.

Definition of Religion Re-examined

These three sides, though they have much in common, should nevertheless be kept carefully distinct in the study of religion. I know it may be said, in fact, it has been said, that the definition of religion which I laid down in my former course of lectures is too narrow and too arbitrary. In one sense, every definition may be said to be arbitrary, for it is meant to fix the limits which the definer, according to his own *arbitrium*, wishes to assign to a certain concept or name. Both in including and excluding, the definer may differ from other definers, and those who differ from him will naturally call his definition arbitrary, and either too narrow or too wide.

I thought it right, for instance, to modify my first definition of religion as "the perception of the Infinite," by restricting that perception to such manifestations as are able to influence the moral conduct of man. My first definition was not wrong, but it was too wide. It cannot be denied that in the beginning the perception of the Infinite had often very little to do with moral ideas, and I am quite aware that many religions enjoin even what is either not moral at all, or even immoral. But if there are perceptions of the Infinite unconnected as yet with moral ideas, we have no right to call them religious till they assume a moral character, that is, till they begin to react on our moral nature. They may be called philosophical, metaphysical, even mathematical, but they form no part of what we call religion. The objection, that some religions actually sanction what is immoral, is purely forensic.

If some religions sanction what is immoral, or what seems to us immoral, this would only serve to prove all the more strongly the influence of religion on the moral conduct of man. We are told, for instance, that "the pre-historic Hebrews killed their first-born in sacrifice to their god. Abraham came very near doing the same thing. Jephtha killed his daughter, and David killed the murderers of the son of Saul, and kept them hanging in the air all summer long, to remind his God that Ishbosheth was avenged. If you catch a Yezidee in the act of stealing, he will tell you that theft is a part of his religion. If you catch a Thug in the act of assassination, he will tell you that murder is to him a religious rite. If you reprove the Todas of the Nilgheris Hills for living in polyandry, they will tell you that this is the very ground-work of their religion. If you reprove the Mormons for living in polygamy, they will remind you that this is the Biblical chart of their faith."[18]

Now suppose that all this were true, would it not prove the very opposite of what it is meant to prove? If religion can induce human beings to commit acts which they themselves, or which we at least, consider doubtful, or objectionable, or altogether criminal, surely it shows that religion, even in this extreme case, exercises an influence on the moral character of man such as probably nothing else could exercise.

From the moment, therefore, that the perception of something supernatural begins to exercise an influence on the moral actions of man, be it for good or for evil, from that moment, I maintain, and from that moment only, have we a right to call it religious.

We must be careful to keep within the limits of a definition which we have once accepted. The definition which I gave of religion, that it consists in a perception of the Infinite, under such manifestations as are able to influence the moral conduct of man, is not too narrow. It is wide enough, at all events, to tax the powers of any single student of the history of religion.

The Meaning of the Infinite

When I said that religion is the perception the Infinite, I took great pains to explain that this perception is to be taken as the true source of religion, as that without which religion would be impossible, or at least inconceivable. But as little as the source is the whole river, is the source of religion the full stream of religion. When Locke said, *Nihil est in intellectu quod non ante fuerit in sensu,* he did not mean that *sensus* was the whole of *intellectus.* He only meant that nothing could be in the intellect that had not come from sensuous perception. I meant the same when I said, *Nihil est in fide quod non ante fuerit in sensu.* I meant that nothing could be in our faith or in our religion that had not come from the perception of the Infinite, but I did not mean that this perception of the Infinite was the whole of religion. As our sensuous percepts grow into concepts and into all that belongs to conceptual thought, our perceptions of the Infinite also are the living germs only which produce in time that marvellous harvest which we call the religions of the world. And if I limited the area of these perceptions of the Infinite to that narrower field which is distinguished by its moral colouring, that field is still of an enormous extent, and will require better and stronger labourers to reap than it has hitherto found.

As to the name which I chose for what forms the real object of all religious perceptions, namely, the Infinite, I know quite well that it may be criticised. But has any one been able to suggest a better name? I wanted a name as wide as possible. I might have chosen Unknowable as equally wide. But to speak of a perception of the Unknowable seemed to me a contradiction in terms. To know has many meanings, and in one of its meanings we may say, no doubt, that the Infinite is the Unknowable. We cannot know the Infinite as we know the Finite, but we can know it in the only way in which we can expect to know it, namely, behind the Finite. In perceiving anything limited, we also perceive what limits it, but to call this Unlimited or Infinite the Unknowable is to do violence to the verb to know.

I am quite aware that what other philosophers have called the Absolute was probably meant by them for what I call the Infinite. I likewise admit that what theologians mean by the Divine is in reality the same. Even the Transcendent might have answered the same purpose. But all those terms had a history. The Absolute reminds us of Hegelian ideas, the Divine is seldom free from a certain mythological colouring, and the Transcendent has its own peculiar meaning in the school of Kant. Infinite, therefore, seemed less objectionable than any of those terms, and submitted more readily to a new definition. It had likewise the advantage of having the term finite for its opposite. If some critics have proclaimed their inability to perceive any difference between infinite and indefinite, I can quite sympathise with them, for I see none whatever. The only distinction which usage would seem to sanction is that indefinite is generally applied to knowledge, infinite to the object of knowledge. We might then say that our knowledge of the infinite must always be indefinite, a proposition to which few critics would demur.

I did not wish, however, to monopolise the word religion in the sense which I assigned to it in my lectures. I simply wished to delimit the subject of these lectures, and to state once for all what segment of human thought would fall within our field of observation. If others define religion in a different sense, we shall know what to expect from them. All that I object to is an undefined use of that word. If Cicero, for instance, defines religion in one place as *cultus pius deorum,* he may be quite right from the Roman or from his own point of view, and we should be forewarned as to what to expect from him, if he were to lecture on religion.[19] Or, if Dr. Robertson Smith, in a recent course of excellent lectures on the ancient Semitic religions, assures us that with the Semites, religion consisted primarily of institutions, such as sacrifices, ablutions, fastings, and all the rest, and not of what was believed about gods or God, we shall know in what sense he uses religion. In modern times also there are many people who hold that religion consists chiefly in ceremonial acts, such as going to church, kneeling, making the sign of the cross, and other ritual observances.

But though I quite admit the right of Cicero or anybody else to define religion in his own sense, and to treat of religion as mere cult, or as mere mythology, I hold as strongly as ever that neither cult nor mythology is possible without a previous elaboration of the concepts and names of the gods. Cult is one of the many manifestations of religion, but by no means the only one, nor a necessary one. The same applies to myths and legends. They are the parasites, not the marrow, of religion.

Besides, there are myths and legends altogether unconnected with religion, and there are solemn acts which have nothing to do with the gods. We saw how some ceremonies and myths connected with Agni, Fire, were religious in their origin, and ceased to be so, while others, purely secular in their origin, assumed in time a religious character. It is often difficult to draw a sharp line between what is no longer and what is not yet religious, but our definition of religion will generally help us in trying to discover whether there are any elements in a ceremonial act or in a mythological tradition which draw their origin, however distantly, from an original perception of the Infinite, and influence, directly or indirectly, the moral conduct of man.

The Religious Element

Let me give you, in conclusion, one more illustration of the difficulties we have to contend with in trying to determine whether certain acts and certain sayings may be called religious or not.

We are told by an excellent Arabic scholar, Baron von Kremer, a member of the Imperial Academy at Vienna, that at Vienna, which is as advanced and refined a capital as any in Europe, you may still see people, when walking in the streets, picking up any bits of bread lying on the pavement and placing them carefully where poor people, or, at least, birds or dogs, may get at them.

Is that a religious act? It may be or it may not. It may be a mere inculcation of the old proverb, "Waste not, want not." But as soon as the bread is called the gift of God, the reluctance to tread it under foot may become religious.

Manzoni, the Arabian traveller, tells us that the Kabili, the agricultural Arab, takes the greatest care not to scatter a crumb of bread. When he sees a piece of bread lying in the street, he lifts it, kisses it thrice, praises God, and puts it where no one can step on it, and where it may be eaten, if only by a dog.[20]

Here we see religious elements entering in. Yet, though the Kabili shows his reverence for bread, though he calls it 'aish, that is, life, as we call it the staff of life, no one would say that bread had become a divine being in Arabia, still less, as some of our friends would say, that it had become a fetish, or a totem.

If, therefore, we find that similar reverence is shown to fire or water, we have not therefore to admit at once that they have thereby been raised to the rank of divine beings. If bread was called life, so was fire. The founder of a new sect among the Ojibways addressed his disciples in the following words: "Henceforth the fire must never go out in thy hut. In summer and winter, by day and by night, in storm and in calm weather, remember that the life in thy body and the fire on thy hearth are the same thing, and date from the same time."[21]

Here fire and life are identified, but the fire within the body is no more than what we should call the warmth of the body, and to say that this warmth is the same as the fire on the hearth implies as yet no kind of divine worship for either the one or the other.

The same respect which is paid to bread, is also paid to other kinds of food. Thus Mohammed forbad to use even the stone of a date for killing a louse, and in another place he is reported to have said: "Honour the palm, for she is your aunt."[22]

In the case of bread, therefore, and also in the case of corn and dates and other kinds of food, we can well understand that they should have been treated with reverence as the gift of Allah, or of any other god, provided always that an acquaintance with such divine beings existed beforehand. Without such previous knowledge, nothing, whether a ceremony or a myth, can be called religious. It seems to me, therefore, that we are perfectly justified in treating that previous knowledge by itself, and to reserve to it exclusively the name of religion. There was religion before sacrifice; there was religion before myth. There was neither sacrifice nor myth before religion, in the true sense of that word. Nothing is more interesting than to find out how sacrifice and myth sprang from the same field as religion. But they did not spring from that field until it had been touched by those rays of light which transform the finite into the infinite, and which called into life the unnumbered seeds that lay hidden in the ground, the seeds of tares as well as of wheat, both growing together until the harvest.

Notes

1. Réville, *Religions des Peuples non-civilisés*, i, p. 144.
2. Pâraskara, *Grihya-sûtras*, I. 11; and Sânkhâyana, I. 18.
3. *Zeitschrift der Deutschen Morgenländischen Gesellschaft*, ix, p. 1 seq.
4. See also Oldenberg, *S. B. E.*, xxix, p. 236.
5. Réville, loc. cit., i, p. 221.
6. Castrén, *Finnische Mythologie*, p. 57.
7. Müller, *Urreligionen*, p. 652.
8. τὸ πῦρ καθαίρει καὶ τὸ ὕδωρ ἁγνίζει. See also Vasishtha, XII. 15, 16.
9. *Corpus Scriptorum Histor. Byzant.*, pars i, p. 381, ed. Bonn; Castrén, loc. cit., p. 57.
10. Hartung, *Religion der Römer*, i, 46, 199; ii, 152.
11. On running through and jumping over the fire, see Grimm, *Deutsche Mythologie*, pp. 592–593; Ovid, *Fasti*, iv. 727 seq.; Müller, *Amerikanische Urreligionen*, p. 56.
12. Grimm, *Deutsche Mythologie*, p. 574.
13. *Description of the Western Islands*, p. 113; quoted by Borlase, *Antiquities of Cornwall*, p. 130.
14. Philostrat., *Heroic.*, p. 740; Welcker, *Trilogie*, p. 247; Grimm, loc. cit., pp. 577, 580.
15. Plutarch, *Aristides*, c. 20; L. v. Schroeder, in Kuhn's *Zeitschrift*, xxix, p. 198.
16. See M. Kovalevsky, *Tableau des origines et de l'évolution de la famille*, 1890, p. 80. He also quotes Geiger, *Ostiranische Kultur*.
17. Müller, loc. cit., pp. 368, 387–388; Brinton, *Myths of the New World*, p. 147.
18. *The Open Court*, No. 112, p. 1883.
19. Cicero, *De Nat. Deorum*, i. 42: "Superstitio in qua est timor inanis Deorum, religio quæ Deorum cultu pio continetur."
20. R. Manzoni, *El-Yemen, Tre Anni nell' Arabia felice*, Roma, 1884, p. 82.
21. Müller, loc. cit., p. 55.
22. Kremer, loc. cit., p. 4.

CHAPTER 15
DISCOVERY OF THE SOUL IN MAN AND NATURE

from Gifford Lectures on Anthropological Religion (1891)

The Three Stages of Early Psychology

We saw in our last lecture how man came first to speak about a *soul*, or, more correctly, about a *breath*. We saw that there was nothing altogether unreasonable in such a name. In fact, whenever we examine that autobiography which man has left us in his language, we shall always find some good sense, something reasonable, even in what seems at first sight most unreasonable or foolish.

If we only bear in mind, what is now a fact doubted by no one, that every word in every language had originally a material meaning, we shall easily understand why that which at the dissolution of the body seemed to have departed and which we consider the most immaterial of all things, should have been called at first by the name of something material, viz., the airy breath. This was the first step in human psychology.

The next step was to use that word *breath*, not only for the breath which had left the body, but likewise for all that formerly existed in the breathing body, the feelings, the perceptions, the conceptions, and that wonderful network of intellectual threads which constituted the man such as he was in life. All this depended on the breath. It certainly was seen to have departed at the same time as the breath.

The third step was equally natural, though it soon led into a wilderness of imaginations. If the breath, with all that belonged to it, had departed, then it must exist somewhere after its departure, and that somewhere, though utterly unknown and unknowable, was soon painted in all the colours that love, fear, and hope could supply.

These three consecutive steps are not mere theory; they have left their foot-prints in language, and even in our own language these foot-prints are not yet altogether effaced.

Let us look at Greek, as we find it in the Homeric poems. At present, I do not mean to speak of what the poet himself may have thought about the soul, about its work during life, and its fate after death. We shall have to speak of that hereafter. What we are now concerned with is what the language which Homer had inherited had to say to him on this subject.

The Original Meaning of Psyche

The most common word for soul in Greek is *psyche* (ψυχή). This psyche meant originally breath. When a man dies, his psyche, his very breath, is said to have passed through the bar of his teeth, the ἕρκος ὀδόντων. Here we see the first step. This word ψυχή, as you know, assumed afterwards every possible kind of meaning. Even in this passage we might translate it by life, or by soul, without destroying the sense. But we can clearly see that what passed through the ἕρκος ὀδόντων was originally meant for the actual breath.

The Psychological Terminology of Homer

Much has been written by Greek scholars about the exact meaning of psyche in every passage where that word occurs in Homer. I am not going to enter on that subject beyond stating my conviction that it is a mistake in poems, such as the *Iliad* and *Odyssey,* to look for a consistent use of words. It would be difficult even in modern poetry to find out what Shakespeare, for instance, thought about the soul, by collecting and comparing all the passages in his plays in which that word occurs. Poets are not bound by logical definitions, and if they used all their words with well-defined meanings, I doubt whether they could have written any poetry at all. They use the living language in which the most heterogeneous thoughts lie imbedded, and whatever word serves best for the moment to convey their thoughts and feelings, is welcome.

In the Homeric poems this difficulty is increased tenfold. Whatever may be thought about the final arrangement of these poems, no one can now hold that they were all originally the outcome of one mind. Nor must we forget that in epic poems different characters may be made to speak very different thoughts, and use the same words in very different meanings, as they best suit the circumstances in which they are uttered.

The Meaning of αὐτός

I shall give you one instance only to show what happens, if we try to interpret Homer as we should interpret Aristotle's treatise on the soul. You remember how it is said in the beginning of the *Iliad,* that Achilles sent the souls of the heroes to Hades, but he gave themselves, αὐτούς, a prey to the dogs. It has been inferred from this and similar passages that Homer looked upon the body as constituting the true self of man. But this is to forget the requirements of poetry. Homer here wanted to bring out the contrast between the souls that went to Hades, and the corpses suffering the indignity of being devoured by dogs. "They themselves" means here no more than "they themselves, as we used to know them in life."

How free Homer feels in the use of such words, we can see from another passage. In the *Odyssey,* we read that Odysseus saw Herakles, or his *eidôlon,* that is, his psyche, in Hades, but he himself, αὐτός, he adds, rejoices among the immortal gods (xi. 601).

In one passage, therefore, αὐτός means the body, or even the corpse, in another, the soul, and to attempt to reconcile the two by any theory except a poet's freedom of expression, would lead, and has led, to mere confusion of thought.

I shall attempt no more than to give you the general impression which a study of the Homeric poems has left on my mind, as to what was thought about the soul, if not by Homer himself, at least by those whose language he used.

Psyche and Menos

What strikes me as most characteristic of *psyche* in the Homeric Greek, is that, whether it means breath, or life, or soul, it is never localised in any organ of the body. It is not in the heart, or in the breast, or in the phrenes, where thought resides. It is in the whole body (σῶμα), yet different from the body.

The Homeric language clearly distinguishes between *psyche* and *menos*, including under the latter name all that we should call mind.

But the most important distinction between psyche and menos, or any other name for mind, seems to me this: that the psyche is something subjective, while all other names express originally rather acts or qualities.

Thymós

Take, for instance, *thymós* (θυμός [soul]). Of course, the Greeks had no recollection of the etymological meaning of that word, as little as we have of our own word soul. But there can be no doubt that *thymós* is derived from *thyein*, to move violently, from which we have also *thyella,* storm. In Sanskrit we have exactly corresponding to *thymós, dhûma*; but this has retained the material meaning of smoke, literally what moves about quickly. *Dhûli,* also, the Sanskrit word for dust, meant originally what is whirled about.

The Greek thymós, therefore, meant originally inward commotion—you remember how in Tamil the soul was called the dancer—and we find in consequence that it is chiefly used with reference to the passions. But though originally thymós meant simply what moves within us, it afterwards comprehended both feelings and thoughts, and has often to be translated by mind in general. It seems to me that it was only *after* it had assumed this meaning, that it could also be used in the sense of life. For if it was said that one man had torn out or destroyed another man's thymós, that was tantamount to his having taken his life.[1] Or when it is said that the thymós left the bones (λίπε ὀστέα θυμός), we know that what is meant is that his mind, and therefore his breath, or his life, had left his body. But it is important to remember that we never hear of a thymós continuing by itself after death, like the psyche, which shows, as I said, that originally the thymós was really an activity, and not, like psyche, a something active.

Phrenes

Another important word which Homer had to use is *phrenes*. It means literally the midriff or diaphragm, which holds the heart and lungs, and separates them from the lower viscera. It is therefore much the same as *stêthos* (στῆθος), the chest, as the abode of the heart. We find it used of animals as well as of men, as when the lion is said to have a stout heart in his chest, ἐν φρεσί (*Iliad,* xvii. 111). But it soon drops its material meaning, and is considered as the seat of all inward acts, both of feeling and of thought. The work of the menos, mind, of noûs, thought, and boulê, will, takes place within the phrenes, just as much as it takes place within the thymós. The Homeric Greek rejoiced, perceived, remembered, reasoned, ἐνὶ φρεσίν, as we should say, in the breast or in the heart. When we meet with such expressions as κατὰ φρένα καὶ κατὰ θυμόν, we should not try to distinguish between the two, as thought and feeling, but translate "in the heart and in the thought," the heart being the locality, the thought the activity. We find a similar juxtaposition in κραδίη καὶ θυμός.

But it is important to observe that the psyche, the soul, is never spoken of as dwelling within the phrenes, the breast, or within the heart (ἦτορ), nor is the psyche ever spoken of by Homer as the abode of the mind.

It has been pointed out that while phrenes in the plural is often used in its purely physical meaning, as we use the words breast and heart, the singular phrên has always reference to the mind or the intellect.

The derivation of phrên and phrénes is not clear in Greek, but there can be little doubt that its meaning, like that of all words, was originally material. It meant the actual diaphragm; then, what was enclosed in the diaphragm, particularly the heart, and lastly what took place within the breast or the heart. To suppose that it was derived from *phroneîn*, to think, and meant originally thought, and afterwards only the seat of thought, the chest or the heart, is to invert altogether the natural order of things. It was only after phrénes had become familiar in the sense of mind in general, that we can account for a large number of derivatives in Greek, such as ἄφρων, πολύφρων, φρονεῖν, and all the rest.

Soul and Ghost

It would be easy to follow the same process in other languages, but the result would always be the same.[2]

It is unfortunate that our own words, soul and ghost, are not quite clear in their etymology. It is most likely, however, that soul, the German *Seele*, the Gothic *saivala*, meant originally, like the Greek θυμός, commotion, and that it is connected with the names for sea, the Gothic saiv-s.[3] And I feel inclined now to trace the English *ghost*, the German *Geist*, which, following an idea of Plato's,[4] I formerly thought connected with *yas*, to boil, with *yeast*, and German *Gischt*, back to the Sanskrit *hîd*, to be angry, *heda*, anger, so that it meant originally heat or commotion.

Differentiation of Meaning

This linguistic process which led to the formation of words for the different phases of the intellectual life of man is full of interest, and deserves a far more careful treatment than it has hitherto received, particularly at the hands of the professed psychologist. What is quite clear is that all these words begin as names of material objects and processes, such as heart, chest, breath, and commotion, just as the names of the gods began with the storm-wind, the fire, the sun, and the sky. At first every one of these words was capable of the widest application. But very soon there began a process of mutual friction and determination, one word being restricted idiomatically to the vital breath or the life, shared in common by man and beast, other words being assigned to the passions or the will, to memory, to knowledge, understanding, and reasoning. This process of widening and narrowing the meaning of words goes on for ever; it goes on even now, and can only be stopped by that dictatorial definition of terms which is so offensive to the majority of mankind, and yet is the *sine quâ non* of all accurate thought. Our own language is over-burdened with an abundance of undefined names, such as soul, mind, understanding, intellect, reason, thought, to say nothing of breast, and heart, and brain, of passion, desire, and will. Who is to define all these words, and to keep one distinct from the other? There is here a real Augean stable to be cleared out, and until it is cleared out by a new Hercules, all philosophy will be in vain.

The Agent

For the purposes of Anthropological Religion we wanted to know how man, for the first time, came to speak and think about a soul as different from the body. We have now seen that the way which led to the discovery of a soul was pointed out to man as clearly as was the way which led him to the discovery of the gods. It was chiefly the breath, which almost visibly left the body at the time of death, that suggested the name of breath, and afterwards the thought of something breathing, living, perceiving, willing, remembering, and thinking within us. The name came first, the name of the material breath. By dropping what seemed material even in this airy breath, there remained the first vague and imperfect concept of what we call the soul.

This something breathing, living, perceiving, and thinking, or, as we may now say again, this postulated agent of the acts of breathing, living, perceiving, and thinking, was recognised as within the body during life, and as without the body after death. It went by the same name, being called psyche in Greek, while inside the living body, and likewise psyche, after having departed from the dead body.

In all this there is nothing strange, nothing which we cannot follow and understand, nothing, or almost nothing, that we cannot make our own. There is one step, no doubt, which we find it difficult to take. We may admit the agent within, we may admit the persistence of that agent after it has left the body, but we should probably consider its identification with the breath as too material. We stand here before the old problem whether the human mind can ever conceive anything, entirely divested of all material attributes. Certain it is, that language cannot express anything except with names taken from material objects. This is a fact to be pondered on by all philosophers, aye, even by those who do not claim that proud title. Language, no doubt, can go on and negat[e] all that is material. One of the poets of the Veda, when speaking of the Supreme Being, says "that it breathed without air" (Rv. X. 129, 2), *ânît avâtám;* and, if we want to follow his example, we may say of the soul after death, that it is a breath without air. Language will perform wonderful feats in that way. But the ancients evidently thought they had gone as far as they could, when they spoke of the soul after death as a breath, that is, as a breathing and as a breather, and with all due respect for modern metaphysical phraseology, I doubt whether, if we keep to positive terms, we shall ever find a better word for the agent within us than breath or *psyché.*

Different Origin of Other Names for Soul

But even though the process which led the ancients to a belief in a soul and in souls may not be quite perfect in the eyes of modern metaphysicians, to study it in the annals of language has one great advantage. It teaches us again and again that the first framers of our language and our thoughts, even though they were not philosophers by profession, were on the whole reasonable beings, men not very different from ourselves, though living in a very different atmosphere. We cannot protest too strongly against what used to be a very general habit among anthropologists, namely, to charge primitive man with all kinds of stupidities in his early views about the soul, whether in this life or in the next.

Shadow

When we are told, for instance, that there was another road also on which man was led to his first discovery of the soul, by recognising it in his own shadow, we simply cannot follow. When man had once realised the idea of a soul and found a name for it, he might liken that soul to many things, and we shall see that he did liken it to many things, such as a bird, a butterfly, a cloud of smoke, and also a shadow. But all this is poetical metaphor, and carefully to be distinguished from that process which we have hitherto examined. There we saw that breath, the actual breath, was identified with life, the individual breath with the living individual, and the departed breath with the departed individual. But to suppose that any human being should ever have mistaken the shadow of his body on the wall as the true agent within, and as that which would remain after his body had been burnt, or buried, or devoured, is more than we can father even on the most primitive savage. It is said that a savage does not know what a shadow is, and how it arises. I doubt it. He might fell a tree, he would never try to fell the shadow of a tree. Anyhow the very word, shade or shadow, shows that the primitive Aryan savage at least knew, even before the Âryas separated, that shade was simply a covering, whether the covering from the sun-light, given by the branches of a tree, or a covering of any light by an opaque body. The name is derived from a root meaning to cover. We can understand how the name of breath became the name of many things dependent on breath, from the breathing of the living man to the soul of the departed. But that any person should ever have looked on his outside shadow which came and went, and could be produced at a moment's notice, as something by which he lived in this life, or something by which he would live in the next, is more than we can take in and digest, more than we ought to charge even against the most primitive savage.

The name of shade did not help the birth of the concept of soul, but the soul, having been conceived as breath flown from the body, was afterwards, for one of its qualities, its thinness or impalpableness, likened to a shadow. So long as the soul was in the body, no likeness was required, and shadow would have been the very last to answer the purpose. After death, however, such a simile became quite natural. The soul was supposed to be like the body, hence it was often called an *eidôlon* or likeness, and what likeness was so like as the shadow which bore the very outlines of the human body, and seemed always to be doing exactly what the man himself was doing? If, as we are told, shadows on a wall suggested to the early artist the first idea of a portrait, what better name could poetical imagination suggest for a disembodied soul than shadow?

We can watch this process in many languages. Thus *ata*, which in Maori means spirit and soul, is clearly identical with *ata*, shadow, image, with the Tahitian *ata*, cloud, with the Marquesan *ata,* likeness. In Mangaian *ata* has actually come to signify the essence of a thing, a concept which many people would consider far beyond the reach of these uncivilised people.

The Maori *wairua* also, which means a spirit, and the human soul, is clearly connected with *wairua,* a shadow, a reflection. In Hawaian *waitua* is a ghost or spirit of a person seen before or after death, separate from the body.

This will show how important the distinction between *radical* and *poetical* metaphor[5] really is for a right appreciation of the thoughts of primitive man in their historical development. He, poor primitive man, can no longer defend himself, but his descendants ought all the more to stand up for his good name. If Mr. Herbert Spencer is right that our common ancestor could never have mistaken a dead for a living thing, how can he believe that a mere shadow was mistaken by him for his own living soul, whether before or after death?

And here language comes again to our help. Though shadow becomes at a very early time a recognised name for the souls after death, its original character as a poetical metaphor is not yet quite forgotten, for instance, by Homer. No doubt, the dead are to him *skiaí* (σκιαί), shades, quite as much as *psychaí,* souls. But in certain passages we can still discover traces of the poetical metaphor. When Odysseus tried to lay hold on the psyche of his dead mother, then thrice it flew away from his hand, "like unto a shadow or even a dream" (σκιῇ εἴκελον ἢ καὶ ὀνείρῳ). And when she explains to him that this is the state of mortals after death, "that their nerves no longer hold the flesh and bones, for these the strong force of fire has consumed, what time their thymós first left the white bones," she adds, "but their psyche flying flits about, *like a dream.*"

All these were originally poetical comparisons. The souls were like shades, like dreams, like smoke (ἠΰτε καπνός); but we never hear of shades, dreams, or smoke leaving the body.[6] That applies to breath only, to psyche, and these two processes must therefore be kept carefully distinct, if we wish to gain a true insight into the working and growing of the human mind in its earliest phases.[7]

Dreams

I have still to say a few words with regard to another theory, according to which the idea of a soul in man is supposed to have been suggested for the first time by dreams and apparitions. What I said with regard to the theory that the soul was originally conceived as a shadow, applies with even greater force to this theory. Before primitive man could bring himself to imagine that his soul was like a dream or like an apparition, it is clear that he must have already framed to himself some name and concept of soul. All the illustrations which have been collected in order to prove that man's first conception of a soul was derived from what he saw in dreams and apparitions, leave no doubt on that point. They all presuppose some knowledge of the soul. When Mr. St. John tells us that the Dayaks think that in sleep the soul sometimes remains in the body, and sometimes leaves it and travels far away,[8] it is clear that they must have elaborated their concept of soul, quite independently of its travels in sleep. They might indeed have likened it to a dream, but they could not have received the first intimation of a soul from hypnotic apparitions.

It is quite true that the separation of subjective and objective impressions is much less fully carried out by uncivilised than by civilised nations, by uneducated than by educated persons. But with regard to dreams, the first impression, whether with civilised or uncivilised people, is that they are not like ordinary objective impressions. As soon as a man wakes even from the most vivid dream, he knows that it was *only* a dream. When, as Sir George Grey describes it, a savage jumps up to get rid of a nightmare, catches a lighted brand from the fire and flings it with many imprecations in the direction where the apparition was seen, he knows, as soon as he is fully awake and has quite shaken off his dream, that the spirit he saw in his dream is not like a real person whom he can lay hold of, punish, and kill. As soon as he is awake, he feels relieved. It was only a dream, he says, it was a nightmare, or whatever name suggests itself; and he comforts himself in his fright by saying: "the fellow only came for a light, and having got it, having been hit by the lighted brand, he will go away."

If people once possess the idea of something within themselves different from the body, even though they call it as yet a commotion or a mover or a dancer only, and if they have once brought themselves to believe that after death, though the body may perish, that which was in the body has not perished, then visions, whether by day or by night, will no doubt help to strengthen their belief in departed spirits, though, alas!, that belief would soon vanish like a dream, if it had nothing but dreams to depend on.

Once given the name and concept of soul, and of departed souls, there would be no limit to poetical metaphor. They might be likened to birds flying away, to smoke vanishing in the air, to shadows that can be seen, but cannot be touched, to dreams that come when least expected, but can never be called back. How far this popular poetry may be carried, may be seen from many popular sayings current among uncivilised and civilised people.

Superstitious Sayings about Shadows

Thus Bastian states that the Benin Negroes regard men's shadows as their souls, that is to say, that they speak of the souls as shadows. Nothing can be more natural. But if he adds that the Wanika are actually afraid of their own shadows, that depends very much on the authority of the interpreter. In the presence of white people, the Wanika may have seemed afraid of many things, even of their own shadows. The reason given that "possibly they think, as some other negroes do, that their shadows watch all their actions and bear witness against them," shows that the explanation was a mere surmise, and all depends on who "the other negroes" were meant to be. If Crantz tells us that the Greenlanders believe a man's shadow to be one of his two souls, is it not clear that they must previously have possessed an independent idea at least of one or the other of the two souls?

Most likely, however, the fact that bodies threw no shadows in the dark, was quite sufficient to suggest the expression, that during sleep and in the darkness of the night a man's soul left his body, just as the shadow did. When speaking of inseparable friends, we may say even now, one follows the other like his shadow, or that one never leaves the other, like his shadow. Even when their relations are less friendly, we speak of one man being shadowed by another. Why then should not the same simile have suggested itself to early thinkers, that in sleep the soul left the body as the shadow leaves the body during night?

There is another popular saying among the Zulus, that a corpse throws no shadow.[9] Is this really to be taken as a myth of observation, as Mr. Tylor would call it? Did any human being ever persuade himself that a corpse, when carried on the bier, threw no shadow, while the bier and those who carried it were followed by their shadows? Mr. Herbert Spencer, in spite of his repeated warnings against taking savages for fools, thinks it was so. I can see in it nothing but a perfectly intelligible process of language. People who had adopted *shades* as one of many idiomatic names for the souls of the departed, might very naturally say that the shade had left the corpse, or that the corpse was without its shade. Fond of riddles, as ancient people are, they might even ask, "What is there in the world without a shadow," and the answer would be, "a corpse." When Eastern nations say now, "May thy shadow never grow less," they know perfectly well that a shadow never grows less by itself. What they mean is perfectly understood, namely, "Mayest thou thyself never grow less."

There are many things which half-educated people repeat and which they pride themselves on believing, though they would often laugh at others for believing that they believe them. Think with how serious and almost solemn a face young ladies will tell ghost-stories in these days. Even though they have never seen a ghost themselves, they are fully convinced that their friends have, and it would amount almost to rudeness to doubt their good faith. However, I ought not to restrict these remarks to young ladies, for I see in the newspapers that the young men at Oxford have just carried a resolution in their Debating Society that they believe in ghosts. And yet they do not really believe in ghosts. They do not even believe that they believe in ghosts, unless they use "to believe" in a very peculiar sense. They like to make believe that they believe in ghosts. To believe is always supposed to be more proper than not to believe. But ask them to bet one shilling on the due apparition of a ghost, and, if I know them well enough, they will decline. It is exactly the same with savages. They also are proud to believe or to profess to believe what ordinary people are not able to believe. The Zulus, for instance, not only profess to believe that a corpse throws no shadow, they look equally serious when they assure their European questioners that "as a man approaches his end, his shadow shortens, and contracts into a very little thing."[10] However, when Bishop Callaway spoke to his Zulu friend seriously, and asked him whether he really believed that the shadow thrown by his body, when walking, was his spirit, he soon collapsed, and falling back on his popular idiom, declared, "No, it is not your *itongo* or spirit (evidently understanding me to mean by my spirit an ancestral guardian spirit watching over me, and not my own spirit), but it will be the *itongo* or ancestral spirit for your children when you are dead." This is hardly more than if we were to say that after death a man's spirit would be to his children a mere shadow, or like a shadow.

Unless we study the wonderful ways of language, we shall never understand the wonderful sayings of men, particularly during the earlier phases of human speech. Here Mr. Herbert Spencer has deprived himself of a microscope that would have disclosed to him again and again perfectly organic thought, while he can see nothing but incoherent specks.

The Ci-près in Language

He often accuses savages of what he calls erroneous classing. He wonders that the Esquimaux should have taken *glass* for *ice*, and quotes this as an instance of their erroneous classing. I do not believe it for a moment. Do we not ourselves call glass *crystal*, and κρύσταλλος meant ice before it came to mean rock-crystal. This is not a case of erroneous classing; it is simply and solely a necessity of language. When we become acquainted with a new subject, such as glass, we have either to invent an entirely new name, and that, particularly in the later periods of language, becomes almost an impossibility, or we must be satisfied with what lawyers call a *ci-près*, and in the case of glass the most natural *ci-près*, or the nearest likeness, seemed to be ice.

Thus again, when we say, "the wall sweats," it is not because we really assume that water comes out of the wall, after a frost, as perspiration comes out of our skin. It is simply a case of poetical metaphor, without which half of our language would become impossible.

I do not believe that because the Orinoco Indians call the dew "the spittle of the stars," they believe that the stars spit during the night. We speak ourselves of cuckoo-spittle, even though we know perfectly well that it is no more than a small grub enclosed in a slimy substance. Nor is it more than a poetical metaphor when we say, it spits with rain.

The Infinite in Man

A student of language knows that all these expressions are not only perfectly natural, but simply inevitable. And the same applies to the words for soul. The soul was discovered in the breath, and hence it was called breath, or *psyché*, by people who at first really believed that that which left the body at death and continued to exist was the breath. This something, called breath, or psyche, was afterwards likened to many things, if they possessed certain attributes which seemed compatible with the nature of the psyche or soul. As the souls after death were supposed to fly away, they were called winged, or even birds, not because they were really taken for birds, or birds for them, but simply and solely because they were supposed to pass through the air, like winged creatures. Even biologists may sometimes speak of angels' wings, but do they believe that vertebrate beings can have wings as well as arms?

The souls were called shades, not because they were ever supposed to be nothing but images thrown on the wall, but because they shared one attribute in common with the shadows, namely, that of being without a body, and almost unsubstantial. Another name, *eidôla*, meant really not much more than shadow. It meant likeness, such as the outline of a man's shadow, or his image reflected in the water. And if the souls were called dreams, this was again because they shared in common with the visions of dreams their unsubstantial nature, their withdrawing themselves from the touch and the embraces of their friends.

When mythology steps in with its irrepressible vagaries, or, what is even more serious, when art invests these unsubstantial similitudes with a substantial form, no doubt the souls often become in the popular mind actual shadows and dreams, birds and winged angels. But though such expressions may satisfy the human heart in moments of grief or hope, though they may inspire the poet with his happiest strains, the serious thinker knows that they are no more than relics of ancient poetry. The soul is not a bird, the soul is not a shadow, the soul is not a dream, not even the shadow of a dream. Here we have the same No, no, which in the Upanishads we saw applied to God. But when the ancients called the soul breath, they really meant what they said; at least, they meant it as much as when they spoke of Dyaus or the sky, meaning not the material sky, but the agent in the sky. No doubt, as the sky was recognised afterwards as only a vesture of God, the breath also was conceived, as early as the time of the Vedic poets, as "breathing without air." On this point we are not wiser than the most primitive savage. We retain his words, however knocked and battered during ages of intellectual toil and moil, and we shall have to retain, whether we like it or not, some of his thoughts also. If breath sounds too material to our ears, we may like the Latin word better, and translate breath by spirit. But so long as we think in human language, we shall never arrive at a truer expression than breath or spirit, unless we rise to a higher octave of thought altogether, and agree to call it the Infinite in Man, as we recognised in the gods of nature the ancient names for the Infinite in Nature.

Why a Belief in a Soul is Necessary

It has been asked what our belief in a soul can have to do with religion and with a belief in God, and what room there is for anthropological by the side of physical religion. To judge from many works on religion, and, more particularly, on the origin of religion, it might seem indeed as if man could have a religion, could believe in gods and in One God, without believing in his own soul, without having even a name or concept of soul. It is true also that our creeds seldom enjoin a belief in a soul as they enjoin a belief in God; and yet, what can be the object, nay, what can be the meaning of our saying, "I believe in God," unless we can say at the same time, "I believe in my soul."

This belief in a soul, however, exactly like the belief in gods, and, at last, in One God, can only be understood as the outcome of a long historical growth. It must be studied in the annals of language, in those ancient words which, meaning originally something quite tangible and visible, came in time to mean something semi-tangible, something intangible, something invisible, nay, something infinite in man. The soul is to man what God is to the universe, and as it was the object of my last course of lectures to follow in the ancient languages and religions of the world the indications of man's progress towards a knowledge of or a faith in God, it is my object now to discover, if possible, in the same historical archives, some evidence that may be left there of man's progress towards a knowledge of or a faith in his own soul. The search for that evidence may often prove tedious, and its interpretation by no means so satisfactory as when we had to deal with the history of man's belief in God. Yet the subject itself is so important that we must not allow ourselves to be discouraged. It is a first attempt, and first attempts, even though they fail, encourage others to try again.

The Soul in Man and the Soul in Nature

The problem which we have to face in trying to discover how the agent within us was first discovered, was first named and conceived, is really in many respects the same as the problem the solution of which we had to study when treating of Physical Religion. There we saw how the agent without, or, at first, the many agents behind the phenomena of nature, had to be named by the names of visible outward phenomena. There, too, if you remember, the question arose whether what had been called *Dyaus,* the sky, or in Chinese, *Tien,* the sky, was the actual, visible, blue sky, or something else, the Agent in the sky. That postulated Agent was actually called in later times the *psyche,* the soul of the sky.

In exactly the same way, we find that the question was asked whether what had been called *prâna* in Sanskrit, or *psyche* in Greek, or *spirit* in Latin, was the actual, visible, warm breath, or something else, the agent in the breath.

We, at our time in the history of the world, may smile at such questions. We know that if we speak of heaven, we do not mean the blue sky. We know that when we speak of our Father in heaven, we do not mean our Father in the clouds. We know that if we speak of the dead as ascending into heaven, as dwelling in heaven, we mean more than a mere ascension into the higher strata of our terrestrial atmosphere. We live in post-Copernican times. Still we must remember that what was once the language of the childhood of the world, will remain for ever the language of the childhood of every generation. A child will always look to the blue sky as the abode of his Father in heaven. A child will always lift his hands and his eyes upward, when praying to God. And no child could conceive the return of the spirit to God who gave it, the return of the Son to the Father, except under the image of an ascent through the clouds. Some bear these fetters of language longer than others. Some bear them all their life, without even being aware of them. Who would blame them, if only they would not grudge to others the freedom for which they have often paid a very heavy price!

In the same way, the soul or the spirit will with many people always remain a breath, an airy breath, for this is the least material image of the soul which they can conceive, just as the sky was the least material image of the deity which many of the ancient nations could conceive.

If we only remember this, we shall better understand how old age is able to use, and, from an historical point of view, to use honestly, the language of childhood, though with a deeper and truer meaning. An old man who prays, "Our Father which art in heaven," is not necessarily a hypocrite. It is a study of the ancient religions of the world that best enables us to see behind the imagery of their language, and the outward show of their sacred customs and ceremonies, something which we can at least understand, something with which to a certain extent we can even sympathise, something that is true, though expressed in helpless and childish words.

But there is another lesson also which an historical study of the origin and growth of the words for soul and for God may help to impress upon our mind. In teaching us how the concept of God arises of necessity in the human mind, it teaches us at the same time that nothing can satisfy the human mind but what we mean by an agent, that is, a real, self-conscious agent, or as we express it, in more anthropomorphic language, a personal soul and a personal God.

We have to guard here against two misunderstandings. There are theologians, even Christian theologians, who hold that the concept of God was the result of a special disclosure, and made to Jews and Christians only. Such assertions can only be silenced by facts, such as I gave in great abundance in my lectures on Physical Religion, though one would have thought that some of these orthodox sceptics would on this point have yielded more ready submission to the express teaching of St. Paul.

But there are other philosophers who hold that the concept of God, though, like the concept of soul, it may be the result of a long-continued historical development which can be traced in all languages and all religions, is nevertheless a name only, which we may retain for old association's sake, but which denotes merely the unity of nature and no more.

This has been repeated again and again, and yet a little reflection would have shown that this whole argument rests simply on a mistake in language and thought. If people prefer to call the agent of their own acts and the agent of the acts of nature a mere unity, modern languages allow such a licence. But we must remember that unity is an abstract term, and that we can never have abstract terms without concrete objects from which they are abstracted. Unity is nothing, if it is not a predicate.

We may predicate unity either when what is substantially one has become differentiated, or when what is substantially different has become combined. The latter is clearly impossible in our case when, if we may trust our reason at all, our reason postulates a self-conscious agent for everything that deserves to be called an act. To speak of an act that acts itself, or of an agent not different from his act, is not to speak, but only to use words. In the former acceptation of unity, we may predicate it of that which is one and the same in different acts, whether the soul in man, or God in nature; but in predicating unity we cannot predicate it except of a unit. We cannot define that *unit,* whether in ourselves or in nature, beyond saying what we mean by it, namely, a self-conscious agent, such as we know our self to be, apart from all other qualifications, and such as we require the self of nature to be, apart from all phenomenal attributes.

Now then we see clearly how closely what I call Anthropological and Physical Religion hang together. The former teaches us how we have come to discover an agent within, and to call that agent soul or person, or ego or self, but not simply a cause, still less a mere unity. The latter shows us how we have come to discover an agent without, and to call that agent soul or person or ego or self, but not simply a cause, still less a mere unity. If in religious language we prefer the name of God, we may do so, but we must not leave out any of the elements of which it is composed. If soul is nothing except it be a self, a self-conscious agent, or a person, God would be nothing, unless He was at least a self, at least a self-conscious agent, or a person, in the highest sense which that word conveys to ourselves.

Notes

1. The same process of thought accounts for such expressions as "ἀπὸ γὰρ μένος εἵλετο χαλκός" ["for the bronze (knife) had robbed them (the sacrificial lambs) of their strength"—The Editor], *Iliad,* iii. 294.
2. Tylor, *Primitive Culture,* ii, p. 388; *Hibbert Lectures,* p. 89 seq.
3. *Science of Language,* vol. 1, p. 522.
4. "Θύσις καὶ ζέσις τῆς ψυχῆς" ["raging and boiling of the soul"—The Editor], *Kratylos,* 419e.

5. *Science of Language,* vol. 2, pp. 456, 479.
6. See on this subject an interesting paper by Mr. C. F. Keary, "The Homeric Words for Soul," in *Mind,* XXIV, Oct. 1881. Though I differ from his conclusion, the facts collected by him have proved very useful.
7. See Dr. Codrington, as quoted in *Hibbert Lectures,* p. 91.
8. Tylor, *Early History of Mankind,* p. 7.
9. Callaway, *Religious System of the Amazulus,* p. 91.
10. Ibid., p. 126.

CHAPTER 16

WHAT WAS THOUGHT ABOUT THE DEPARTED

from *Gifford Lectures on Anthropological Religion* (1891)

The Soul *Minus* the Body

We saw in a former lecture how the name and concept of soul arose. It was a perfectly simple process; what may almost be called a mere process of subtraction. There was man, a living body, acting, feeling, perceiving, thinking, and speaking. Suddenly, after receiving one blow with a club, that living body collapses, dies, putrefies, falls to dust. The body, therefore, is seen to be destroyed. But there is nothing to prove that the agent within that body, who felt, who perceived, who thought and spoke, had likewise been destroyed, had died, putrefied, and fallen to dust. Hence the very natural conclusion that, though that agent had departed, it continued to exist somewhere, even though there was no evidence to show *how* it existed and *where* it existed.

Continuance of Feelings towards the Dead

We next examined the different ways in which some of the principal nations of antiquity treated the dead bodies of their friends, and we could clearly perceive that they were all suggested by feelings of either love or fear, not directed towards the material remains which had either been destroyed by fire or hidden in the earth, but directed towards something else which they called the souls, or the spirits, or the shades.

If a father had been loved and revered while living among his children, it would have been against all the tendencies of the human heart if those sentiments had suddenly ceased at death. And if an enemy had died or had actually been killed, it seemed by no means unnatural that feelings of hatred and fear should be entertained against him, even after his death.

However, whether it was due to the fact that among uncivilised races the sentiment of love was less prevalent, or whether the terrible appearances which often accompanied death told on the survivors, certain it is that in ancient times the feelings towards the departed consisted more largely of fear or awe than of tenderness and love.

The Zulus, however, when giving an account of their ancestral worship to Bishop Callaway, laid great stress on their feelings of love and gratitude towards the deceased. Their father whom they knew is the head by whom they begin and end in their prayer, for they know him best, and his love for his children; they remember his kindness to them whilst he was living; they compare his treatment of them whilst he was living, support themselves by it, and say: "he will still treat us in the same way now he is dead."

It was this feeling of love that led to a continuance of acts of kindness towards the departed. Not only was their body carefully disposed of, but their memory was cherished, and at the ordinary meals of the family and at festive gatherings some share of food and drink was often thrown into the fire of the hearth, as a gift to them. This led to what I believe to have been a subsequent belief namely, that, in some way or other, the souls were really able to enjoy these gifts, nay, that they had a right to them, and would resent their withdrawal.

The feelings of fear would dictate almost the same acts, though the motive would be rather a wish to propitiate than a desire to benefit the departed. The thought that the souls of the dead had the power of injuring the survivors seems to have been very common, and the great attention paid to a proper disposal of the dead bodies, was due in no small degree to a desire to pacify the departed spirits, and to give them what was considered their due (τὰ νόμιμα).

I hardly think, however, that, as has lately been suggested, we can ascribe the wide-spread custom of burning the dead to an apprehension, lest, if only buried, their ghosts might return. There are many much more natural motives that would have suggested cremation rather than burial. Nor do I know of any evidence that people who burnt the dead bodies thought that they were safe thereby against the mischievous power of the ghosts of the departed, or that they were freed from the obligation of honouring and appeasing the ancestral spirits by commemorations and sacrificial offerings.

The Germs of Ancestral Religion

Two results would almost inevitably follow. The honours paid to the dead at the time of the funeral would assume a solemn and sacred character, and their continuance at certain times and seasons would become part and parcel of the religious life of the people. Secondly, the souls themselves, to whom these honours were paid, would soon assume a more and more exalted character, and occupy in the minds of their worshippers a place second only to that which had been assigned to the gods.

This process would naturally assume different forms among different people, but its general character would be the same. A second religion would arise, a second class of no longer human, and soon of half-divine beings would be believed in, and the powers ascribed to them and the worship paid to them would become almost, if not entirely, identical with those of the gods.

The First Ancestor

It has been asserted that in some cases this worship of the departed formed all that can be called religion among certain tribes. Religion has sometimes been called a *retrogressio in infinitum,* and by a very simple *retrogressio in infinitum* certain tribes were supposed to have argued that as they had known a father, a grandfather, and a great-grandfather, there must have been earlier fathers, and fathers of fathers, all to be honoured and to be feared, all in their way powerful, able to punish and able to reward, till at last human reason wanted a rest, and postulated at the beginning of all things one father, the father of all fathers, and, very soon, the maker of all things. You see that this process of reasoning is perfectly natural, however startling its result may seem to us. When a missionary tells us for the first time that a savage believes that the world was made by his grandfather, we find it difficult to believe such a statement. Nevertheless the train of thought that leads from a real father to a father of all fathers, is not so very different from that which leads our own minds from the conception of one cause to the conception of a cause of all causes. The Zulus, for instance, believe in *Unkulunkulu,* the great-grandfather, and this great-grandfather is regarded by some of the Zulu tribes as the creator and ruler of the world.

Were there Races without Physical Religion?

Some students assure us that there are races whose religion consists entirely of ancestral worship. This question, whether there are and whether there have been races whose whole religion consists in worship of ancestors, can only be solved by a careful examination of those very troublesome accounts of travellers of which I had to speak before. As far as my own studies go, I have not succeeded in discovering one single race believing in ancestral souls only, and not in gods. Among civilised and literary nations, whose history must always form the starting-point and the foundation of any truly scientific research, there is no trace of such a state of things. That does not prove its entire impossibility, by no means. But it will certainly make the true historian very careful, before he draws his general conclusions from the fragmentary accounts of travellers among Zulus and Australians, rather than from the sacred literature of the great nations of the world.

Are there Races Whose Religion is Exclusively Ancestor-Worship?

I have been looking out for many years, wherever there was any likelihood of meeting with a religion that consisted entirely of ancestor-worship. There is no *à priori* reason why races should not exist now who, even in the eyes of careful observers, might seem to worship nothing but their ancestors. Even where a worship of nature-gods existed, it is quite possible that it might have vanished and have been superseded by a worship of ancestral spirits.

Religion of Zulus

The best-known of the races who were formerly considered as having no religion except ancestor-worship are the Zulus, and the other so-called Bântu tribes of South Africa. I do not wonder that this mistake should have arisen. These tribes are migratory, they are not held together by a common priesthood, and, like most savages, they are very unwilling to allow themselves to be examined on religious topics. It was quite natural therefore that some missionaries, when they were told that these Kafirs worshipped *Unkulunkulu,* the great-grandfather, should have stated that the whole religion of these natives consisted in worship of grandfathers and ancestors. When at a later time some missionaries, such as Dr. Colenso or Dr. Callaway, acquired a more accurate knowledge of the dialects spoken by the Bântu tribes, and were able to carry on rational discussions with some old chiefs and priests, they were able to give a much better account. But they also were under the impression for a long time that the Zulu religion was exclusively ancestral.

We owe the best description of the Zulu religion to Bishop Callaway, in his *Religious System of the Amazulus,* a book which certainly leaves us under the impression that, at present, the Zulu religion consists almost exclusively of worship of ancestors. Their ancestral spirits are called *Amatongo,*[1] and at their head stands *Unkulunkulu.*

This Unkulunkulu, their great primal ancestor, has come to be regarded as the Creator, though other authorities deny that the Zulus have any idea of creation. The Rev. J. Macdonald, for instance, assures us that "they hold that the earth and the heavenly bodies have always been as we see them now, and that they will thus always continue, unless some terrestrial catastrophe should set the whole on fire, or, in some other way, disperse everything."[2] But even when Unkulunkulu is regarded as the Creator, he is not exactly an object of worship, awe, or reverence with the Zulus. He is often supposed to have given place to newer men of more recent times. The great objects of fear and reverence to which they pray and sacrifice are the spirits of dead ancestors for the last few generations.

"Ancestor-worship," as the same intelligent observer, the Rev. J. Macdonald, remarks,

> is not only professed by them, but they actually regulate their conduct by it. If a man has a narrow escape from accident and death, he says, 'My father's soul saved me,' and he offers a sacrifice of thanksgiving accordingly. In cases of sickness, propitiatory sacrifices are offered to remove the displeasure of the ancestors, and secure a return of their favour. Should any one neglect a national custom in the conduct of his affairs, he must offer sacrifice to avert calamity as the consequence of his neglect. When offering propitiatory sacrifices, the form of prayer used by the priest is: 'Ye who are above, accept our offering and remove our trouble.' In free-will offerings, as in escape from danger or at the ripening of crops, the prayer takes the following form: 'Ye who are above, accept the food we have provided for you; smell our offering now burning, and grant us prosperity and peace.'[3]

Many missionaries who did not consider that ancestor-worship could be considered as religion, declared in consequence that the South African tribes had no religion at all, and no belief in God. Even Dr. Callaway, though fully aware of the religious character of ancestral worship, seems to have been doubtful for a time whether the Bântu tribes had any knowledge of divine beings, apart from their ancestral spirits.

All observers, however, agree now that these tribes believe in other spirits also, besides those of men. They speak of water or river spirits, whom they describe as dwarfs or fairies. These are called *Incanti,* and are always mischievous.[4]

They have also a number of superstitions about thunder, lightning, rain, the rainbow, eclipses of sun and moon, and other physical phenomena. But all this would not yet prove that they believed in any of the gods of physical religion, or in a Supreme Being above the ancestral spirits.

However, in his last paper, "On the Religious Sentiment amongst the Tribes of South Africa," Dr. Callaway reports the statement made to him by an intelligent Gqika chief. "We used not to say," he told the bishop, "*Utikxo* for God, but *Ukqamata.*

When men feared anything they used to say, 'May Ukqamata help us.' Here then," the bishop writes, "we have hidden in the language of the people a word which shows that long ago, before the word Utikxo was introduced amongst them for God, they had a name representative to them of the Supreme—a being not like Unkulunkula amongst the Zulus, who is sometimes represented as having begun, died, and passed away; but one who is seen now with them, and to whom they constantly appeal in time of necessity, much in the same way as the devout amongst ourselves appeal to God."

Other people whom the bishop asked told him that Ukqamata is a living spirit, but that they knew not where it dwells; and if asked where it dwells, they would answer, "It goes beside me, and yet I see it not." And they said, "Spirits go out of men to go to Ukqamata, to the place where they dwell with him. The corpse does not go to Ukqamata, it is the spirit only which goes to him; the corpse remains in the earth."

Dr. Callaway then continues: "I have been aware that apparently apart from, above, and beyond their mere ancestral worship, ... the Kafir races ... universally speak of a Great *Itongo,* and appeal to it, pretty much in the same way as these frontier Kafir tribes are said to appeal to Ukqamata." The bishop also states that the Kafirs seem quite aware that the various names applied to the Supreme Spirit are but various names of the same Being. They admitted that they had borrowed Utikxo as the name of the Supreme Spirit from their neighbours, the Hottentots.

In spite of much confusion in these statements, one thing is clear, that the Kafir races of South Africa are not exclusively worshippers of ancestral spirits, but that they recognise or remember a Supreme Spirit, standing above their ancestors, and exercising a personal power over nature.[5]

Religion of the Niassans

Another race which quite recently revived my hopes of finding a religion consisting exclusively of ancestor-worship are the Niassans. An interesting description of the religion of the Niassans was published lately by Kramer in the *Tijdschrift vor Indische Taal-, Land-, en Volkenkunde,* deel xxxii, 1890.

These Niassans, who live in a solitary island west of Sumatra, have more than a hundred idols. Formerly they had less, now they go on adding to their number. Priests and priestesses make a living by serving these idols.

Their really important and permanent idols are the images of ancestors and house-idols. The ancestors are represented as human figures, about six to eight inches high, and carefully carved. Poor people, however, have to be satisfied with a piece of wood, with holes for eyes and mouth.

The house-idols are in the shape of children, and in the houses of rich chiefs these also are carefully executed.

After a man has been buried, the priest covers his grave with a mat, and seeks till he find a six-legged spider under it. That is taken as the soul of the departed, deposited in a reed, and placed by the side of the image.

Such images, however, are made of those only who have left male descendants. They occasionally borrow them from one another.

The Niassans expect all blessings from their ancestors, who likewise protect them against all dangers. But for that purpose it is necessary that they should receive constant offerings. No event of any importance takes place without some communication being made or some honour shown to the ancestral images. On some occasions their names have to be repeated. In fact their whole life seems to be under the sway of their ancestors. It is true they believe in evil spirits also (Bechus), and even in a devil (Bela). But these might be traced back to a belief in hostile ancestors, or ancestors of hostile tribes. At all events, they would not prove the existence of anything like a belief in nature-gods. What are called their *hazimas* (p. 492), and what many people would call fetishes, are really nothing but amulets, stones, teeth, pieces of lead, &c., which they wear as a protection against evil spirits. They are often supposed to have fallen from the sky, to have been the head of a serpent, or to consist of condensed stormwind.

However, within and above all this chaos of ancestral spirits, ghosts, and fetishes, there suddenly appears our old friend, the sun. Yes, solar worship even among the Niassans! The owner, the lord and master of all men is Lature, and he dwells in the sun. As we are the possessors of our pigs, the Niassans say, Lature is the possessor of all men. Nay, they are proud to call themselves the pigs of the sun. Sacrifices are offered to the sun-god that he may grant a long life to his pigs.

But though we look in vain for a religion consisting of ancestor-worship only, we often find that in the same religion the worship of ancestral spirits and the worship of the gods of nature exist side by side, and, what is important, we find that they are never confounded, but kept carefully distinct even in the terminology that is applied to them (pp. 478, 490).

Worship of Gods and Worship of Ancestors Kept Distinct

Professor Ch. de la Saussaye, in his *Manual of the Science of Religion,* has pointed out that in Greece, for instance, "other names are applied to the altars, sacrifices, and offerings connected with the dead than those used in the worship of the Olympian gods. The altar is called ἐσχάρα, not βωμός; the offering of the sacrifice ἐναγίζειν, ἐντέμνειν, not θύειν; the libations themselves χοαί, not σπονδαί" (p. 113). This is the rule, but there are exceptions.

Dr. Rohde also, in his *Psyche,* remarks that "the Greeks sacrificed to the gods by day, to the heroes in the evening or by night, not on high altars, but on a low sacrificial hearth, which was close to the ground, and sometimes hollow. Black-coloured animals of the male sex were killed for them, and their heads were not, as in the case of victims intended for the gods, turned towards the sky, but pressed down to the ground. The blood was allowed to run on the ground or on the sacrificial hearth, as a blood-feast (αἱμακουρία) to the heroes; their body was burnt entire, so that no man might eat of it. Sometimes cooked viands were offered to the heroes, and they were invited to partake of them" (p. 140).

Much the same applies to Sanskrit. When offerings are made to the gods, the Brâhmanic thread has to hang from the left shoulder, under the right arm (*upavîtin*). When the offering is intended for the departed spirits, the same thread has to be hung on the right shoulder, and under the left arm (*prâkînavîtin*).[6]

The departed spirits are called Pitris, fathers, the gods, Devas, the bright.

The exclamation used in sacrificing to the gods is *svâhâ;* in sacrifices to the departed it is *svadhâ.*

This shows among two nations, so widely separated as Greeks and Hindus, but whose language is known to us accurately, a clear recollection of the different meaning with which from the very first, gods and ancestors were worshipped by the Âryas.

Let us now see more in detail what some of the Aryan nations thought on this subject, and what place in their religion they assigned respectively to the worship of the gods and to the worship of ancestral spirits.

It is easy to see that when we speak of worship of ancestors and worship of gods, the meaning of worship must be different according to the different nature of those to whom it is addressed. The worship of the dead began with acts of kindness shown to the departed from the day of their death to the day of their funeral; the worship of gods was inspired by a feeling of awe, and arose from a sense of what was due to higher powers. In the end these two kinds of worship may have become almost identical, but in their first motives they stand wide apart.

Before we proceed further, however, we must first try to find out by what process of reasoning, or, it may be, unreasoning, people came to believe anything about the dead beyond their mere existence. The fact of their existence, as we saw, was proved, if it required any proof at all, by an argument as irresistible to-day as it was thousands of years ago. In the absence of all proof to the contrary, the agent in man was believed to exist on the same ground on which the agents in nature were believed to exist. There can be no action without an agent. Agni, the agent of fire, was not believed to be destroyed and annihilated, although the individual fire in which he appeared might be extinguished. And the soul, the agent in man, could not be believed to be destroyed and annihilated, although the individual body in which it appeared had fallen to dust and ashes.

But the human mind, and more especially the human heart, is not satisfied with this general belief in the mere existence of souls. It wants to know, what it cannot possibly know, *where* and *how* the soul exists after the body has left it, or rather, after it has left the body. Here we enter into the domain of mythology as different from that of religion. A study of the imaginations of ancient people with regard to the state of the soul after death may be useful for a knowledge of the character of different nations, possibly for a knowledge of human character in general. But it can no longer influence our own convictions. What Plato says in support of his belief in one God "who holds in His hands the beginning, middle, and end of all that is, and moves according to His nature in a straight line," is said for us quite as much as for the ancient Greeks, and even the name of Zeus, if we but know its true meaning, need not offend us. But the mythological stories told of Zeus and the other Olympian gods, are nothing to us. They were very little even to Plato.

It is the same with the mythology concerning the souls of the departed. That these souls exist is as true for us as it was for Plato. But where they exist and how they exist is a question the various answers to which may form an important subject of study to the historian, but can hardly, if at all, influence our own conviction that here, as in so many other things, we must learn to wait, and for a while to remain ignorant.

Such ignorance, however, was difficult to brook, and we find, I believe, among all nations some attempts, however futile, at lifting the veil and catching a few glimpses of the life of the souls after death.

Where Do the Departed Exist?

With people who burn their dead, there can be little doubt that the body, as such, has come to an end, when there is nothing left of it but dust and ashes. Their views about the abode of the departed souls are therefore generally less coarse and material than the views of people who bury their dead near their own abodes, sometimes beneath their own houses; or of the Egyptians and other races who mummify the bodies and keep them daily before their eyes.

When the Greeks say that the likenesses of men, the *eidôla*, mere shadows, went to the realm of *Hades*, we have only to restore the meaning of *Hades*, namely *Aides*, the Invisible, and we could ourselves use their language, that the likenesses of the departed go to the realm of the Invisible. It is only a more poetical expression for what we express in more homely terms, when we say that the soul has departed and has become invisible. You know, however, how very soon this invisible world, this house of the Invisible, becomes a Hades, with terrible rivers to be crossed in Charon's boat;[7] with a three-headed watch-dog, with judges judging the souls, and punishments and tortures inflicted on the wicked.

All this is mythology, and cannot affect us.

With people who buried their dead new ideas sprang up, as we see, for instance, among the Jews, who placed the abode of the dead below the earth. Their Sheol was indeed a lower world, and the same idea gained ground among other nations also. So long as the souls of the departed were supposed to exist in a place separated from the seats of the gods, they were naturally considered as *Inferi*, living below, in opposition to *Superi*, the gods, living above. From this simple distinction have sprung in time all the horrors of the infernal regions, which are supposed to have influenced the lives of men more powerfully than any other article of religious faith.

Before this definite localisation of the departed took place, we find another very general impression prevailing among civilised and uncivilised nations, that the departed went to the West. The Hervey islanders believed that when a man died his spirit returned to *Avaiki*, the original home of their ancestors in the region of sunset. Sometimes this region is called *te-po*, the night, i.e., the place where the sun hides itself at night, or, in other words, the West. Dr. William Wyatt Gill thinks that this is due to the fact that the Eastern Polynesians came originally from the West, from Avaiki (= Hawaiki, Hawai'i, Savaiki, Savai'i, all forms of the same name).[8] This may be so. But Avaiki is conceived also as a vast hollow beneath the earth. In it there are many regions, bearing separate names, but all to be regarded as parts of spirit-land. And in this sense Avaiki is clearly the West, as the land of departed spirits.

It required but little poetical imagination to speak of the sun as dying every day and vanishing in the west, and no expression seemed more natural than to speak of man's day having closed, of his life having set, of his soul following the path of the sun to the abode of the Blessed. The sun was even conceived as the first who found the way to a realm beyond, and under various names he soon assumed the character of the Lord of the Departed, and then, by a natural reaction, was claimed as a man, as the first of men who had lived and died.[9] Sometimes, however, as in the Veda, that realm of Yama, the first of the departed, and the Pitris, the Fathers, is placed in the highest heaven, an expression which almost ceases to be local, and assumes something of an ethical character.

Nor is this more exalted view as to the abode of the departed confined to civilised races. Though the Zulus, for instance, localise some of their spirits in caverns, on the roofs of houses, and other places, yet their general idea seems to be that at death the spirit goes upwards to the spirit-land. This is best shown by their own usual form of prayer, which is: "Ye who are above, ye who have gone before."[10]

So much with regard to the ideas as to *where* the departed existed after death.

How Do the Departed Exist?

The answer to the question *how* they existed was suggested, in the first place, by moral sentiments. We know of several nations who, though they believed in the existence of the departed after death, did not believe that they were liable to punishments or rewards for what they had done in this life.

Belief in Punishments and Rewards

But suppose a crime had been discovered after the death of the man who had committed it, how could the thought be suppressed that he would be punished in the next world? And if people stood at the funeral pile of a father or a mother or a friend whose kindness they had not been able to requite during life, what was more natural for them than to hope and to believe that their love would be rewarded in the next world? And this did not remain a fond hope. It was soon looked upon as a necessity and a certainty, for without it there would be no justice in the world, and that there is justice in the world is an ineradicable belief of the human heart.

I know that this expression, "an ineradicable belief of the human heart," gives great offence to certain philosophers. They either deny the existence of such beliefs *in toto,* or they try to account for them as the result of repeated experience. But in our case, how could it be said that a belief in universal justice arises from repeated experience? Surely, no one would say that our experience teaches us again and again that the good are rewarded and the bad punished in this life. One might even go so far as to say that it is the repeated experience of the very contrary, namely, of the misfortunes of the good and the triumphs of the bad, that provokes an appeal to and a belief in a higher justice.

Plato

Plato, in a famous passage in the *Laws,* says: "Now we must believe the legislator when he tells us that the soul is in all respects superior to the body, and that even in life what makes each one of us to be what we are is only the soul; and that the body follows us about in the likeness of each of us, and therefore, when we are dead, the bodies of the dead are rightly said to be our shades or images; for the true and immortal being of each one of us, which is called the soul, goes on her way to other gods, that before them she may give an account—an inspiring hope to the good, but very terrible to the bad, *as the law of our fathers tells us*" (p. 959).

I quote this passage, not because it is Plato's, but because we see from it how even Plato submits to the authority of this νόμος πάτριος, this law of our fathers, which, without vouchsafing any proof, declares that justice will prevail.

The Law of Cause and Effect

But though I call this an ineradicable belief of the human heart, I do not mean to say that no proof can be produced for it. I only mean that no proof is required, until doubt has first been thrown on it. When this belief in justice has been challenged, it requires but little reflection to see that it is but another form of the old law of causality which underlies the whole of our thoughts, nay, without which no thought whatever would be possible. It may be said that this law of causality is an abstraction, which must of necessity belong to a much later phase in the history of the human mind. So it does. But it exists, nevertheless, even in the most ancient times, and in the laws of our fathers; nay, it exercises its influence on the thoughts of men, even though they have as yet no name for it.

We saw in our last course of lectures how the same hidden law of causality necessitated a belief in agents, behind the acts of nature. In its most general form this belief might be called the inevitable result of the simple and universal proposition, *Ex nihilo nihil fit.* Applied to the phenomena of nature, it would mean, "Nothing can be done without a doer."

Now the ancient belief in justice is likewise but another version of the same rule, namely, *Ex aliquo fit aliquid.* If we call the former the law of causality, we may call the latter the law of effect, which denies in the most absolute way that anything can be annihilated, can be without an effect.

Karma na kshîyate

There is a saying among the old laws of our fathers in India, which as applied to our moral actions expresses this truth in the simplest and strongest words, *karma na kshîyate,*[11] "a deed does not perish," that is to say, whatever guilt or merit there is in a human action, man will not come out of it till he has paid or received the uttermost farthing.

This idea of karma, which forms the foundation of the system of Buddhist morality, does not belong to the Buddhists only. In the Upanishads, karma has become already a technical term, and its power and influence must help to account for many things that otherwise seem unaccountable. Thus in a dialogue between Yâgñavalkya and Gâratkârava Ârtabhâga, the latter asks:

> 'When the speech of this dead person enters into the fire, breath into the air, the eye into the sun, the mind into the moon, the hearing into space, into the earth the body, into the ether the self, into the shrubs the hairs of the body, into the trees the hairs of the head, when the blood and the seed are deposited in the water, where is the man himself?'
>
> Yâgñavalkya answers: 'Give me thy hand, O friend. We two alone shall know of this; let this question of ours not be (discussed) in public.' Then these two went out and argued, and what they said was karma, what they proclaimed was karma, viz., that man becomes good by good karma, and evil by evil karma. (Brih.-Âr. Up. III. 2).

In the ancient codes of law the same idea occurs again and again. In the Code of Vishnu, for instance, it is fully placed before us among the words of comfort to be addressed to the mourners at funeral ceremonies. Here we read: "Every creature is seized by Kâla (time) and carried into the other world. It is the slave of its actions.

Wherefore then should you wail?"; [and] "As both his good and bad actions will follow him like associates, what does it matter to a man whether his relatives mourn over him or no[t]?" (XX. 28, 31).

You see that here again there is no doubt in the mind of the old lawgiver on this subject. It is simply stated as a fact, that his good and his bad actions follow a man into the next world, and that he is the slave of his acts, that is, that he has to bear their consequences.

The same idea meets us again, though in a more mythological dress, in the Kaushîtaki Upanishad (I. 4), where the good and evil deeds are represented as following the departed till he approaches the hall Vibhu and the glory of Brahma reaches him. Then he shakes them all off, and his good deeds go to his beloved, his evil deeds to his unbeloved relatives.

Still more mythological is the account given of the soul after death in the Avesta. Here the soul is represented as being met by a beautiful maiden, white-armed, tall, and noble, as if in her fifteenth year, and when he asks her who she is, she replies: "O thou youth, of good thoughts, good words, and good deeds, of good religion! I am thy own conscience" (Vîstâsp Yast, 56).

If, then, I call this belief in rewards and punishments "an ineradicable belief of the human heart," I mean that, if divested of its many mythological disguises, it is but one of the many paraphrases of the law of cause and effect, a law which for beings, such as we are, is irresistible, which requires no proof, nay, which admits of no proof, because it is self-evident.

Are the Departed Conscious of What Passes on Earth?

These two beliefs, the belief in the continued existence of the soul after death, and that in its liability to rewards and punishments, seem to me as irresistible to-day as they were in the days of Plato. We cannot say that a belief in rewards and punishments is universal. We look for it in vain, for instance, in the Old Testament or in Homer. But when that belief has once presented itself to the human mind, it holds its own against all objections. It is possible, no doubt, to object to the purely human distinction between rewards and punishments, because, from a higher point of view, punishment itself may be called a reward. Even eternal punishment, as Charles Kingsley used to say, is but another name for eternal love, and the very fire of hell may be taken as a childish expression only for the constant purification of the soul. All this may be conceded, if only the continuity of cause and effect, between this life and the next, is preserved.

But when we come to the next question, whether the Departed, as has been fondly supposed, are able to feel, not only what concerns them, but likewise what concerns their friends on earth, we may call this a very natural deduction, a very intelligible hope, we may even admit that no evidence can be brought forward against it, but beyond that we cannot go.

Still, if we have followed the thoughts of the early Greeks so far, that they did not, like Homer, believe the departed souls to be simply senseless, ἀφραδέεις, or, like the Jews, simply to sleep and be at rest, but capable of suffering from punishments, and rejoicing in rewards, we can understand at least, even though we cannot follow them, when they go a step beyond and hold that the departed souls could be cognisant of and take pleasure in the honours rendered to them by their relatives and friends on earth. It is curious to find that Plato admits that the souls of the dead have the power after their death of taking an interest in human affairs (*Law,* xi. 927). It is true he does not attempt any proof of this belief, but he appeals to ancient tales and traditions, and to lawgivers who tell us that these things are true, nay, he actually condescends to use the old, though by no means extinct argument, that these ancient lawgivers would have been utter fools, if they had said such things, without knowing them to be true.

Sometimes we meet with a still lower, though likewise quite intelligible view, that the Departed could actually be pleased by food and drink and such other things as they enjoyed during their life on earth. Here, no doubt, the evidence of the senses would soon lead to a wholesome scepticism. Gifts thrown into the fire and burnt in it, might be supposed to reach the Departed, as the smoke of sacrifices was believed to reach the nostrils of the gods. But other gifts, left untouched on the graves, could hardly be supposed to have benefitted the Departed. However, even the savages of Rarotonga have found a way out of these difficulties. It is true, they say, that the visible part of the food is eaten by the rats, but the gods come at dusk and feed on the *essence* of these offerings. This is not bad for a Rarotongan casuist.

We have thus seen how the *psyché,* the breath, the *anima,* had in the eyes of the ancients, whether civilised or uncivilised, become endowed with all that was implied by the *thymós* or *animus,* the mind, and man's fancy was thenceforth set free to finish the picture with such colours, and such light and shade, as best suited the taste of poets, artists, philosophers, and last, not least, of priests.

The poets would speak of the souls flitting about in the air like birds, or hiding beneath the earth like serpents. This was purely symbolical language. But as soon as artists began to speak the same language, we find in sculpture and paintings the souls of the departed represented as small winged beings, others also as serpents, dwelling near their graves. Such symbolical representations are apt to become myths and to be believed in their literal sense by a portion at least of the people. But we must take care not to see in them proofs of a former serpent-worship, as little as of bird-worship. Their true explanation is much more simple and natural.

Even philosophers remain poets. The Emperor Hadrian, though initiated in Greek philosophy and in the Greek mysteries, addressed his soul, when dying, in words half-childish, half-poetical:

> Animula, vagula, blandula,
> Hospes comesque corporis,
> Quae nunc abibis in loca?
> Pallidula, rigida, nudula,
> Nec ut soles dabis jocos.

> Whither—thou wandering, fondling sprite,
> The body's mate and guest—
> Soon must thou fly?

> Wan, robeless, homeless, formless mite!
> Thy mirth and wonted jest
> With thee shall die.

(Translated by the Hon. Lionel A. Tollemache, in *Safe Studies,* p. 396)

Influence of Priests

Lastly, we must not forget, when we sometimes wonder at the elaborate and extravagant offerings made to the Departed, that there was somebody else to be fed besides the spirits. The profession of a priest is a very old one, and we find it almost everywhere. Now it stands to reason that if people want to have priests, they must feed them, or, as we call it by a slightly changed term, they must fee them. I mean that *fee* is the same word as the Latin *pecus,* and meant originally an animal, given by way of payment. What is called a sacrifice to the gods and to ancestral spirits was almost always an offering to the priest also. And if it did not mean a fee, it always meant a feast, more particularly in the case of funeral functions. A single case will serve to make this clearer than any general theories.

The Rev. J. Macdonald tells us that "should a Zulu dream the same dream more than once, he consults the magicians, who profess to have much of their own revelations through dreams. ... If a dreamer sees a departed relative, the magician says oracularly, 'He is hungry.' A beast is then killed as a quasi-sacrifice. The blood is carefully collected and placed in a vessel at the side of the hut, farthest from the door. The liver is hung up in the hut and must not be eaten, until all the flesh of the animal has been used. During the night the spirit is regaled and refreshed by the food thus provided, and eats or withdraws the *essence* that goes to feed and sustain spirits. After a specified time all may be eaten except the portions the magician orders to be burned, generally bones and fat."[12]

Is not all this simple human nature? The man is disturbed by a repeated dream. He asks the priest. The priest says, "the spirit is hungry," the fact being that he himself is hungry. He advises the killing of an animal. The essence of the animal goes to the spirit, the fat and bones are burnt, and the rest is eaten by the hungry priest and his friend, who, after a good feast, has probably another terrible dream, and thus provides fresh employment to the priest.

It is easy to see how rapidly a stone will roll downhill after the first impulse has been given. It is that first impulse which interests the psychologist, and that must, if possible, be accounted for. To discover the secret springs of the first movements of the human mind, when brought face to face with the problems of nature and the problems of its own existence, has been my chief object throughout all these lectures. Our progress, I know, has sometimes been slow and tedious. But it seems to me that to discover the sources of religion in the darkest antiquity is a task worthy at least of the same devotion and the same perseverance as to discover the sources of the Nile in darkest Africa.

Notes

1. *Amatongo* is the plural of *itongo.*
2. *Journal of the Anthropological Institute,* Nov. 1890, p. 128.
3. Ibid., p. 122.
4. Ibid., p. 124.
5. Callaway, *Religious System of the Amazulus,* p. 117, "The Lord of Heaven."
6. Âpastamba Paribhâshâ-sûtras, Sûtras 58 and 59; *Zeitschrift der D.M.G.,* 1850, p. lv.
7. Charon, and the obolos to be put into the mouth of the departed, are later. See Hermann, *Griechische Privatalterthümer,* p. 368.

8. *Transactions of the Australian Association for the Advancement of Science,* Melbourne, 1890, p. 636.
9. Rv. X. 14, 2; Ath. XVIII. 3, 13; Kaegi, pp. 69, 159.
10. Macdonald, loc. cit., p. 121.
11. It occurs in Vasishtha XXII. 4, in Gautama XIX. 5, in Baudhâyana III. 10, 4.
12. *Journal of the Anthropological Institute,* Nov. 1890, p. 121.

CHAPTER 17

THE DIVINE AND THE HUMAN

from *Gifford Lectures on Anthropological Religion* (1891)

The principal object of the study of *Anthropological Religion* in its historical development is to learn how man has been searching for the god-like element in human nature, just as a study of *Physical Religion* showed him to us as bent on discovering something divine or infinite in that objective nature by which he found himself surrounded. I have tried therefore to place before you the various attempts by which the human mind, whether in India, Greece, and Rome, or in Palestine and Egypt, or even in countries not yet illuminated by the rays of civilisation, arrived at the discovery of something more than human in human nature, of something immortal in mortal man, at a belief in a soul, and in the divine kinship of that soul. It required an effort, perhaps the greatest effort of which human nature is capable, to bring the two concepts of the human and the divine, which for a time seemed diametrically opposed to each other, into one focus again. It is the history of these efforts, and, at the same time, the justification of these efforts, that forms the second great division of *Natural Religion,* nay, of all religion; for what would religion be without this second article of faith: "to believe in my own soul and its divine sonship."

Worship of the Departed Leads to the Recognition of the Divine in Man

We saw that one of the most powerful helps to bring the too widely distant concepts of the divine and the human together again, was derived from ancestor-worship. The worship of the spirits of the departed which, under various forms, was so widely spread over the ancient world, could not but accustom the human mind to the idea that there was something in man which deserved such worship. The souls of the departed were lifted higher and higher, till at last they reached the highest stage which existed in the human mind, namely that of divine beings, in the ancient sense of that word. The Romans had their *Divi Manes,* their divine ancestral spirits. In their ancient laws it was laid down that the rights of these gods should be sacred, *Deorum Manium jura sancta sunto;* and that our friends, when dead, should be held as gods. *Sos* (i.e., *suos*) *leto dato divos habento.*[1]

In Greece the ancestral spirits of families became θεοί πατρῷοι, paternal gods, and the same name was given to the ancestors of a race and the founders of towns.

Among the Vedic Indians the Pitris or fathers became the companions of the Devas; in later times they are placed even above the Devas.[2]

In ancient Persia the Fravashis, the Fravashis of the faithful, the awful and over-powering Fravashis, helped Ahura-mazda in all his works.

Apotheosis

This idea of apotheosis or deification of man, as it meets us in many parts of the world, may seem very strange to us. It would be more than strange, it would be an idea simply impossible, unless there had been preparatory steps leading up to it. It is to all intents and purposes a transition *in alterum genus*. Nay, if there are two *genera*, which seem completely to exclude one another, they are those of gods and men. Gods might well have been defined as beings who, whatever else they may be, are not men; men as beings who, whatever else they may be, are not gods. Yet from very early times we saw how both Greeks and Romans had accustomed their minds to the idea, that a man may become a god. That gods also may assume the form of men, and even appear to man disguised in human shape, is more intelligible. For after all, the gods, we are told, can do all things. But to conceive that human nature could ever be changed into divine nature, requires an effort that seems at first beyond the powers at all events of those whose idea of deity was represented by beings such as Zeus, Apollo, and Athene.

Let us try whether we may still discover how the Greeks, who must always remain a representative race in the evolution of human thought, were helped over this difficulty, how they came to reconcile their reason, which was always so keen and vigorous, to what must at first have appeared to them a contradiction in terms, an apotheosis of man, or an apanthroposis of God. Nowhere else, not even in India, can we discover the vestiges of this transition more clearly than in Greece. I propose therefore to examine it more carefully, so that it may serve as an illustration of similar processes which have taken place in other countries also, but have not left us there the same complete record which we find in the literature of Greece.

Heroes as Διογενεῖς

We saw how the Greeks were led to a belief in Zeus, as the supreme deity of nature. When Zeus had once been recognised as the father of gods and men, it was natural and intelligible that the ancient, powerful kings of Greece should have been called, in a special sense, the offspring of Zeus. When we find these kings called Διογενεῖς, this need not at first have implied much more than what was meant by our "Kings by the grace of God."

The transition from this to actual sonship was very easy. If Aeacus, for instance, was born in Aegina or of Aegina, and if he was also a βασιλεὺς Διογενής, a Zeus-born king, it was almost inevitable that in more or less poetical language Aegina should be represented as his mother, and Zeus as his father. And while in this case we see a mortal woman married to a god, there are other cases where a goddess is married to a mortal man. Thus Achilles is represented as the son of the goddess Thetis and of Peleus.

In all these cases, whenever there was some Olympian blood in the veins of human heroes, they were supposed to be half-way between the gods and men, and in many cases they were believed not to die like ordinary mortals, but to live in the Isles of the Blest, in the Elysian Fields, or even with the Olympian gods. Menelaos, though only the son-in-law of Zeus, is told that it is not decreed for him by the gods to die, but that the immortals will send him to the Elysian plain, because he is the son-in-law of Zeus.[3]

This apotheosis or deification of heroes, though purely mythological in its origin, may have helped towards the discovery of something divine in man. Only we must always bear in mind that these heroes of the Theban and Trojan wars were exceptional beings. In some cases it is almost certain that they were not real men, but purely mythological creations. If Dionysos, for instance, because he is the son of a mortal woman, is represented to us as a hero, who afterwards became a god, even an Olympian god,[4] we must remember that Dionysos was probably an ancient deity, before he was represented as a human hero. He never died. Helena also was certainly a goddess,[5] before she was represented as the wife of Menelaos, and the cause of the Trojan war. As to Herakles, even the ancients admitted that there was a mortal and an immortal Herakles. Herodotus tells us that the one, the Olympian, received divine honours, the other ancestral honours only (ii. 44).

The introduction of these new beings, half human, half divine, could not but modify the whole character of the religion of the Greeks. Religion is, no doubt, a very general term, and it is difficult to lay hold of it when, as in ancient Greece, it existed as yet in the minds of individuals, or families, or villages only. On this account Hesiod's works, containing the first attempt at a systematic treatment of Greek religion, deserve to be considered by students of religion quite as carefully as the works of Homer. It may be useful, therefore, before we proceed further, to cast a glance at his *Theogony,* and at what he considered the objects of religious belief and worship to have been among the Greeks of his own time.

Whether the system of Greek religion which we find in Hesiod, was entirely his own work or the work of the Greek people at large, is difficult to say. It is quite certain, no doubt, that neither Homer nor Hesiod "made" the gods of the Greeks, but it is truer of Hesiod than of Homer that he reduced them to something like order.

Hesiod admitted gods and men, but he also admitted intermediate beings. He seems to have been the first to use the word and the concept of ἡμίθεος, half-god, which was probably unknown to Homer (*Iliad,* xii. 23). He applies it to those very heroes of the Trojan and Theban wars, of whom we have been speaking. But he never admits that an ordinary man could become a god or even a half-god. Between gods and men he placed the heroes, ἥρωες, or half-gods ἡμίθεοι, and the Daimones, δαίμονες.

Heroes

It is quite true, as I pointed out, that Hesiod's heroes were exceptional beings, not ordinary mortals. Yet the fact that these heroes were believed to have lived on earth, like ordinary mortals, and had, after their death, been transferred either to the isles of the blessed, or had been admitted to the society of the Olympian gods, made it easier in later times to imagine that a real mortal, distinguished above the rest, might share the same honours as those who were the direct descendants of the gods (ἐκ θεῶν γεγονότες, Isocr. 4, 84). We read of such men as receiving not only the usual ancestral worship, but as worshipped, like the heroes, and, in some cases, as worshipped like the gods. We read of the Chersonitae, for instance, sacrificing to Miltiades, the victor of Marathon, 490 B.C., after his death, as to their founder (Herod, vi. 38). In Greece, where the patriotic feeling was so strong, nothing was more natural than that national honours should be paid to those who had died for their country. Pausanias tells us that the Marathonians worshipped all who had fallen in the battle of Marathon, calling them Heroes, and especially Marathon, from whom their Demos had its name (i. 32, 4).

Plutarch relates that after the battle of Plataea, 479 B.C., the inhabitants performed funeral sacrifices (ἐναγίζειν) every year to the Hellenes who had fallen and been buried there (*Arist.* 21). And they do so still, he continues:

> On the 16th of the Attic month Maimakterion they have a great procession, preceded at day-break by a trumpeter who gives the sign to attack. Then follow carriages with myrtle and garlands, a black bull, and free-born boys, carrying libations of wine and milk in amphoras, and oil and ointments in jugs. No slave is allowed to take any part, because those Greeks once died for freedom. At last follows the Archon of Plataea, walking through the town to the graves. He is never allowed to touch a sword or to wear any but white vestments, but on this occasion he is clothed in a red chiton, carrying an urn which he has taken from the archive, and girded with a sword. He takes water from a spring, washes the funeral columns himself, and anoints them with sweet oil. After he has killed the bull on the altar, and prayed to Zeus and to the chthonic Hermes, he invites the brave who have died for Hellas to the meal and the partaking of the blood. He then mixes a goblet of wine, and while pouring out the libation, he says: 'I drink to men who died for the liberty of the Hellenes.' This festival is kept by the Plataeensians to the present day.

Thucydides tells us that "Brasidas, who died after gaining the victory at Amphipolis (422 B.C.), was buried in the city with public honours in front of what is now the Agora. The whole body of the allies in military array followed him to the grave. The Amphipolitans enclosed his sepulchre, and to this day they sacrifice to him as to a hero, and also celebrate games and yearly offerings in his honour. They likewise made him their founder, and dedicated their colony to him" (v. 11).

Many similar cases might be quoted, all showing how the distance which originally separated men from the heroes as well as from the gods, became smaller and smaller in the eyes of the Greeks. When Demosthenes speaks of those who have fallen for their country, he says in so many words:[6] "How should one not consider men of such merit as blessed, men who have received such recognition and honour? One may fairly say of them that they are the assessors of the lower gods, and in the isles of the blessed placed in the same rank as the brave men of former times."[7]

Plutarch tells us that Lysander (died 394 B.C.) was the first to whom the Greeks actually erected altars and offered sacrifices as to a god (*Vita Lys.* 18). In later times even bodily strength, athletic power, nay, mere physical beauty, might raise a man to the rank of heroes. Herodotus tells us that Philip of Croton, a contemporary of Lysander, obtained, on account of his beauty, what no one else had obtained, for the Egestans built a Heroion on his tomb, and honoured him by sacrifices (θυσίῃσι).

This custom was continued till at last ἥρως became the received name for any departed spirit. On Boeotian tombstones nothing is more common than the inscription, ἥρως χαῖρε, hail, O hero!

If then we find that an ordinary mortal could be honoured like a hero, and a hero could be worshipped like a god, it is easy to see that the mere likeness would soon be forgotten, that the mortal would be honoured as a hero, and the hero worshipped as a god. And thus a transition was effected half-unconsciously between the human and the divine, and the existence of something divine in man would be admitted almost by necessity.

Daimones

The conception of Daimones led by another road to the discovery of something divine in man. While the heroes were originally human, the Daimones were in the beginning purely divine beings with nothing human about them. Plutarch tells us that Hesiod was the first who divided all rational beings into four classes, namely *gods, daimones,* many and good, *heroes,* including the half-gods, and *men.*[8]

It may be doubtful whether Hesiod had this fourfold division always clearly before his mind, and whether it ever formed the recognised foundation of the religion of the Greeks. His identifying the daimones with the men of the golden race seems at all events purely arbitrary. This name of Daimones is in fact one of the most difficult names in the religious phraseology of the ancient Greeks. We translate the word by demons,[9] which gives of course a very imperfect idea of its import. At first δαίμων clearly meant a god, the same as θεός. Thus we read in Homer of Athene returning from the earth to the Daimones on Olympus. Δαίμων is used more especially when a god appears as an active power. It then comes very near to the Latin *numen,* a divine power, as distinguished from the more personal *deus.* Thus Helios, the sun, is called by Aristophanes a great demon among gods and men.

But we saw already how the same word δαίμων was used in another sense too. It was applied by Hesiod to the first or the golden race of men. This has caused much confusion.

Hesiod's Four Races

Hesiod speaks of four races of men who lived before the present race, the *golden,* the *silver,* and the *copper* race, and lastly, the race of *heroes* or *half-gods.*

The first race, the golden, was made by the Olympian gods under Kronos. They lived a life of perfect happiness; and, without suffering from the evils of old age, they passed away as in a sleep. But after they had been buried, Zeus changed them into Daimones, and they now roam about on earth, clothed in air, and watching over laws and crimes.

The second or silver race was given up to folly, and neglected the worship of the gods. After passing away, they became the Blessed under the earth (ἐπιχθόνιοι μάκαρες θνητοὶ καλέονται).

Then Zeus made the third or copper race, in the days when iron was not yet known. It is said that these men were made out of ash-trees (ἐκ μελιῶν), and it is curious to observe that the Germans also had a tradition that men were made out of such trees. They were warriors of enormous strength. After a time they died and passed away.

After these three races, the golden, the silver, and the copper races, we should expect the present or iron race. But here Hesiod interpolates the race of the heroes—the heroes of the Theban and Trojan wars—who are called by him ἡμίθεοι or half-gods. These, after they died, did not, as Homer represents them, descend into the house of Hades, but, according to Hesiod, Zeus allowed them to live in the isles of the blessed, ruled by Kronos as their king.

Daimon therefore meant originally god, and was probably on Greek soil as old a word as theós. Its etymology is unfortunately uncertain, but its meaning differs from that of theós chiefly by its impersonal and unmythological character. Daimones were divine agents, at first, without any proper names, without temples and sacrifices. Afterwards they were represented as the followers (Plat., *Legg.*, viii, p. 848 D), the messengers, and the servants of the gods (Plut., *Def. Orac.*, 13).

As these divine powers watch over man, it was supposed after a time that every man had his own daimon,[10] likewise every family, and every town. At first, these daimones were kind, tutelary beings. Hence the ἀγαθὸς δαίμων, the good genius, of a man. But we also hear of an evil daimon, who is accountable for the misfortunes that befall a man. It is difficult for us to connect any clear ideas with these daimones who determined the fates of man. But how familiar this idea was to the Greeks, we see from such expressions as δαιμόνιος. A δαιμόνιος ἀνήρ was meant for a man possessed, strange, luckless, wretched, sometimes also for an inspired and marvellous person.[11] Our [word] awful comes often very near to δαιμόνιος.

The Daimonion of Sokrates

We shall now better understand what Sokrates really meant by his Daimonion, and why he held that a belief in this Daimonion involved a belief in gods and even in sons of gods or heroes. This is what Plato puts into the mouth of Sokrates in his *Apology:*

> "You say that I teach and believe in δαιμόνια, new or old, no matter for that.... But if I believe in δαιμόνια, I must believe in δαίμονες; is not that true? ... Now, what are δαίμονες?, are they not either gods or the sons of gods? Is that true ?"
>
> "Yes, that is true."
>
> "But this is just the ingenious riddle of which I was speaking: the δαίμονες are gods, and you say first that I do not believe in gods, and then again that I do believe in gods; that is, if I believe in δαίμονες. Or, if the δαίμονες are the illegitimate sons of gods, whether by the nymphs or by any other mothers, as is thought, that, as all men will allow, necessarily implies the existence of their parents. You might as well affirm the existence of mules, and deny that of horses and asses.... But no one who has a particle of understanding will ever be convinced by you that the same man can believe in δαιμόνια and θεῖα and yet not believe that there are δαίμονες, θεοί and ἥρωες."
> (*Apol.* 27 D)

This Daimonion or this Daimon became under various names more and more important in later philosophical systems, particularly in the widespread system of the

Stoics. With them the rational part of the human soul was considered as a part of the divine reason, given to every man as the god within, or as his δαίμων.

All this is clear and intelligible. The only disturbing element in the history of the word δαίμων is the idea of Hesiod that the men of the golden race were raised by Zeus to the rank of daimones. It is hardly ever mentioned as an article of national belief by other representative thinkers of Greece, and it utterly perverts the original character of the daimones which is divine from the beginning and remains divine to the end, when it has been recognised as the Divine, dwelling in man.

The Three Roads Leading to the Discovery of Something Divine in Man

We have thus discovered three roads on which the Greeks were conducted to the discovery of something more than human, something superhuman, something divine or infinite in man. The most important road was that of ancestor-worship, beginning with the honours paid to departed parents, grandparents, and great-grandparents, then leading on to the worship of the ancestors of a family, of a clan, of a town, and of a state, and ending in the recognition of a world of spirits, not far removed from the world of the gods.

The second road started from a kind of mythological belief in human heroes, as the offspring of Zeus. Afterwards ordinary mortals also were raised to the same level, and thus another approach was made to the discovery of something divine, or, at least, god-like in man.

The third road started from a belief in divine powers, called Daímones. These spirits were supposed to watch over the destiny of a man, then to become his destiny. A man being possessed by his daimon, was at last identified with it, and the divine in man was thus once more recognised as the δαιμόνιον of Sokrates and other philosophers.

Nearness, likeness, and oneness with the Divine are the three goals which the human mind reached in Greece. In each case we see that a belief in nature-gods is pre-supposed, nay, that without that belief anthropological religion would be simply impossible.

Plato on Gods, Daimones, Heroes, and Ancestral Spirits

We cannot find a better summing-up of the last results of Greek religion than what is given us by Plato: "First," he says, "comes a belief in God, in that God who, as the old tradition declares, holds in His hand the beginning, middle, and end of all that is, and moves according to His nature in a straight line (*rita*) towards the accomplishment of His end. Justice always follows Him, and is the punisher of those who fall short of the divine law. Every man therefore ought to make up his mind that he will be one of the followers of God—and he who would be dear to God, must, as far as possible, be like Him and such as He is" (*Laws,* 716 seq.).

Now this may seem a very philosophical religion, but this belief in God, quite apart from a belief in the many Olympian gods, can be discovered in Homer quite as much as in Plato. In the *Iliad,* Diomedes says: "Let all flee home, but we two, I and Sthenelos, will fight till we see the end of Troy: *for we came with God*"(ix. 49).

In the *Odyssey*, the swineherd says to Ulysses: "Eat and enjoy what is here, for God will grant one thing, but another He will refuse, whatever He will in His mind, *for He can do all things*" (xiv. 444; x. 306).[12]

And Plato himself, after he has thus spoken of God, continues: "This is the conclusion, which is also the noblest and truest of all sayings, that for the good man to offer sacrifices to the gods, and hold converse with them by means of prayers and offerings and every kind of service, is the noblest and best of all things, and also the most conducive to a happy life, and very fit and meet." He then continues:

> Next after the Olympian gods, and the gods of the state, honour should be given to the gods below.
>
> Next to these gods, a wise man will do service to the daimones or spirits, and then to the heroes, and after them will follow the sacred places of private and ancestral gods, having their ritual according to law.
>
> Next comes the honour of living parents, to whom, as is meet, we have to pay the first and greatest and oldest of all debts. ... And all his life long a man ought never to utter an unbecoming word to them; for of all light and winged words he will have to give an account; Nemesis, the messenger of Justice, is appointed to watch over them. When they are angry and want to satisfy their feelings in word or deed, he should not resist them; for a father who thinks that he has been wronged by his son may be reasonably expected to be very angry.
>
> At their death, the most moderate funeral is best. ... And let a man not forget to pay the yearly tribute of respect to the dead, honouring them chiefly by omitting nothing that conduces to a perpetual remembrance of them, and giving a reasonable portion of his fortune to the dead.

Whatever in this account of Greek religion in its widest sense may be ascribed to Plato personally, one thing seems very clear, that at his time a belief in the Olympian gods, and a belief in the spirits of the departed, existed peaceably side by side, and that funeral ceremonies, and a continued commemoration of the dead were considered essential elements of a truly religious life, quite as much as the sacrifices and praises of the great gods of nature.

Our Own Problems

And now you will perceive how near these problems which occupied the ancient world, approach the problems of our own time. We need not look upon the struggles through which, as we saw, the Greek mind passed in its search for the divine in man, τὸ θεῖον ἐν ἀνθρώπῳ, as the immediate antecedents of our own struggles. These searchings after truth, whether in Greece and Rome, or in India and Persia, need not be more to us than parallel instances, historical lessons, showing us how others toiled at the same task at which we ourselves are toiling still. But even thus they are full of import, for, after all, the same heart beats in every human breast.

Belief in Immortality in the Old Testament

But there is one religion, which forms not only a parallel, but is in one sense the real antecedent of our own religion, supplying the historical background of our faith, and still feeding the thoughts of many Christians. I mean, of course, the Jewish.

Now it is well known that in no religion is the abyss which separates the divine from the human greater than in that of the Jews. An idea such as that expressed in a verse ascribed to Hesiod, ὡς ὁμόθεν γεγάασι θεοί θνητοί τ᾽ ἄνθρωποι,[13] could hardly be expressed in Hebrew.

The question therefore which we have to answer is, whether the ancient Jews ever bridged over that abyss, whether they also discovered something divine in man, whether they believed in a life after death or in the immortality of the soul. This question has been discussed by the most learned theologians for many centuries, and, strange to say, they have not arrived at a unanimous conclusion yet. Unfortunately this question, like many others connected with religion, has been dragged out of the quiet field of historical research into the noisy arena of theological gladiatorship. Some theologians considered it orthodox to deny the existence of a belief in immortality in the Old Testament, because, if the Jews had believed in the immortality of the soul, St. Paul could not have said that it was "Christ who abolished death and brought life and immortality to light by the gospel" (2 Tim. 1.10). Others equally eager to appear orthodox held the opposite opinion, because Christ Himself had appealed to the Old Testament, in which God calls Himself "the God of Abraham, and the God of Isaac, and the God of Jacob, and God is not the God of the dead, but of the living" (Matt. 22.31).

But quite apart from the delusive influence which theological prepossessions always exercise on historical research, the reason why these answers have hitherto proved so unsatisfactory must be sought for in the indefinite character of the question itself. First of all, the Old Testament is a very general and bewildering term. The Old Testament, as is now generally admitted, comprehends a number of books, written by different authors and at different times. Secondly, even immortality, though it seem at first sight a very simple and clear term, may mean many things and requires therefore a very careful definition.

Meaning of Immortality

If immortality is meant for no more than a continuance of existence, if by a belief in immortality on the part of the Jews is meant no more than that the Jews did not believe in the annihilation of the soul at the time of death, we may confidently assert that, to the bulk of the Jewish nation, this very idea of annihilation was as yet unfamiliar. Dr. H. Schultz, to whose learned work, *Alttestamentliche Theologie*, I feel greatly indebted, seems to think that it is the more highly developed races only who cannot conceive a personal being as absolutely coming to an end. But the fact is that the idea of absolute annihilation and nothingness is hardly ever found except among people whose mind has received some amount of philosophical cultivation, certainly more than what the Jews possessed in early times (p. 698). The Jews did not believe in the utter destruction of the soul, but, on the other hand, their idea of life after death was hardly that of life at all. It was existence, without life. Death was considered by them, as by the Greeks, as the greatest of misfortunes. To rejoice in death is a purely Christian, not a Jewish idea. It may be that expressions such as "Abraham was gathered unto his people" (Gen. 25.8), or the words of David, "I shall go to my son, but he will not return to me" (2 Sam. 12.23), show that a certain degree of personal identity was supposed by the Jews also to remain after death. This is indicated also by the recalling of the ghosts of the dead, as in the case of Samuel. The very name given to these ghosts, *elohim*, seems to show that they were supposed to have retained their personal character, while the strict injunctions against such superstitious customs serve only to prove that they existed. But beyond this, all is vague and dark.

Sheol

The place in which the departed were believed to dwell was called by the Jews Sheol, which seems originally to have meant no more than grave or cave. Assyrian scholars derive it from an Accadian word, *shual,* but it can hardly be considered as a foreign word in Hebrew. Sheol is not meant for an individual grave (*bor, keber*), though its first idea may have been borrowed from it, but for a vast space in the interior of the earth. It is the land of shadows, of the Rephaim, where there is no work, nor device, nor knowledge, nor wisdom. The dead lie down and are at rest. They have fewer interests than even the Greek shades in Hades. All distinctions are gone, though sometimes kings are mentioned as sitting there on thrones (Jes. [*sic*] 14.9). But though the Jews believed that the souls continued to exist in Sheol, they did not believe that the wicked would there be punished and the good rewarded. All rewards and punishments for virtue or vice were confined to this world, and a long life was regarded as the surest proof of the favour of Jehovah. It may, no doubt, be taken as a sign of wonderful humility that, with two exceptions, Enoch and Elijah, the greatest of the Jewish saints and heroes should have been satisfied with this meagre reward. But it was their conception of God, as infinitely removed from this world, that made a belief in true immortality almost impossible for them, and excluded all hope for a nearer approach to God, or for any share in that true immortality which belonged to Him and to Him alone.

Controversy as to the Jewish Belief in Immortality

To most students of religion it seems, indeed, as if a religion without a faith in immortality was impossible. Many years ago (1865), and in opposition to the very highest authorities, I ventured to say:

> Without a belief in personal immortality, religion surely is like an arch resting on one pillar only, like a bridge ending in an abyss. We cannot wonder at the great difficulties felt and expressed by Bishop Warburton and other eminent divines, with regard to the supposed total absence of the doctrine of immortality or personal immortality in the Old Testament, and it is equally startling that the Sadducees who sat in the same council with the high-priest, openly denied the resurrection. However, though not expressly asserted anywhere, a belief in personal immortality is taken for granted in several passages of the Old Testament, and it is difficult to think of Abrahamn or Moses as without a belief in life and immortality.[14]

For this passage I was severely taken to task by one of the greatest classical scholars of our age, Professor K. Lehrs. In his *Populäre Aufsätze aus dem Alterthum,* he wrote: "No one would be more disgusted than Lessing by the words boldly uttered by M.M., that, though not expressly asserted anywhere, a belief in personal immortality is taken for granted in several passages of the O.T." (p. 304). This is as strong in language as weak in argument. Lehrs knew that Lessing had denied the possibility of any religion existing without a belief in a future life, future rewards and punishments. He likewise knew, or ought to have known, how disparagingly Kant[15] and Schopenhauer[16] had spoken of the Jewish religion on account of the very absence of that belief.

It was not likely, therefore, that I should have ventured to say what I said without having what I believed to be some good reason on my side. If, therefore, Professor Lehrs felt disgusted, the easiest way to get over his disgust would have been to read once more the book of which he spoke so rashly.

I quite agree with him, and other scholars, that a belief in immortality does not appear on the surface of the Old Testament, as it appears in the New Testament, and in the sacred books of other religions. We should look in vain in the Old Testament for such an utterance as "He that believeth in me, though he were dead, yet shall he live." We could not match the words of a Vedic poet, who exclaims: "Who will restore me to the great Aditi, that I may see father and mother." No heroine in the Old Testament would have said what Antigone is made to say: "Departing hence, I strongly cherish the hope that I shall be fondly welcomed by my father, and by my mother, and by my brother."

But though a belief in immortality does not pervade the whole organism of the Old, as it pervades that of the New Testament, I still hold that we can catch, by what I ventured to call a kind of microscopic analysis, hidden germs, at least, of that belief in many passages of the sacred writings of the Jews. I shall mention a few of these passages, in order to show that I did not, like Professor Lehrs, speak at random.

I am well aware that certain passages which have been most frequently quoted as showing a belief in immortality in the Old Testament, have to be surrendered. The first is the passage from Job 19.25. This was formerly translated: "For I know that my redeemer liveth, and that he shall stand at the latter day upon the earth." Most scholars, however much they differ on the exact interpretation of these difficult lines, agree now that they cannot mean what they seemed to mean to former translators, and that what is meant by a redeemer is a vindicator, here God Himself, who will stand up for the innocence of Job. Professor Schultz translates: "For I know my vindicator lives, and a revenger (Bluträcher) will rise up on the dust (the grave), and, after this my skin has been devoured, and I am denuded of my flesh, I see God (i.e., the revenger on the dust), Him whom I see (fighting) for me, and my eyes see Him—no longer angry; the heart in my bosom fails. If ye say, how shall we persecute him, and that the root of the matter is found in me, be ye afraid for yourselves of the sword, for sword-guilt in anger—that ye know (the Almighty)."[17]

Another passage, Psalm 16.10, which was formerly translated: "Thou wilt not leave my soul in hell," is rendered in the Revised Version by "Thou wilt not leave my soul to Sheol," that is, "Thou wilt not surrender my soul to Sheol," or "Thou wilt not let me die yet."

As passages, however, which seem to me to contain a silent recognition of something divine, and therefore immortal in man, I should mention Gen. 1.26, "And

God said, Let us make man in our image, after our likeness." And again, Gen. 2.7, "And the Lord God formed man of the dust of the ground, and breathed into his nostrils the breath of life; and man became a living soul." Now a God-like being that lives by the breath of God, cannot have been conceived at the same time as something totally different from God. Nay, we are told distinctly that when the dust returns to dust, the spirit or breath returns to God who gave it.

A verse like that in Psalm 8.5, "For thou hast made man but little lower than the Elohim (angels)," could hardly have been written by one who believed man to be but dust, and no better than the beasts of the field.

Again, when we read in Gen. 3.22, that the Lord sent Adam and Eve forth from the garden of Eden, we are told that this was done, lest he put forth his hand and take also of the tree of life, and eat, and live for ever. This seems to imply that man, if he had remained in the garden of Eden, would have been capable of eternal life.

The two cases of Enoch and Elijah are, no doubt, too exceptional to prove a belief in immortality, but they show, at all events, that the mind of the Hebrews was familiar with the idea of a human being returning to God, without suffering the penalty of death.

I do not mean to say that passages such as these prove that the Jews had anticipated the Christian belief in the immortality of the soul. I only maintain that they contain the germs of such a belief, and that they are incompatible with a belief in the utter annihilation of the soul, which Lehrs, Schopenhauer, and others would force on the religion of the Old Testament.

But it would be as easy, nay, much easier, to collect a number of passages from the Old Testament which seem to contain a distinct denial of immortality. To those who believe in a complete unity of the Old Testament, who ignore its composite character and its historical growth, and who look upon the whole of it as miraculously revealed, such contradictions must be perplexing, and can only be removed by a great effort of special pleading. In the eyes of the historian, however, they only serve to confirm the truly historical character of that collection of ancient and modern books. Thus the author of Psalm 39.14, says, "Before I go hence and am no more." We read in the Book of Job 7.8–10, "The eye of him that seeth me shall behold me no more: thine eyes shall be upon me, but I shall not be. As the cloud is consumed and vanisheth away, so he that goeth down to Sheol shall come up no more. He shall return no more to his house, neither shall his place know him any more."

Job 14.7, "For there is hope for a tree, if it be cut down—But man dieth and wasteth away, yea, man giveth up the ghost, and where is he? As the waters fall from the sea, and the river decayeth and drieth up; so man lieth down and riseth not. Till the heavens be no more, they shall not awake, nor be roused out of their sleep."

If we turn to Ecclesiastes, his utterances become more and more despairing. "That which befalleth the sons of men befalleth beasts; even one thing befalleth them: as the one dieth, so dieth the other; yea, they have all one breath; so that man hath no preëminence above a beast" (3.19).

Such sceptical utterances, however, could not but provoke resistance. In the later history of the Jews, whether from their own heart's desire, or from their intercourse with foreign nations, we find that a belief in a life after death became more and more prominent.

Thus, while at the very dawn of Christianity the Sadducees openly denied that "there was any resurrection or spirit or angel," the Pharisees, according to Josephus, believed that "the souls have an immortal strength in them, and that in the underworld they will experience rewards or punishments, according as they have lived well or ill in this life."

Still there always remained in the Jewish mind the idea of the unapproachable majesty of Jehovah. The souls might have their rewards and punishments in the lower world, but true immortality, a communion of the soul with God, was beyond the horizon of the Jewish mind. The idea of anything approaching apotheosis was, and remained to the last, blasphemy in the eyes of the Jews. Adam, though created by Jehovah, is never called the son of God, in a genealogical sense, except once in the Gospel of St. Luke. Here the genealogy of Joseph is traced back to David, and that of David to Enos, Seth, and Adam, "who was the son of God." If we may recognise Rabbinical influences in this genealogy, they are at all events very modern, and we know that at the very same time the fact that Jesus called Himself the Son of God was enough to condemn Him to death.

Reaction

But it is exactly among the Jews, where the two ideas of the Divine and the Human had been most widely wrenched apart, that we witness the strongest reaction. The desire for nearness to God, likeness to God, oneness or atone-ment with God, may be suppressed for a time. It may be silenced by the awe with which the majesty of the Divine fills the human heart. But it is always there. Though the Jew lies prostrate before Jehovah, yet his soul always panted for Him, as the hart panteth after the waterbrook; and it was, after all, the Jew who, in the great history of the world, was destined to solve the riddle of the Divine in man.

But how was it to be solved? Not one of the three roads that led the Greek to the discovery of the Divine in man was open to the Jew.

The first road which led man through the worship of ancestral spirits to the recognition of something Divine in man was barred to the Jew. The ghosts of his departed friends were in Sheol, and to offer sacrifices to them was unlawful.

The second road was too mythological. The Jew knew no Διογενεῖς, no sons of Jehovah, like Herakles, or Dionysos, or Menelaos. Abraham was the friend, Moses was the servant, David was the anointed of God, and if they were never raised to a divine, or even half-divine rank, much less could ordinary mortals hope for such an elevation.

And yet there are passages, scattered about in the Old Testament, in which some idea of a divine sonship seems very clearly expressed. Thus we read in Deuteronomy, 14.1, "Ye are the children of the Lord, your God." But the full meaning of these passages seems never to have been realised. "Father and son" were used in a poetical, often also in a moral sense, as when we read in 2 Sam. 7.14, the words meant for David: "I will be his father, and he shall be my son. If he commit any iniquity, I will chasten him with the rod of men, and with the stripes of the children of men." But we never see it used in the Old Testament as St. John used it, when he wrote, "Behold, what manner of love the Father has bestowed upon us, that we should be called the sons of God" (1 John 3.1).

The third road also was not likely to tempt the Jewish mind. Their belief in angels might have helped them, as the belief in daimones helped the Greek, in their first faltering steps towards that goal. But though the Jewish angels of the Lord might indeed "encamp round about them that fear him, and deliver them" (Psalm 12.7), they would never become indwelling spirits, like the daimonion of Sokrates, and never point the way to the discovery of the Divine in human nature.

Christianity, the Revelation of the Divine Sonship of Christ and Man

And yet it was the soil of Jewish thought that in the end gave birth to the true conception of the relation between the Divine in nature and the Divine in man. In what I am going to say, I shall pay little regard to the miraculous events in which the birth of that concept is supposed to have been manifested. What are those miraculous wrappings to us? When the Divine in the outward world has once been fully recognised, there can be nothing more or less divine, nothing more or less miraculous, either in nature or in history. Those who assign a divine and miraculous character to certain consecrated events only in the history of the world, are in great danger of desecrating thereby the whole drama of history, and to make it, not only profane, but godless.

It is easy to call this a pantheistic view of the world. It is pantheistic, in the best sense of the word, so much so that any other view would soon become atheistic. Even the ancient Greeks suspected the ubiquity or omnipresence of the Divine, when, as early as the time of Thales, they declared that *all* is full of the gods. The choice here lies really between Pantheism and Atheism. If anything, the greatest or the smallest, can ever happen without the will of God, then God is no longer God. To distinguish between a direct and indirect influence of the Divine, to admit a general and a special providence, is like a relapse into polytheism, a belief in one and many gods.

What we call Christianity embraces several fundamental doctrines, but the most important of them all is the recognition of the Divine in man, or, as we call it, the belief in the divinity of the son. The belief in God, let us say in God the Father, or the Creator and Ruler of the world, had been elaborated by the Jews. It was ready to hand. Greeks and Romans, most of the civilised and uncivilised nations of the world, had arrived at it.

But when the Founder of Christianity called God His Father, and not only His Father, but the Father of all mankind, He did no longer speak the language of either Jews or Greeks. To the Jews, to claim divine sonship for man would have been blasphemy. To the Greeks, divine sonship would have meant no more than a miraculous, a mythological event, such as the birth of Hercules. Christ spoke a new language, a language liable, no doubt, to be misunderstood, as all language is; but a language which, to those who understood it, has imparted a new glory to the face of the whole world. It is well known how this event, the discovery of the Divine in man, which involves a complete change in the spiritual condition of mankind, and marks the great turning-point in the history of the world, has been surrounded by a legendary halo, has been obscured, has been changed into mere mythology, so that its real meaning has often been quite forgotten, and has to be discovered again by honest and fearless seeking. Christ had to speak the language of His time, but He gave a new meaning to it, and yet that language has often retained its old discarded meaning in the minds of His earliest, nay, sometimes of His latest disciples also.

The Divine sonship of which He speaks was not blasphemy, as the Jews thought; it was not mythology, as so many of His own followers imagined, and still imagine. Father and son, divine and human, were like the old bottles that could hardly hold the new wine; and yet how often have the old broken bottles been preferred to the new wine, that was to give new life to the world.

The Words Father and Son

Let us first examine the words father and son a little more closely. They seem the best-known words of our language, and yet it would be difficult to find two words more full of mystery, even in their every-day acceptation. Again, nothing seems at first more natural than to apply these words to God and man. In many, if not in most religions, man has addressed God as his father, and has looked upon himself as his son. The expression has become so familiar that we hardly feel that it is, and can only be, a metaphor. And yet it is really the boldest metaphor in the whole of human language. The two words must be almost completely emptied of their contents, before they become fit to express the relation between God and man. Such is human language. We cannot help it, only we should not forget it.

Parable of the King's Son

But now, let us go a step further. It can easily be seen that true sonship depends mainly on knowledge. A man may be the son of a king, but if he is brought up by an old shepherd with his other children, he is a shepherd-boy, and not a prince. And yet as soon as he discovers and knows that the king is his father, and not the shepherd, he at once becomes a prince, he is a prince, he feels himself a prince, the son of a king. It is in the same way that man must discover that God is his father, before he can become a son of God. To know is here to be, to be to know. No mere miracle will change the shepherd-boy into a prince; no mere miracle will make man the son of God. That sonship can be gained through knowledge only, "through man knowing God, or rather being known of God" (Gal. 4.9), and till it is so gained, it does not exist, even though it be a fact.

If we apply this to the words in which Christ speaks of Himself as the Son of God, we shall see that to Him it is no miracle, it is no mystery, it is no question of supernatural contrivance, it is simply clear knowledge; and it was this self-knowledge which made Christ what He was, it was this which constituted His true, His eternal divinity. This is not *Apotheosis*, a word which by its formation seems to imply a removal from the level of humanity to that of deity; it is rather, if I may be allowed to coin a new word, *Anatheosis*, a taking back, or a taking up of the human into the divine nature. What can be clearer than the words of Christ Himself: "No man knoweth the Son, but the Father; neither knoweth any man the Father, save the Son, and he to whomsoever the Son will reveal Him" (Luke 10.22).

But we must remember that though Christ uses the homely words of father and son, He Himself warns His disciples against the wrong use of these words. "Call no man your father upon the earth," He says, "for one is your father, which is in heaven" (Matt. 23.9). Can anything be clearer and stronger? Instead of saying, as we should say, "Call not God father, because father means your father upon the earth," He says, "Call no man father, for father has now assumed a new and higher meaning, and can no longer be used in its old familiar sense."

The Position of Christianity in the History of the World

If we have learnt to look upon Christianity, not as something unreal and unhistorical, but as an integral part of history, of the historical growth of the human race, we can see now how all the searchings after the Divine or the Infinite in man, which we watched in our former lectures, were fulfilled in these simple utterances of Christ. His preaching, we are told, brought life and immortality to light. Life, the life of the soul, and immortality, the immortality of the soul, were there, and had always been there. But they were brought to light, man was made fully conscious of them, man remembered his royal birth, when the word had been spoken by Christ.

The Second Birth

This was called a new birth, and it was so as much as it would be a new birth to the shepherd-boy, when he knew that he was a born prince. This expression, a new birth, or a second birth, which so staggered Nicodemus, is a very familiar expression in Sanskrit. One feels surprised at first when one sees *Dviga,* twice-born, as the regular name for the higher castes in India. The Brâhmans themselves, to judge from the various explanations they give of that title, seem to have forgotten its true meaning. But its original conception can hardly have been different from that of a new birth, that is, the recognition of something superhuman, divine, and immortal in man, which marks to every man the beginning of a new life.

For we must never forget that it was not the principal object of Christ's teaching to make others believe that He only was divine, immortal, or the Son of God. He wished them to believe this for *their own* sake, for *their own* regeneration. Thus we read, "As many as received him, to them gave he power to become the sons of God, even to them that believe in his name: which were born, not of blood, nor of the will of the flesh, nor of the will of man, but of God" (John 1.12). The same doctrine is repeated again and again. "He that believeth in the Son, has everlasting life" (John 3.36; 6.47). "He that heareth my word, hath everlasting life" (John 5.24). "He giveth life unto the world" (John 6.33). Can we doubt what was the meaning of that life? It was immortality, an immortality which need not wait for death, but has its surety even in this life. "If a man keep my saying, he shall never see death" (John 9.52). "Whosoever liveth and believeth in me, shall never die" (John 11.26). "As thou, Father, art in me, and I in thee, that they also may be one in us" (John 17.21).

It might be thought, at first, that this recognition of a divine element in man must necessarily lower the conception of the Divine. And so it does in one sense. It brings God nearer to us; it brings the Divine from the clouds to the earth. It bridges over the abyss by which the Divine and the human were completely separated in the Jewish and likewise in many of the Pagan religions. It rends the veil of the temple.

This lowering, therefore, is no real lowering of the Divine. It is an expanding of the concept of the Divine, and at the same time a raising of the concept of humanity, or rather a restoration of what is called human to its true character, a regeneration, or a second birth, as it is often called by Christ Himself: "Except a man be born again, he cannot see the kingdom of God" (John 3.3).

Christ's Teaching and Its Later Interpretation

The endless theological discussions which, beginning from the first, and not yet ended in the nineteenth century, were meant to define the words of Christ and to draw new limits between the Divine and the human, have fortunately no interest for us. The amount of learning spent on these speculations is incredible. The Church has been rent asunder by them. Hundreds, nay thousands, who thought too freely and spoke too boldly on this subject were sent to prison and to the stake; and, when all other arguments failed, the argument of the biggest battalions has often been invoked for a final solution. These are sad chapters in the history of the world, written in blood and tears. They would never have been written if the Church had been satisfied with the words of Christ. We have not many of His own words. We cannot even be certain that we always have them exactly as He spoke them. Christ never wrote, He never composed a treatise on the true relation between the Divine and the human, either with regard to Himself or with regard to humanity at large. His utterances were always short and complete. Nothing can be left out in them, and nothing ought to be added. A truth does not gain by many words. Often it is completely strangled by them. You have only to read some of those heavy folios of the so-called Fathers of the Church, or of the great theological authorities of the Middle Ages, and you will be appalled at the havoc which, with all their logic and all their piety, these learned Rabbis have wrought in the simple words of Christ. I do not mean to say that His words are not full of meaning, or that they would not supply texts for thousands of sermons and commentaries. But we must not forget that they were meant to be what they are, that they were meant to say exactly what they say, neither more nor less, and that to add one jot or tittle to them, is often to destroy them altogether. As early a witness as Justin Martyr, when speaking of the teaching of Christ, says: "His speeches were brief and cut short, for He was not a sophist, but His speech was the power of God" (*Apol.*, i. 14).

It is quite true that to the student of history it is of the deepest interest to discover the antecedents and parallels of these short utterances, to watch the previous struggles of the human mind, while searching for the true expression of these nascent truths. But when that expression has at last been found, it ought to suffice. The historian may descend once more into the shaft from which the ore has been raised, and examine once more the ore from which the gold has been extracted. But it is the small ounce of precious gold, purified, weighed, and coined, that is wanted for our daily life, and to tamper with it, or to mix it up once more with the slags from which it has been extracted, or cast it back into the shaft from which it has been raised, would be sheer madness. And yet that is what so many theological writers, both in ancient and modern times, are constantly doing. Christ, when speaking of Himself and His relation to God, expressed all He wished to say in a few words. "I and my

Father are one" (John 10.30), and "My Father is greater than I" (John 14.28). And when addressing His disciples, and through them the whole of mankind, His words are again as short and telling as words can be: "As thou, Father, art in me, and I in thee, they also may be one in us" (John 17.21).

And as if to protest once more against the too human interpretation of such purely symbolical words as father and mother, son and brother, you remember the words which sounded so startling to many ears. When His mother and His brethren were seeking for Him, "He answered, saying, 'Who is my mother, or my brethren?' And he looked round about on them which sat about him, and said: 'Behold my mother and my brethren! For whosoever shall do the will of God, the same is my brother, and my sister, and mother'" (Mark 3.32–35; Luke 8.21).

These utterances are very short, but why will people imagine that short utterances contain less than long treatises? There are subjects on which but little *can* be said. When treating of Physical Religion, we found that after all had been said in different sacred books about God in nature, there remained in the end but that one short name, "I Am that I Am." In treating of Anthropological Religion we arrive at the same result. We see how, starting from different points, the deepest thinkers in every part of the world suspected in man something more than the body, something not mortal, soon something immortal; something not merely human, soon something superhuman, divine, and infinite. They called it by ever so many names, the breath, the soul, the spirit, the self; but in the end no words seemed to express the relation between the Divine in man and the Divine in nature better than that of father and son, though even that expression had to be carefully guarded against mythological corruption.

Objections Considered

I know full well the objections that will be raised against the line of argument which I have followed in this course of Lectures on Anthropological Religion.

It will be said on one side that I have deserted the impartial standpoint from which the student of the Science of Religion should never flinch, and that my chief object has been to magnify Christianity, by showing that it is the fulfilment of all that the world has been hoping and striving for. In one sense that is true. But if I hold that Christianity has given the best and truest expression to what the old world had tried to express in various and less perfect ways, I have at least given the facts on which I rely. If my facts can be proved to be wrong, my conclusions will fall; and if any better expression can be given to what the witness within calls the truth, I should be most ready to accept it. Nor should I ever wish to convey the impression that, because the teaching of Christ is true, therefore all the teachings of other religions are false. It has been, on the contrary, my constant endeavour to show how much truth there is in other religions, nay, to use the words of St. Augustine, that there is not one which does not contain some grains of truth. But because in Christian countries Christianity has often been exalted by exaggerated and meaningless praise, there is no reason why we should be ashamed to claim for it that place which, not only the voice of our own heart, but the voice of history, assigns to it among all the religions of the world. At some times silence may be the truest homage, but if there is a time to be silent, there is also a time to speak.

When in my last course of Lectures on Physical Religion I endeavoured to trace the various roads which led to the discovery and the naming of the Infinite in nature, I did not say that the different names and concepts of the gods or agents in nature which we find in non-Christian religions were all wrong. On the contrary, I tried to show in all of them the earnest endeavour to feel after God, if haply they might find Him. I still hold therefore that the whole of Physical Religion may best be summed up in the words: "I believe in God the Father Almighty, maker of heaven and earth." That language is religious, and, if you like, mythological, but it is nevertheless the highest expression that human language can devise.

In the present course of lectures I have likewise endeavoured to describe the various attempts at discovering something infinite in man, and I have shown that a belief in something within us, different from the body, is universal. I hold to this till one single language can be produced which is without a name for soul. There is not the same jealousy here. On the contrary, there is great readiness to accept the different names for soul and the different forms of belief in its perpetuity or immortality, as supports of our own belief in something immortal within us. I go even further, and am quite willing to admit that there are certain philosophies which have entered more deeply into this problem of the Divine in man or the immortality of the soul than any religion. But philosophy, we must remember, is not religion. Philosophy is for the few, religion for the many, nay for all, and the question which concerns us is whether any religion has discovered a truer expression for the relation between man and God than Christianity. Here also the words, "I believe in Jesus Christ, the Son of God," may, if properly understood, serve to sum up nearly all that has been thought on the Divine element in human nature, on the Infinite in man.

But while on one side I shall incur the displeasure of those who carry impartiality to the brink of injustice, I expect even stronger objections from the opposite side. So far from accepting the position which I have assigned to Christianity in the historical growth of religion, many theologians will hold that Christianity stands altogether outside the stream of history, and beyond the reach of any comparison with other religions. The true divinity which, as I tried to show, Christ claimed for Himself and for His brethren, would not satisfy them at all. They want, not a real, but a miraculous divinity, a divinity not very different, in fact, from that which, soon after his death, was ascribed to Plato, as the son of Apollo, or which was claimed for other founders of religions. If people are satisfied with such a belief, it probably contains all that they require, because it is all they can as yet comprehend. Nor do I deny that they have a warrant for their belief in some of the earliest documents of the Christian Church. But the very fact that by the side of the three Synoptical Gospels we find the Gospel according to St. John, should teach us that there is a natural progress and easy transition from the one to the other, and that the same lesson may be conveyed to some in parables, to others in all plainness of thought and speech.

I am prepared for these objections from two opposite quarters, and while I never notice mere abuse, I shall always feel most grateful if any of my opponents will point out when my facts are wrong historically, or my deductions faulty logically. I know but too well how easy it is to err in treating of the origin and history of religions, and no pioneer need be ashamed if he has sometimes missed the right road. But I may repeat at the end of this course what I said when I began it: "Do you think it is possible to lecture on religion, even on natural religion, without giving offence either on the right or on the left? And do you think that a man would be worth his salt who, in lecturing on religion, even on natural religion, were to look either right or left, instead of looking all facts, as they meet him, straight in the face to see whether they are facts or not, and, if they are facts, to find out what they mean?"

It is possible that to some philosophers the subject which I have treated under the name of Anthropological Religion may seem to form no part of religion at all. It is true that "I believe in the existence of the soul" forms as yet no part of any creed; and yet what would it profit a man if he believed all the creeds, and did not believe in his own soul? But if I should not have succeeded in showing how the belief in something infinite, immortal, and divine in man forms an essential part of all religions, I may, in conclusion, appeal to three authorities in my support.

The first is Lord Gifford himself. His opinions have a right to be considered by those who have been entrusted with carrying out the intentions of his Will.

Lord Gifford, in his remarkable essay on Substance, says: "God must be the very substance and essence of the human soul. The human soul is neither self-derived nor self-subsisting. It would vanish if it had not a substance, and its substance is God" (p. 207).

What Lord Gifford asserts simply as a fact, is expressed in more diffident language by one who had all his life been almost face to face with the workings of God in nature, and to whom, if to any one, the heavens had declared not only the glory, but the eternal wisdom of God. Kepler, the discoverer of the three laws on which our planetary system is founded, declared "that it was his highest wish to find *within* the God whom he had found everywhere *without*."[18]

The third witness is Kant. He expresses the same thought, though again from a different point of view.

Before the tribunal of his own critical philosophy both the Divine in nature and the Divine in man were treated as transcendent, and as beyond the reach of our categories, while all the arguments for the existence of God, the cosmological, the teleological, and the ontological, were summarily dismissed. And yet he says, in a passage that has often been quoted:

> Two things fill the mind with ever new and growing admiration and awe, the more frequently and the more intensely we ponder on them: the starry firmament above me, and the moral law within me. Neither of them is hidden in darkness, so that I need look for it or could only suspect it in what is beyond. I see them both before me, and I connect them directly with the consciousness of my own existence. The former begins with the place which I myself occupy in the external world of sense, and enlarges my connection here into the infinitely great, with worlds beyond worlds, and systems of systems, nay, also into the unlimited time of their periodical motion, their beginning and their continuance. The latter begins with my own invisible self, my personality, and places me in a world of true infinitude, perceptible to the understanding only, with which I know myself to be connected, not, as before, by a casual bond only, but by a general and necessary union.[19]

The divine presence which Kant beheld in the starry firmament is the Infinite in nature. The divine presence which he perceived in his conscience or in his own invisible self is the Infinite in man.

The historical development of our belief in the Infinite in nature I tried to explain in my lectures in 1890. The gradual growth of a belief in something infinite, immortal, and divine in man, formed the subject of my present course. If life and strength be spared I hope to treat of the true nature of the soul, and of the relation between the Infinite in man and the Infinite in nature in my next and, what will be, I am sorry to say, my last course of lectures before the members of the University and the citizens of this busy town of Glasgow.

Notes

1. Cic., *De Leg.*, 2, 9, 22.
2. *India, What can it teach us?*, p. 372.
3. *Odyssey*, iv. 555.
4. Diod., *Siculus*, 4, 15; Eurip., *Bacchae, 767.*
5. Isocrates, 10, 61.
6. Demosth., epitaph. 1399, § 36, p. 590, t. ii. Bekk.
7. Lehrs, *Populäre Aufsätze*, p. 332.
8. See Plut., *Def. Orac.*, 10.
9. See H. Spencer, *Sociology*, p. 323.
10. Phokylides, *Fragm.*, 17B.
11. Lehrs, *Aufsätze aus dem Alterthum*, p. 146.
12. *Lectures on the Science of Language*, vol. 2, p. 460.
13. Hesiod, *Opera et Dies*, line 108; roughly translated: "how gods and mortal men have come from the same place of origin"—The Editor.
14. *Chips from a German Workshop*, vol. 1, p. 45.
15. "Kant meint, da ohne Glauben an ein künftiges Leben gar keine Religion gedacht werden könne, so enthalte das Judenthum, als solches genommen, gar keine Religion."
16. Schopenhauer, *Paral.* i, p. 137, writes: "Die eigentliche Judenreligion, wie sie in der Genesis und allen historischen Büchern bis zum Ende der Chronica dargestellt und gelehrt wird, ist die roheste aller Religionen, weil sie die einzige ist, die durchaus keine Unsterblichkeitslehre, noch irgend eine Spur davon hat."
17. Schultz, loc. cit., p. 705: *Voraussetzungen der christlichen Lehre von der Unsterblichkeit*, Göttingen, 1861, pp. 219–223.
18. *Philosophy and Theology*, by J. H. Stirling, p. 32.
19. *Works*, ed. Rosenkranz and Schubert, vol. 8, p. 312 seq.

CHAPTER 18

THE PARLIAMENT OF RELIGIONS
IN CHICAGO, 1893[1] (1894)

There are few things which I so truly regret having missed as the great Parliament of Religions held in Chicago as a part of the Columbian Exhibition in 1893.[2] Who would have thought that what was announced as simply an auxiliary branch of that exhibition could have developed into what it was, could have become the most important part of that immense undertaking, could have become the greatest success of the year, and I do not hesitate to say, could now take its place as one of the most memorable events in the history of the world?

As it seems to me, those to whom the great success of this œcumenical council was chiefly due, I mean President Bonney and Dr. Barrows, hardly made it sufficiently clear at the beginning what was their real purpose and scope. Had they done so, every one who cares for the future of religion might have felt it his bounden duty to take part in the congress. But it seemed at the first glance that it would be a mere show, a part of the great show of industry and art. But instead of a show it developed into a reality, which, if I am not greatly mistaken, will be remembered, aye, will bear fruit, when everything else of the mighty Columbian Exhibition has long been swept away from the memory of man.

Possibly, like many bright ideas, the idea of exhibiting all the religions of the world grew into something far grander than its authors had at first suspected. Even in America, where people have not yet lost the faculty of admiring, and of giving hearty expression to their admiration, the greatness of that event seems to me not yet fully appreciated, while, in other countries, vague rumours only have as yet reached the public at large of what took place in the religious parliament at Chicago. Here and there, I am sorry to say, ridicule also, the impotent weapon of ignorance and envy, has been used against what ought to have been sacred to every man of sense and culture; but ridicule is blown away like offensive smoke; the windows are opened, and the fresh air of truth streams in.

It is difficult, no doubt, to measure correctly the importance of events of which we ourselves have been the witnesses. We have only to read histories and chronicles, written some hundreds of years ago by eyewitnesses and by the chief actors in certain events, to see how signally the observers have failed in correctly appreciating the permanent and historical significance of what they saw and heard, or of what they themselves did. Everything is monumental and epoch-making in the eyes of ephemeral critics, but History must wait before she can pronounce a valid judgement, and it is the impatience of the present to await the sober verdict of History which is answerable for so many monuments having been erected in memory of events or of men whose very names are now unknown, or known to the stones of their pedestals only.

But there is one fact in connexion with the Parliament of Religions which no sceptic can belittle, and on which even contemporary judgement cannot be at fault. Such a gathering of representatives of the principal religions of the world has never before taken place; it is unique, it is unprecedented; nay, we may truly add, it could hardly have been conceived before our own time. Of course even this has been denied, and it has been asserted that the meeting at Chicago was by no means the first realisation of a new idea upon this subject, but that similar meetings had taken place before. Is this true or is it not? To me it seems a complete mistake. If the religious parliament was not an entirely new idea, it was certainly the first realisation of an idea which has lived silently in the hearts of prophets, or has been uttered now and then by poets only, who are free to dream dreams and to see visions. Let me quote some lines of Browning's, which certainly sound like true prophecy:

> Better pursue a pilgrimage
> Through ancient and through modern times,
> To many peoples, various climes,
> Where I may see saint, savage, sage
> Fuse their respective creeds in one
> Before the general Father's throne.

Here you have no doubt the idea, the vision of the religious parliament of the world; but Browning was not allowed to see it. *You* have seen it, and America may be proud of having given substance to Browning's dream and to Browning's desire, if only it will see that what has hitherto been achieved must not be allowed to perish again.

To compare that parliament with the council of the Buddhist King Asoka, in the third century before Christ, is to take great liberties with historical facts. Asoka was no doubt an enlightened sovereign, who preached and practised religious toleration more truly than has any sovereign before or after him. I am the last person to belittle his fame; but we must remember that all the people who assembled at his council belonged to one and the same religion, the religion of Buddha, and although that religion was even at that early time (242 B.C.) broken up into numerous sects, yet all who were present at the Great Council professed to be followers of Buddha only. We do not hear of Gainas nor Agîvikas or Brahmans, nor of any other non-Buddhist religion being represented at the Council of Pâtaliputra.

It is still more incongruous to compare the Council of Chicago with the Council of Nicæa. That council was no doubt called an œcumenical council, but what was the οἰκουμένη, the inhabited world, of that time (A.D. 325) compared with the world as represented at the Columbian Exhibition of last year? Nor was there any idea under Constantine of extending the hand of fellowship to any non-Christian religion. On the contrary, the object was to narrow the limits of Christian love and toleration by expelling the followers of Arius from the pale of the Christian church. As to the behaviour of the bishops assembled at Nicæa, the less that is said about it the better; but I doubt whether the members of the Chicago council, including bishops, archbishops, and cardinals, would feel flattered if they were to be likened to the fathers assembled at Nicæa.

One more religious gathering has been quoted as a precedent of the Parliament of Religions at Chicago; it is that of the Emperor Akbar. But although the spirit which moved the Emperor Akbar (1542–1605) to invite representatives of different creeds to meet at Delhi, was certainly the same spirit which stirred the hearts of those who originated the meeting at Chicago, yet not only was the number of religions represented at Delhi much more limited, but the whole purpose was different. Here I say again, I am the last person to try to belittle the fame of the Emperor Akbar. He was dissatisfied with his own religion, the religion founded by Mohammed; and for an emperor to be dissatisfied with his own religion and the religion of his people, augurs, generally, great independence of judgement and true honesty of purpose. We possess full accounts of his work as a religious reformer, from both friendly and unfriendly sources; from Abufazl on one side, and from Badáoní on the other.[3]

Akbar's idea was to found a new religion, and it was for that purpose that he wished to become acquainted with the prominent religions of the world. He first invited the most learned ulemahs to discuss certain moot points of Islam, but we are told by Badáoní that the disputants behaved very badly, and that one night, as he expresses it, the necks of the ulemahs swelled up, and a horrid noise and confusion ensued. The emperor announced to Badáoní that all who could not behave, and who talked nonsense, should leave the hall, upon which Badáoní remarked that in that case they would *all* have to leave.[4] Nothing of this kind happened at Chicago, I believe. The Emperor Akbar no doubt did all he could to become acquainted with other religions, but he certainly was not half so successful as was the president of the Chicago religious congress in assembling around him representatives of the principal religions of the world. Jews and Christians were summoned to the imperial court, and requested to translate the Old and the New Testament. We hear of Christian missionaries, such as Rodolpho Aquaviva, Antonio de Monserrato, Francisco Enriques and others; nay, for some time a rumour was spread that the emperor himself had actually been converted to Christianity.

Akbar appointed a regular staff of translators, and his library must have been very rich in religious books. Still he tried in vain to persuade the Brahmans to communicate the Vedas to him or to translate them into a language which he could read. He knew nothing of them, except possibly some portions of the Atharva-veda, probably the Upanishads only. Nor was he much more successful with the Zend-Avesta, though portions of it were translated for him by one Ardshiv. His minister, Abufazl, tried in vain to assist the emperor in gaining a knowledge of Buddhism; but we have no reason to suppose that the emperor ever cared to become acquainted with the religious systems of China, whether that of Confucius or that of Lao-tse. Besides, there was in all these religious conferences the restraining presence of the emperor and of the powerful heads of the different ecclesiastical parties of Islam. Abufazl, who entered fully into the thoughts of Akbar, expressed his conviction that the religions of the world have all one common ground.[5] "One man," he writes, "thinks that he worships God by keeping his passions in subjection; another finds self-discipline in watching over the destinies of a nation. The religion of thousands consists in clinging to a mere idea; they are happy in their sloth and unfitness of judging for themselves. But when the time of reflection comes, and men shake off the prejudices of their education, the threads of the web of religious blindness break, and the eye sees the glory of harmoniousness. But," he adds, "the ray of such wisdom does not light up every house, nor could every heart bear such knowledge. Again," he says, "although some are enlightened, many would observe silence from fear of fanatics, who lust for blood though they look like men. And should any one muster sufficient courage, and openly proclaim his enlightened thoughts, pious simpletons would call him a madman, and throw him aside as of no account, whilst the ill-starred wretches would at once think of heresy and atheism, and go about with the intention of killing him."[6]

This was written, more than three hundred years ago, by a minister of Akbar, a contemporary of Henry VIII; but if it had been written in our own days, in the days of Bishop Colenso and Dean Stanley, it would hardly have been exaggerated, barring the intention of killing such "madmen as openly declare their enlightened thoughts"; for burning heretics is no longer either legal or fashionable. How closely even the emperor and his friends were watched by his enemies we may learn from the fact that in some cases he had to see his informants in the dead of night, sitting on a balcony of his palace, to which his guest had to be pulled up by a rope! There was no necessity for that at Chicago. The parliament at Chicago had not to consider the frowns or smiles of an emperor like Constantine; it was encouraged, not intimidated, by the presence of bishops and cardinals; it was a free and friendly meeting, nay, I may say a brotherly meeting, and what is still more—for even brothers will sometimes quarrel—it was a harmonious meeting from beginning to end. All the religions of the world were represented at the congress, far more completely and far more ably than in the palace at Delhi, and I repeat once more, without fear of contradiction, that the Parliament of Religions at Chicago stands unique, stands unprecedented in the whole history of the world.

There are, after all, not so many religions in the world as people imagine. There are only eight great historical religions which can claim that name on the strength of their possessing sacred books. All these religions came from the East; three from an Aryan, three from a Semitic source, and two from China. The three Aryan religions are the *Vedic*, with its modern offshoots in India, the *Avestic* of Zoroaster in Persia, and the religion of *Buddha*, likewise the offspring of Brahmanism in India. The three great religions of Semitic origin are the *Jewish*, the *Christian*, and the *Mohammedan*. There are, besides, the two Chinese religions, that of *Confucius* and that of *Lao-tse*, and that is all; unless we assign a separate place to such creeds as Gainism, a near relative of Buddhism, which was ably represented at Chicago, or the religion of the Sikhs, which is after all but a compromise between Brahmanism and Mohammedanism.

All these religions were represented at Chicago; the only one that might complain of being neglected was Mohammedanism. Unfortunately the Sultan, in his capacity as Khalif, was persuaded not to send a representative to Chicago. One cannot help thinking that both in his case and in that of the Archbishop of Canterbury, who likewise kept aloof from the congress, there must have been some unfortunate misapprehension as to the real objects of that meeting. The consequence was that Mohammedanism was left without any authoritative representative in a general gathering of all the religions of the world. It was different with the Episcopalian Church of England, for although the archbishop withheld his sanction, his church was ably represented both by English and American divines.

But what surprised everybody was the large attendance of representatives of all the other religions of the world. There were Buddhists and Shintoists from Japan, followers of Confucius and Lao-tse from China; there was a Parsee to speak for Zoroaster, there were learned Brahmans from India to explain the Veda and Vedânta. Even the most recent phases of Brahmanism were ably and eloquently represented by Mozoomdar, the friend and successor of Keshub Chunder Sen, and the modern reformers of Buddhism in Ceylon had their powerful spokesman in Dharmapâla. A brother of the King of Siam came to speak for the Buddhism of his country. Judaism was defended by learned rabbis, while Christianity spoke through bishops and archbishops, nay, even through a cardinal who is supposed to stand very near the papal chair. How had these men been persuaded to travel thousands of miles, to spend their time and their money in order to attend a congress, the very character and object of which were mere matters of speculation?

Great credit no doubt is due to Dr. Barrows and his fellow labourers; but it is clear that the world was really ripe for such a congress, nay, was waiting and longing for it. Many people belonging to different religions had been thinking about a universal religion, or at least about a union of the different religions, resting on a recognition of the truths shared in common by all of them, and on a respectful toleration of what is peculiar to each, unless it offended against reason or morality. It was curious to see, after the meeting was over, from how many sides voices were raised, not only expressing approval of what had been done, but regret that it had not been done long ago. And yet I doubt whether the world would really have been ready for such a truly œcumenical council at a much earlier period. We all remember the time, not so very long ago, when we used to pray for Jews, Turks, and infidels, and thought of all of them as true sons of Belial. Mohammed was looked upon as the arch-enemy of Christianity, the people of India were idolaters of the darkest die, all Buddhists were atheists, and even the Parsees were supposed to worship the fire as their god.

It is due to a more frequent intercourse between Christians and non-Christians that this feeling of aversion towards and misrepresentation of other religions has of late been considerably softened. Much is due to honest missionaries, who lived in India, China, and even among the savages of Africa, and who could not help seeing the excellent influence which even less perfect religions may exercise on honest believers. Much also is due to travellers who stayed long enough in countries such as Turkey, China, or Japan to see in how many respects the people there were as good, nay, even better, than those who call themselves Christians. I read not long ago a book of travels by Mrs. Gordon, called *Clear Round.* The author starts with the strongest prejudices against all heathens, but she comes home with the kindliest feelings towards the religions which she has watched in their practical working in India, in Japan, and elsewhere.

Nothing, however, if I am not blinded by my own paternal feelings, has contributed more powerfully to spread a feeling of toleration, nay, in some cases, of respect for other religions, than has the publication of the *Sacred Books of the East.* It reflects the highest credit on Lord Salisbury, at the time Secretary of State for India, and on the university of which he is the chancellor, that so large an undertaking could have been carried out; and I am deeply grateful that it should have fallen to my lot to be the editor of this series, and that I should thus have been allowed to help in laying the solid foundation of the large temple of the religion of the future—a foundation which shall be broad enough to comprehend every shade of honest faith in that Power which by nearly all religions is called *Our Father,* a name only, it is true, and it may be a very imperfect name; yet there is no other name in human language that goes nearer to that for-ever-unknown Majesty in which we ourselves live and move and have our being.

But although this feeling of kindliness for and the desire to be just to non-Christian religions has been growing up for some time, it never before found such an open and solemn recognition as at Chicago. That meeting was not intended, like that under Akbar at Delhi, for elaborating a new religion, but it established a fact of the greatest significance, namely, that there exists an ancient and universal religion, and that the highest dignitaries and representatives of all the religions of the world can meet as members of one common brotherhood, can listen respectfully to what each religion had to say for itself, nay, can join in a common prayer and accept a common blessing, one day from the hands of a Christian archbishop, another day from a Jewish rabbi, and again another day from a Buddhist priest (Dharmapâla). Another fact, also, was established once for all, namely, that the points on which the great religions differ are far less numerous, and certainly far less important, than are the points on which they all agree. The words, "that God has not left Himself without a witness," became for the first time revealed as a fact at this congress.

Whoever knows what human nature is will not feel surprised that every one present at the religious parliament looked on his own religion as the best, nay, loved it all the same, even when on certain points it seemed clearly deficient or antiquated as compared with other religions. Yet that predilection did not interfere with a hearty appreciation of what seemed good and excellent in other religions. When an old Jewish rabbi summed up the whole of his religion in the words, "Be good, my boy, for God's sake," no member of the Parliament of Religions would have said "No"; and when another rabbi declared that the whole law and the prophets depend on our loving God and loving our neighbour as ourselves, there are few religions that could not have quoted from their own sacred scriptures more or less perfect expressions of the same sentiment.

I wish indeed it could have been possible at this parliament to put forward the most essential doctrines of Christianity or Islam, for example, and to ask the representatives of the other religions of the world whether their own sacred books said Yes or No to any of them. For that purpose, however, it would have been necessary, no doubt, to ask each speaker to give chapter and verse for his declaration—and here is the only weak point that has struck me, and is sure to strike others, in reading the transactions of the Parliament of Religions. Statements were put forward by those who professed to speak in the name of Buddhism, Brahmanism, Christianity, and Zoroastrianism—by followers of these religions who happened to be present—which, if the speakers had been asked for chapter and verse from their own canonical books, would have been difficult to substantiate, or, at all events, would have assumed a very modified aspect. Perhaps this was inevitable, particularly as the rules of the parliament did not encourage anything like discussion, and it might have seemed hardly courteous to call upon a Buddhist archbishop to produce his authority from the Tripitaka, or from the nine Dharmas.

We know how much our own Christian sects differ in the interpretation of the Bible, and how they contradict one another on many of their articles of faith. Yet they all accept the Bible as their highest authority. Whatever doctrine is contradicted by the Bible they would at once surrender as false; whatever doctrine is not supported by it they could not claim as revealed. It is the same with all the other so-called book-religions. Whatever differences of opinion may separate different sects, they all submit to the authority of their own sacred books.

I may therefore be pardoned if I think that the Parliament of Religions, the record of which has been assembled in fifty silent volumes, is in some respects more authoritative than the parliament that was held at Chicago. At Chicago you had, no doubt, the immense advantage of listening to living witnesses; you were *making* the history of the future—my parliament in type records only the history of the past. Besides, the immense number of hearers, your crowded hall joining in singing sacred hymns, nay, even the magnificent display of colour by the representatives of Oriental and Occidental creeds—the snowy lawn, the orange and crimson satin, the vermilion brocade of the various ecclesiastical vestments so eloquently described by your reporters—all this contributed to stir an enthusiasm in your hearts which I hope will never die. If there are two worlds, the world of deeds and the world of words, you moved at Chicago in the world of deeds. But in the end what remains of the world of deeds is the world of words, or, as we call it, *history*, and in those fifty volumes you may see the history, the outcome, or, in some cases, the short inscription on the tombstones of those who in their time have battled for truth, as the speakers assembled at Chicago have battled for truth, for love, and for charity to our neighbours.

I know full well what may be said against all sacred books. Mark, first of all, that not one has been written by the founder of a religion; secondly, that nearly all were written hundreds, in some cases thousands, of years after the rise of the religion which they profess to represent; thirdly, that even after they were written they were exposed to dangers and interpolations; and fourthly, that it requires a very accurate and scholarlike knowledge of their language and of the thoughts of the time when they were composed, in order to comprehend their true meaning. All this should be honestly confessed; and yet there remains the fact that no religion has ever recognised an authority higher than that of its sacred book, whether for the past, or the present, or the future. It was the absence of this authority, the impossibility of checking the enthusiastic descriptions of the supreme excellence of every single religion, that seems to me to have somewhat interfered with the usefulness of that great œcumenical meeting at Chicago.

But let us not forget, therefore, what has been achieved by this parliament in the world of deeds. Thousands of people from every part of the world have for the first time been seen praying together, "Our Father, which art in heaven," and have testified to the words of the prophet Malachi: "Have we not all *one* Father, hath not *one* God created us?" They have declared that "in every nation he that feareth God and worketh righteousness is acceptable to Him." They have seen with their own eyes that God is not far from each one of those who seek God, if haply they may feel after Him. Let theologians pile up volume upon volume of what they call theology; religion is a very simple matter, and that which is so simple and yet so all-important to us, the living kernel of religion, can be found, I believe, in almost every creed, however much the husk may vary. And think what that means! It means that above and beneath and behind all religions there is one eternal, one universal religion, a religion to which every man, whether black, or white, or yellow, or red, belongs or may belong.

What can be more disturbing and distressing than to see the divisions in our own religion, and likewise the divisions in the eternal and universal religion of mankind? Not only are the believers in different religions divided from each other, but they think it right to hate and to anathematise each other on account of their belief. As long as religions encourage such feelings none of them can be the true one.

And if it is impossible to prevent theologians from quarrelling, or popes, cardinals, archbishops and bishops, priests and ministers, from pronouncing their anathemas, the true people of God, the universal laity, have surely a higher duty to fulfil. Their religion, whether formulated by Buddha, Mohammed, or Christ, is before all things practical, a religion of love and trust, not of hatred and excommunication.

Suppose that there are and that there always will remain differences of creed; are such differences fatal to a universal religion? Must we hate one another because we have different creeds, or because we express in different ways what we believe?

Let us look at some of the most important articles of faith, such as *miracles, the immortality of the soul,* and *the existence of God.* It is well known that both Buddha and Mohammed declined to perform miracles, nay, despised them if required as evidence in support of the truth of their doctrines. If, on the contrary, the founder of our own religion appealed, as we are told, to His works in support of the truth of His teaching, does that establish either the falsehood or the truth of the Buddhist, the Mohammedan, or the Christian religion? May there not be truth even without miracles? Nay, as others would put it, may there not be truth even if resting apparently on the evidence of miracles only? Whenever all three religions proclaim the same truth, may they not all be true, even if they vary slightly in their expression, and may not their fundamental agreement serve as stronger evidence even than all miracles?

Or take a more important point, the belief in the immortality of the soul. Christianity and Mohammedanism teach it, ancient Mosaism seems almost to deny it, while Buddhism refrains from any positive utterance, neither asserting nor denying it. Does even that necessitate rupture and excommunication? Are we less immortal because the Jews doubted and the Buddhists shrank from asserting the indestructible nature of the soul?

Nay, even what is called *atheism* is, often, not the denial of a Supreme Being, but simply a refusal to recognise what seem to some minds human attributes, unworthy of the Deity. Whoever thinks that he can really deny Deity, must also deny humanity; that is, he must deny himself, and that, as you know, is a logical impossibility.

But true religion, that is, practical, active, living religion, has little or nothing to do with such logical or metaphysical quibbles. Practical religion is life, is a new life, a life in the sight of God; and it springs from what may truly be called a new birth. And even this belief in a new birth is by no means an exclusively Christian idea. Nicodemus might ask, How can a man be born again? The old Brahmans, however, knew perfectly well the meaning of that second birth. They called themselves *Dviga,* that is Twice-born, because their religion had led them to discover their divine birthright, long before we were taught to call ourselves the children of God.

In this way it would be possible to discover a number of fundamental doctrines, shared in common by the great religions of the world, though clothed in slightly varying phraseology. Nay, I believe it would have been possible, even at Chicago, to draw up a small number of articles of faith, not, of course, thirty-nine, to which all who were present could have honestly subscribed. And think what that would have meant! It rests with us to carry forth the torch that has been lighted in America, and not to allow it to be extinguished again till a beacon has been raised lighting up the whole world, and drawing towards it the eyes and hearts of all the sons of men in brotherly love and in reverence for that God who has been worshipped since the world began, albeit in different languages and under different names, but never before in such unison, in such world-embracing harmony and love, as at the great religious council at Chicago.

Notes

1. Substance of a Lecture delivered in Oxford in 1894.
2. Letter to the Rev. John Henry Barrows, D.D., Chairman of the General Committee; Dated, Easter Sunday, April 2nd, 1893:

DEAR SIR,
What I have aimed at in my Gifford Lectures on "Natural Religion" is to show that all religions are natural, and you will see from my last volume on *Theosophy or Psychological Religion* that what I hope for is not simply a reform, but a complete revival of religion, more particularly of the Christian religion. You will hardly have time to read the whole of my volume before the opening of your Religious Congress at Chicago, but you can easily see the drift of it. I had often asked myself the question how independent thinkers and honest men like St. Clement and Origen came to embrace Christianity, and to elaborate the first system of Christian theology. There was nothing to induce them to accept Christianity, or to cling to it, if they had found it in any way irreconcilable with their philosophical convictions. They were philosophers first, Christians afterward. They had nothing to gain and much to lose by joining and remaining in this new sect of Christians. We may safely conclude therefore that they found their own philosophical convictions, the final outcome of the long preceding development of philosophical thought in Greece, perfectly compatible with the religious and moral doctrines of Christianity as conceived by themselves.

Now, what was the highest result of Greek philosophy as it reached Alexandria, whether in its Stoic or neo-Platonic garb? It was the ineradicable conviction that there is Reason or Logos in the world. When asked, Whence that Reason, as seen by the eye of science in the phenomenal world?, they said: "From the cause of all things which is beyond all names and comprehension, except so far as it is manifested or revealed in the phenomenal world."

What we call the different types or ideas, or logoi, in the world, are the logoi or thoughts or wills of that Being whom human language has called God. These thoughts, which embraced everything that is, existed at first as thoughts, as a thought-world (κόσμος νοητός), before by will and force they could become what we see them to be, the types or species realised in the visible world (κόσμος ὁρατός). So far all is clear and incontrovertible, and a sharp line is drawn between this philosophy and another, likewise powerfully represented in the previous history of Greek philosophy, which denied the existence of that eternal Reason, denied that the world was thought and willed, as even the Klamaths, a tribe of Red Indians, profess and ascribe the world, as we see it as men of science, to purely mechanical causes, to what we now call uncreate protoplasm, assuming various casual forms by means of natural selection, influence of environment, survival of the fittest, and all the rest.

The critical step which some of the philosophers of Alexandria took, while others refused to take it, was to recognise the perfect realisation of the Divine Thought or Logos of manhood in Christ, as in the true sense the Son of God; not in the vulgar mythological sense, but in the deep metaphysical meaning of which the term υἱὸς μονογενής [only-begotten son] had long been possessed in Greek philosophy. Those who declined to take that step, such as Celsus and his friends, did so either because they denied the possibility of any Divine Thought ever becoming fully realised in the flesh or in the phenomenal world, or because they could not bring themselves to recognise that realisation in Jesus of Nazareth. St. Clement's conviction that the phenomenal world was a realisation of the Divine Reason was based on purely philosophical grounds, while his conviction that the ideal or the Divine conception of Manhood had been fully realised in Christ and in Christ only, dying on the Cross for the truth as revealed to Him and by Him, could have been based on historical grounds only.

Everything else followed. Christian morality was really in complete harmony with the morality of the Stoic school of philosophy, though it gave to it a new life and a higher purpose. But by means of Christian philosophy the whole world assumed a new aspect. It was seen to be supported and pervaded by Reason or Logos, it was throughout teleological, thought and willed by a rational power. The same Divine presence was now perceived for the first time in all its fullness and perfection in the one Son of God, the pattern of the whole race of men henceforth to be called "the sons of God."

This was the groundwork of the earliest Christian theology, as presupposed by the author of the fourth Gospel, and likewise by many passages in the Synoptical Gospels, though fully elaborated for the first time by such men as St. Clement and Origen. If we want to be true and honest Christians, we must go back to those earliest ante-Nicene authorities, the true Fathers of the Church. Thus only can we use the words: "In the beginning was the Word, and the Word became flesh," not as thoughtless repeaters, but as honest thinkers and believers. The first sentence, "In the beginning was the Word," requires thought and thought only; the second, "and the Logos became flesh," requires faith—faith such as those who knew Jesus had in Jesus, and which we may accept unless we have any reasons for doubting their testimony.

There is nothing new in all this, it is only the earliest Christian theology restated, restored, and revised. It gives us at the same time a truer conception of the history of the whole world, showing us that there was a purpose in the ancient religions and philosophies of the world, and that Christianity was really from the beginning a synthesis of the best thoughts of the past, as they had been slowly elaborated by the two principal representatives of the human race, the Aryan and the Semitic.

On this ancient foundation, which was strangely neglected, if not purposely rejected, at the time of the Reformation, a true revival of the Christian religion and a reunion of all its divisions may become possible, and I have no doubt that your Congress of the religions of the world might do excellent work for the resuscitation of pure and primitive ante-Nicene Christianity.

Yours very truly,
F. Max Müller.

3. Müller, *Introduction to the Science of Religion* [Appendix to Lecture One], p. 209 et seq.
4. Ibid., p. 221.
5. Ibid., p. 210.
6. Ibid., p. 211.

CHAPTER 19
SCIENCE OF RELIGION: A RETROSPECT (1898)

There is one advantage in growing old: one is able to see that the world is growing also. Whether it is growing better or worse may be left an open question, but it certainly is not to-day what we knew it to be, say, fifty years ago.

Another advantage is that from a distance we can better perceive the general drift of a science, the direction in which it really has moved and is moving. We are less distracted by the books that appear from year to year, occupy our attention for a time, and then are forgotten. We are better able, also, to see how books that are now almost forgotten, have, like sunken rocks, determined the undercurrent of the stream of scientific work.

I can well think back at least fifty years, when I attended the first lectures on "Religionsgeschichte"—History of Religions—in the University of Leipzig, and afterwards at Berlin. These lectures were then strictly confined to the Christian and the Jewish religions, and they were generally delivered by the professor of Hebrew, or by some professor belonging to the faculty of theology. Nothing else was thought worthy of the name of religion at that time, not even what existed on the classical soil of Greece and Italy. What we now call the religion of Greeks and Romans was then considered as either mere mythology or as pagan superstition, and lectures on the popular traditions or sacred customs of these two classical countries fell naturally to the share of the professors of Greek and Latin.

As far as I remember, the first German scholar who wrote on the religion of the Romans, as distinct from their mythology, was J. A. Hartung. His valuable book, *Die Religion der Römer*, published in 1836, seems to have attracted little attention outside of Germany, but it certainly marked a new era in the historical study of religion, and is by no means antiquated, even now. Hartung's admirers and followers expected another work from him on the religion of the Greeks, but, unfortunately, he died before it was finished, and what was published after his death from his manuscript, *Die Religion und Mythologie der Griechen*, 1865, is not to be compared to his first book. The first volume, containing what is called a "Natural history of heathen religions," throws out some useful hints on the origin and growth of religion as then understood. It repeats the usual explanations of the origin of mythology. Men, we are told, could not but represent to themselves whatever in nature affected them with pleasure or pain, as itself animated. This was simple Animism, but no attempt was made as yet to explain this animistic tendency in man, and to trace it back to its real source, a peculiarity of ancient language. That the gods were created by men, and therefore reflect in their character the peculiarities of their creators, whether savage or civilised, is likewise admitted, and an important hint is thrown out that religion and language are contemporaneous in their origin, marking the very beginnings of social life—in fact that it is through language and religion that man first became man.

After Hartung's publications, those who had to lecture on the history of religions had to pay more attention to the forms of belief and worship among Greeks and Romans by the side of Christians and Jews; but the idea that pagan religion was of the same kind as the religion of Christians and Jews was hardly hinted at as yet. With the spreading of Semitic studies beyond the narrow sphere of Hebrew, the religion of the Phœnicians in ancient, and of Mohammedans in modern, times, had likewise to be included in the history of religions, while the gradual decipherment of Egyptian hieroglyphics and of cuneiform inscriptions, added new chapters to this ever-increasing subject. The archives of the ancient religion of India and Persia, were likewise opened, and Chinese missionaries added large materials to what was still called the History of Religions, not yet the History of Religion. The accumulation of material had been so sudden and so enormous, that no one ventured as yet on a comprehensive study of all these forms of faith. The professor of Chinese lectured on Confucius and Lao-tse, the professor of Persian on Zoroaster, the professor of Sanskrit on the Vedas and Purânas, the professor of Arabic on Mohammed. This system lasted for some time, and it certainly had one great advantage; no one lectured on any religion unless he knew something of it, and not merely about it, unless he knew at least the language in which its sacred books were written, and was able to appeal to authoritative documents in support of his opinions.

Soon, however, new interests arose. As a comparative study of languages had proved quite a new relationship between the principal languages of Europe and Asia, it was supposed that the same kind of relationship might be discovered between the various religions of the ancient world also. And so it was. As all the Semitic languages had one unmistakable type, and all Aryan languages another, every Semitic religion turned out to possess one physiognomy, every Aryan religion another. Hence, to derive any Aryan religion from a Semitic source was, in ancient times, at least as impossible and unscientific as to derive Greek from Hebrew. Whatever there was of Semitic thought and language in any of the Aryan religions was of necessity borrowed, and could not claim any organic relationship, however interesting it might be for historical purposes. It thus became possible to construct historical pedigrees of the Semitic as well as of the Aryan religions, though, of course, for the earliest periods of their history only.

A new and very critical step was taken soon after. As long as these studies remained almost exclusively in the hands of scholars and historians, they attempted no more than a history of the principal religions of the world. Meiner's *Allgemeine kritische Geschichte der Religionen,* 1806, is a well-known specimen of that class of work. But as facts accumulated, the love of generalisation set in, and instead of religions and their history, we begin to hear of religion as a thing by itself, the same in the South and in the North, the same among savage and highly civilised nations. Philosophers take the place of historians, and undertake to account for the origin, not of such and such a religion, but of religion in general, and even to explain the laws which, they suppose, governed its development. The history of religions was thus supplanted by the history of religion; only it was difficult to say where that religion in general was to be found. A good example of this class of works may be seen in Benjamin Constant's *De la Réligion considérée dans sa source, ses formes, et ses développements,* 1824–31. This represented, no doubt, an advance; but it was a most dangerous advance, because it opened the door to all kinds of theories long before a sufficient number of facts had been accumulated and critically sifted. From an historical point of view, the historical existence of such a thing as religion in general had yet to be proved, while the admission of a common pre-historic religion from which all historic religions were derived, was a mere postulate, pregnant with the most misleading deductions, and hardly preferable to the belief in a primeval revelation, of which so much was written during the eighteenth century.

Preëminent among the leaders of this philosophic and generalising movement stand two names, Schelling and Hegel. They endeavoured to show that there was an intelligible origin, not so much for any individual religion, but rather for religion in the abstract, and that its historical development was determined by certain laws—nay, by logical necessities—so that it could not have been different from what, as history shows us, it has been. No one can deny that this treatment has thrown much unexpected light on many of the phases of religious thought, but it is responsible also for considerable confusion of thought on the subject. Where was this general religion to be found, except in the individual religions; and where could those individual religions be studied, except in their sacred books, many of which were not yet accessible? Thus it happened that not only were many of the facts on which some very large theories had been built up very ill-ascertained, but they had often been adapted to the very theories which they were meant to support; so that we were left with many theories, and with but very few well-established facts.

Neither Schelling nor Hegel could have read a line of the Rig-veda or the Avesta, yet they assigned to each what they supposed its right place in the development of religion. Others compared religions such as Buddhism and Christianity, knowing, no doubt, Christianity in its present form, but hardly anything authentic or chronologically settled of the history of Buddhism. There is, no doubt, such a thing as religion in the abstract, or religion common to all mankind, but have we any right to identify that religion with the few historical religions the history of which is known to us?

Very soon another step followed. If religion was to be studied in the religions of the leading nations of the world, why should it not be studied equally well in the religions of savage, barbarous, and uncivilised tribes? The question was very natural, but the difficulties in this case were enormous. No one without a knowledge of the language spoken by such savage tribes, whether a missionary or a casual traveller, could claim a hearing from serious students. If Schelling did not know either Sanskrit or Zend, what did men like De Brosses know of the language and of the thoughts of the Negroes on the West Coast of Africa, where feitiços (*factitia,* amulets) were supposed to have had their natural home? And yet he had not only traced the origin of the religious views and practices of African Negroes, of which he knew next to nothing, back to a worship of fetishes, but he boldly proclaimed fetishism to be the origin of most, though not yet of all, religions. This last step was left to Bastholm. A more preposterous theory has seldom been promulgated; but, as the idea of religion in general had once been started and accepted, new attempts were made from time to time to find the origin of that general religion in some peculiar variety of religion, particularly if it happened to be prevalent among races upon a very low level of civilisation. Thus totemism, ancestor worship, animism were all tried in turn to serve as keys to the origin of religion. To say that these theories were built up on "scandalously ill-certified facts" is going too far. The stories of savage or barbarous tribes as collected by Klemm, Bastholm, Waitz and Tylor, cannot claim the same authority as the stories collected by Pausanias or by Grimm, much contested as even these have been, but they are by no means to be rejected altogether, and it would be unfair to charge a man such as Waitz, the editor of Aristotle's *Organon*, with having been uncritical in collecting his evidence. On the contrary, it was he who protested against trusting to the unauthenticated reports of travellers and even of missionaries, and who pointed out, for instance, that some of the lowest African idolaters had always possessed, with all their fetishism, a very clear idea of one Supreme Deity.

The mistake common to all these attempts was their treating religion as one, and trying to recognise in the rationale of one the rationale of all religions. We may compare the separate streams of religion one with the other, and it is no doubt this comparative study of religions which has excited the greatest interest of late. It has sometimes been called Comparative Religion; but if we can form no definite idea of religion as such, what shall we think of Comparative Religion? A comparative study of bones is called comparative anatomy, not comparative bones. Why, then, should a comparative study of religions be called comparative religion, and not comparative theology, or a comparative study of religion, or simply the Science of Religion? Most sciences in this age of ours have become comparative even without being called so, and as every science is based on a comparison of facts, the Science of Religion also would naturally include a comparison of religions from their inevitable mythological beginning to their latest philosophical aspirations.

In this comparative theology, however, as much as in comparative philology, the beginning must always be made with comparing homogeneous or organically related religions—Semitic, Aryan, Australian, American or African. It may be instructive also to collect coincidences between religious that cannot possibly have had the same origin. But such casual likenesses can receive a truly scientific value in cases only where religions or languages have been proved to be genealogically or historically connected. There is a large field still open to students of religion: first, in collecting and critically sifting materials; secondly, in discovering coincidences; and thirdly, in finding out, if possible, the reason of such coincidences, whether in the common nature of the human mind, or in the peculiar character of the physical environment which acted on the human mind in different parts of the world and in successive periods of its historical development. If, as is now generally admitted, mythology was the first attempt at a poetical interpretation of the most important phenomena of nature, we can easily see how there was an easy transition from these efforts to know all the causes of things (*rerum cognoscere causas*) to the higher efforts to know the cause of all things. And if we remember that the nature of Aryan speech was such that it could at first express agents only—doers, not things done; rainers, not rain; lighteners, not lightnings—it is not difficult to understand how the agents of the great and constantly present drama of Nature were merged at last in the Supreme Agent, the Author and Ruler of all things. On this point all serious scholars seem to be agreed, however they may differ, and honestly differ, on certain points of detail.

INDEX

8695